Data Analytics and Artificial Intelligence for Predictive Maintenance in Smart Manufacturing

Today, in this smart era, data analytics and artificial intelligence (AI) play an important role in predictive maintenance (PdM) within the manufacturing industry. This innovative approach aims to optimize maintenance strategies by predicting when equipment or machinery is likely to fail so that maintenance can be performed just in time to prevent costly breakdowns. This book contains up-to-date information on predictive maintenance and the latest advancements, trends, and tools required to reduce costs and save time for manufacturers and industries.

Data Analytics and Artificial Intelligence for Predictive Maintenance in Smart Manufacturing provides an extensive and in-depth exploration of the intersection of data analytics, artificial intelligence, and predictive maintenance in the manufacturing industry and covers fundamental concepts, advanced techniques, case studies, and practical applications. Using a multidisciplinary approach, this book recognizes that predictive maintenance in manufacturing requires collaboration among engineers, data scientists, and business professionals and includes case studies from various manufacturing sectors showcasing successful applications of predictive maintenance. The real-world examples explain the useful benefits and ROI achieved by organizations. The emphasis is on scalability, making it suitable for both small and large manufacturing operations, and readers will learn how to adapt predictive maintenance strategies to different scales and industries. This book presents resources and references to keep readers updated on the latest advancements, tools, and trends, ensuring continuous learning.

Serving as a reference guide, this book focuses on the latest advancements, trends, and tools relevant to predictive maintenance and can also serve as an educational resource for students studying manufacturing, data science, or related fields.

Advances in Intelligent Decision-Making, Systems Engineering, and Project Management

This new book series will report the latest research and developments in the field of information technology, engineering and manufacturing, construction, consulting, healthcare, military applications, production, networks, traffic management, crisis response, human interfaces, and other related and applied fields. It will cover all project types, such as organizational development, strategy, product development, engineer-to-order manufacturing, infrastructure and systems delivery, and industries and industry-sectors where projects take place, such as professional services, and the public sector including international development and cooperation, etc. This new series will publish research on all fields of information technology, engineering, and manufacturing including the growth and testing of new computational methods, the management and analysis of different types of data, and the implementation of novel engineering applications in all areas of information technology and engineering. It will also publish on inventive treatment methodologies, diagnosis tools and techniques, and the best practices for managers, practitioners, and consultants in a wide range of organizations and fields including police, defense, procurement, communications, transport, management, electrical, electronic, aerospace, requirements.

Hybrid Intelligence for Smart Grid Systems
Edited by Seelam VSV Prabhu Deva Kumar, Shyam Akashe, Hee-Je Kim, and Chinmay Chakrabarty

Machine Learning-Based Fault Diagnosis for Industrial Engineering Systems
Rui Yang and Maiying Zhong

Smart Technologies for Improved Performance of Manufacturing Systems and Services
Edited by Bikash Chandra Behera, Bikash Ranjan Moharana, Kamalakanta Muduli, and Sardar M. N. Islam

Wireless Communication Technologies: Roles, Responsibilities and Impact of IoT, 6G, and Blockchain Practices
Edited by Vandana Sharma, Balamurugan Balusamy, Gianluigi Ferrari, and Prerna Ajmani

Digital Technology Enabled Circular Economy: Models for Environmental and Resource Sustainability
Edited by Bikash Moharana, Bikash Behera, and Kamalakanta Muduli

Industry 4.0, Smart Manufacturing, and Industrial Engineering: Challenges and Opportunities
Edited by Amit Kumar Tyagi, Shrikant Tiwari, and Sayed Sayeed Ahmad

Data Analytics and Artificial Intelligence for Predictive Maintenance in Smart Manufacturing
Edited by Amit Kumar Tyagi, Shrikant Tiwari, and Gulshan Soni

For more information about this series, please visit: www.routledge.com/Advances-in-Intelligent-Decision-Making-Systems-Engineering-and-Project-Management/book-series/CRCAIDMSEPM

Data Analytics and Artificial Intelligence for Predictive Maintenance in Smart Manufacturing

Edited by
Amit Kumar Tyagi, Shrikant Tiwari, and
Gulshan Soni

CRC Press
Taylor & Francis Group
Boca Raton London New York

CRC Press is an imprint of the
Taylor & Francis Group, an **informa** business

Designed cover image: Shutterstock – Summit Art Creations

First edition published 2025
by CRC Press
2385 NW Executive Center Drive, Suite 320, Boca Raton FL 33431

and by CRC Press
4 Park Square, Milton Park, Abingdon, Oxon, OX14 4RN

CRC Press is an imprint of Taylor & Francis Group, LLC

ISBN: 978-1-032-76952-3 (hbk)
ISBN: 978-1-032-77016-1 (pbk)
ISBN: 978-1-003-48086-0 (ebk)

DOI: 10.1201/9781003480860

Typeset in Times
by Newgen Publishing UK

Contents

Preface

In this smart era, every application is embedded by innovation and technological advancement, so the manufacturing industry stands at the threshold of a profound transformation. The manufacturing industry has always been a platform of innovation and needs solutions to today's problems (with rapidly changing world's need). We meet with the future and need innovation to sort many raised issues like maintenance, cost, etc. for which data analytics and AI are solutions for it. Data analytics and artificial intelligence (AI) or data mining techniques have emerged as essential tools for this decade which invite a new revolution in the way manufacturers manage their operations. This has resulted in a transformation in predictive maintenance (PdM), an indispensable strategy that uses the power of data and AI to optimize maintenance practices and, in doing so, redefine the manufacturing landscape/process. Hence, this book provides detail of various topics such as where it uses data analytics and artificial intelligence techniques in predictive maintenance in the manufacturing industry. This book will start with fundamental concepts, the transformative impact, and reasons why this convergence of technology and industry is not just a trend but a revolution.

As we move on this journey through the realms of data, algorithms, sensors, and real-time decision-making, this book explain essential implications of these emerging (data analytics/data mining techniques) technologies on the future of manufacturing. It starts from data collection and preprocessing to the deployment of required machine learning models in the respective area. Through this book, we will discover how predictive analytics can identify equipment failures, extend the life of machinery, enhance safety, and drive cost-efficiency in manufacturing operations. The manufacturing industry has always been an essential of innovation, and today, it finds itself at a crossroads where tradition meets the future. Data analytics and AI are being used to transform the industry, its machinery, making manufacturing smarter, more efficient, and more sustainable.

In the last, this book can serve as a reference guide to our readers which will inform about the latest advancements, trends, and tools relevant to predictive maintenance (with a global audience and addressing universal challenges faced by manufacturers).

Editors

Amit Kumar Tyagi, PhD, is an Assistant Professor at the National Institute of Fashion Technology, New Delhi, India. Previously he was an Assistant Professor (Senior Grade 2) and Senior Researcher at Vellore Institute of Technology (VIT), Chennai, Tamilandu, India, from 2019 to 2022. He earned a PhD in 2018 at Pondicherry Central University, Puducherry, India. Dr. Tyagi joined the Lord Krishna College of Engineering, Ghaziabad (LKCE), from 2009 to 2010 and from 2012 to 2013. He was an Assistant Professor and Head of Research, Lingaya's Vidyapeeth (formerly known as Lingaya's University), Faridabad, Haryana, India, from 2018 to 2019. His supervision experience includes more than ten master's dissertations and one PhD thesis. He has contributed to several projects such as AARIN and P3-Block to address some of the open issues related to privacy breaches in vehicular applications (such as parking) and medical cyber physical systems (MCPS). He has published over 200 papers in refereed high-impact journals, conferences, and books, and some of his articles were awarded best paper awards. Dr. Tyagi has filed more than 25 patents (nationally and internationally) in the areas of deep learning, internet of things, cyber physical systems, and computer vision. He has edited more than 25 books for IET, Elsevier, Springer, CRC Press, etc. Also, Dr. Tyagi has authored four books on intelligent transportation systems, vehicular ad-hoc network, machine learning, and internet of things, with IET UK, Springer Germany, and BPB India. He is a winner of faculty research awards for 2020, 2021, and 2022 (three consecutive years) given by the Vellore Institute of Technology, Chennai, India. Recently, he was awarded the best paper award for a paper titled "A Novel Feature Extractor Based on the Modified Approach of Histogram of Oriented Gradient", in ICCSA 2020, Italy (Europe). His research focuses on next-generation machine-based communications, blockchain technology, smart and secure computing, and privacy. He is a regular member of ACM, IEEE, MIRLabs, Ramanujan Mathematical Society, Cryptology Research Society, Universal Scientific Education and Research Network, CSI, and ISTE.

Shrikant Tiwari, PhD, is an Associate Professor in the Department of Computer Science and Engineering (CSE), School of Computing Science and Engineering (SCSE) at Galgotias University, Greater Noida, Uttar Pradesh, India. Dr. Tiwari also is a Senior Member of IEEE. He earned a PhD in computer science and engineering at the Indian Institute of Technology (Banaras Hindu University), Varanasi, India, in 2012 and an MTech in computer science and technology at the University of Mysore, India, in 2009. He has authored or co-authored more than 75 national and international journal publications, book chapters, and conference articles. He has five patents filed to his credit. His research interests include machine learning, deep learning, computer vision, medical image analysis, pattern recognition, and biometrics. Dr. Tiwari is a member of ACM, IET, FIETE, CSI, ISTE, IAENG, and SCIEI. He is also a guest editorial board member and a reviewer for many international journals of repute.

Gulshan Soni, PhD, is an Associate Professor and Principal in Charge in the Computer Science Engineering Department at the School of Engineering and Information Technology, Mahaveer Academy of Technology and Science University (MATS University), Raipur, India. He earned a PhD at Pondicherry University, India, along with a BTech at the National Institute of Technology (NIT), Raipur, India, and an ME at the National Institute of Technical Teachers Training and Research (NITTTR), Chandigarh, India. His research interests include wireless sensor networks, wireless body area networks, MAC protocols, and routing protocols, as well as distributed computing. Dr. Soni has published extensively in reputable journals and presented at national and international conferences. With over eight years of teaching experience, he brings valuable expertise to both government and private academic institutions in India.

Contributors

Thilagavathy A
Department of Computer Science and Engineering
R.M.K. Engineering College
Kavaraipettai, India

Ravi Arora
Department of Information Technology
Bharati Vidyapeeth's College of Engineering
New Delhi, India

Shruthi Arunkumar
Velammal Engineering College
Baba Nagar, Chennai, India

M. Dinesh Babu
Rajalakshmi Institute of Technolgoy
Chennai, India

Bhavesh Bhatia
Department of Computer Science
University of Western Ontario
London, Ontario, Canada

Varhsa Bhatia
Defence Institute of Advanced Technology
Pune, Maharashtra, India

Vineet Bhatia
Defence Institute of Advanced Technology (DIAT)
Pune, Maharashtra, India

Nikita Bhatt
Department of Computer Engineering
Chandubhai S. Patel Institute of Technology
Charotar University of Science and Technology
Changa, India

Nirav Bhatt
Department of Artificial Intelligence and Machine Learning
Chandubhai S. Patel Institute of Technology
Charotar University of Science and Technology
Changa, India

T. Chandrasekar
Kalasalingam Academy of Research and Education
Krishnankoil, Tamil Nadu, India

Nalayini C.M.
Velammal Engineering College
Velammal Nagar
Surapet, Chennai, India

Mohit Dayal
Department of Information Technology
Bharati Vidyapeeth's College of Engineering
New Delhi, India

Kesha Miralbhai Desai
Department of Artificial Intelligence and Machine Learning
Chandubhai S. Patel Institute of Technology
Charotar University of Science and Technology
Changa, India

Ritesh Kumar Dewangan
Department of Mechanical Engineering
Rungta College of Engineering and Technology
Raipur, India

Vinita Dewangan
Department of Mathematics
School of Engineering and Information Technology (MSEIT)
Mahaveer Academy of Technology and Science University
(MATS University)
Raipur, India

S Dhanushya
Department of Information Technology
Kongu Engineering College (Autonomous)
Perundurai, Tamil Nadu, India

Ajay Dureja
Department of Information Technology
Bharati Vidyapeeth's College of Engineering
New Delhi, India

Roopa Devi E M
Associate Professor Department of Information Technology
Kongu Engineering College (Autonomous)
Perundurai, Tamil Nadu, India

Gaurav Kumar Gautam
Department of Artificial Intelligence and Machine Learning
Chandubhai S. Patel Institute of Technology
Charotar University of Science and Technology
Changa, India

Anindya Ghatak
School of Computer Science Engineering and Technology
Bennett University
Gautam Buddha Nagar, Uttar Pradesh, India

B. Jaisinghani
Autodesk Inc.
San Francisco, California, USA

T.C. Jermin Jeaunita
Department of Computer Science and Engineering
Loyola Institute of Technology and Science
Thovalai, Tamil Nadu, India

A.V. Kalpana
Department of Data Science and Business Systems
SRM Institute of Science and Technology
Kattankulathur, Tamil Nadu, India

P. Karthikeyan
Department of Computer Science and Information Engineering
National Chung Cheng University
Chiayi, Taiwan

Surinder Kaur
Department of Information Technology
Bharati Vidyapeeth's College of Engineering
New Delhi, India

Sanjeev Kumar Khare
Defence Institute of Advanced Technology
Pune, Maharashtra, India

Pallavi P. Khobragade
Department of Civil Engineering
Rungta College of Engineering and Technology
Raipur, India

G Kiruthika
Department of Information Technology
Kongu Engineering College (Autonomous)
Perundurai, Tamil Nadu, India

Nishant Pushpak Koshti
Department of Artificial Intelligence and Machine Learning
Chandubhai S. Patel Institute of Technology
Charotar University of Science and Technology
Changa, India

N. Krishnaraj
Department of Networking and Communications
Faculty of Engineering and Technology
SRM Institute of Science and Technology
Kattankulathur, Tamil Nadu, India

Bhupendra Kumar
College of Business and Economics
Debre Tabor University
Debre Tabor, Ethiopia

Mahesh Kumar
Department of Information Technology
Bharati Vidyapeeth's College of Engineering
New Delhi, India

M. Maranco
Department of Networking and Communications
Faculty of Engineering and Technology
SRM Institute of Science and Technology
Kattankulathur, Tamil Nadu, India

Bireshwar Dass Mazumdar
Department of Computer Science and Engineering
School of Computer Science Engineering and Technology
Bennett University
Noida, Uttar Pradesh, India

Hiren Mewada
Prince Mohammad Bin Fahd University
Dhahran, Saudi Arabia

Manmohan Mishra
Department of Computer Science Application
United Institute of Management
Prayagraj, Uttar Pradesh, India

Kumudavalli M.V.
Dayananda Sagar College of Arts Science and Commerce
Bengaluru, Karnataka, India

Venkatesha Nayak
Department of Postgraduate Studies in Commerce
University Evening College
Mangalore, Karnataka, India

Timilehin O. Olubiyi
Department of Business Administration and Marketing
School of Management Sciences
Babcock University
Ilishan-Remo, Ogun State, Nigeria

R Shanthakumari
Department of Information Technology
Kongu Engineering College (Autonomous)
Perundurai, Tamil Nadu, India

Vijay Anandh R
Department of Electronics and Communication Engineering
R.M.K. College of Engineering and Technology
Chennai, India

Yadav Krishna Kumar Rajnath
Department of Industrial and Production Engineering
Institute of Engineering and Rural Technology (IERT)
Prayagraj, Uttar Pradesh, India

U. Srinivasulu Reddy
Machine Learning and Data Analytics Lab
Centre of Excellence in Artificial Intelligence
Department of Computer Applications
National Institute of Technology
Tiruchirappalli, India

Adeyemi Omolade S.
Department of Business Administration
Oduduwa University
Ipetumodu, Osun State, Nigeria

Kumar Chandar S
School of Business and Management
CHRIST (Deemed to be University)
Bangalore, Karnataka, India

V. Sathya
SRM Institute of Science and Technology
Chennai, India

Selva Sharmila
Dayanandasagar College of Engineering
Bengaluru, Karnataka, India

Shyla
Department of Information Technology
Bharati Vidyapeeth's College of Engineering
New Delhi, India

S. Singh
Vivekanand College of Arts
Science and Commerce (Autonomous)
Chembur, Mumbai, India

M. Sivakumar
Department of Networking and Communications
Faculty of Engineering and Technology
SRM Institute of Science and Technology
Kattankulathur, Tamil Nadu, India

S. Thenmozhi
Dayanandasagar College of Engineering
Bengaluru, Karnataka, India

Shrikant Tiwari
Department of Computer Science and Engineering
School of Computing Science and Engineering
Galgotias University
Noida, Uttar Pradesh, India

Rati Kailash Prasad Tripathi
Department of Pharmaceutical Sciences
Sushruta School of Medical and Paramedical Sciences
Assam University (A Central University)
Silchar, Assam, India

Amit Kumar Tyagi
Department of Fashion Technology
National Institute of Fashion Technology
New Delhi, India

M Vaidya
Department of Technology
Jodhpur Institute of Engineering and Technology
Jodhpur, Rajasthan, India

Virendra Kumar Verma
Department of Industrial and Production Engineering
Institute of Engineering and Rural Technology (IERT)
Prayagraj, Uttar Pradesh, India

Nilesh Ware
Defence Institute of Advanced Technology
Pune, Maharashtra, India

1 Introduction to Machine Learning Fundamentals

Ritesh Kumar Dewangan, Vinita Dewangan, and Pallavi P. Khobragade

1.1 INTRODUCTION

Machine learning, a subset of artificial intelligence (AI), has become a fundamental aspect of contemporary computing. Its capacity to enable systems to learn from data and independently make informed choices has sparked revolutionary advancements across various sectors. [1], [2].

In recent years, the field has witnessed unprecedented growth, driven by advancements in computational power, data availability, and algorithmic sophistication.

The essence of machine learning lies in its departure from traditional rule-based programming paradigms. Instead of relying on explicitly defined instructions, machine learning algorithms learn iteratively from data, continuously improving their performance as they encounter new examples. This change in perspective has unlocked numerous opportunities for diverse applications across a wide array of fields, spanning from healthcare and finance to transportation and entertainment.

The allure of machine learning lies in its capability to extract insights, patterns, and relationships from vast amounts of data that may be too complex for human analysts to decipher manually. By discerning these patterns, machine learning algorithms can make predictions, identify anomalies, recommend actions, and automate decision-making processes, thus offering valuable assistance to human operators and augmenting their capabilities.

In this chapter, we embark on a journey to explore the fundamental concepts and principles that underpin machine learning. We delve into the core types of machine learning, namely supervised, unsupervised, and reinforcement learning, elucidating their respective methodologies and applications. Additionally, we examine key techniques such as feature engineering, model selection, and evaluation, which play pivotal roles in the development and deployment of machine learning systems.

Moreover, we delve into the ethical considerations inherent in the adoption of machine learning technologies. As these systems exert increasing influence on society and individuals, it becomes imperative to address issues related to privacy, fairness, transparency, and accountability. By navigating these ethical dilemmas with care and foresight, we can ensure that machine-learning technologies serve the collective good while minimizing potential harms and biases.

DOI: 10.1201/9781003480860-1

Essentially, this chapter provides a fundamental roadmap for individuals new to machine learning as well as experienced professionals in the field. By grasping the fundamental principles and methodologies discussed here, readers will be prepared to navigate the intricate realm of machine learning, harness its capacity for innovation and societal advancement, and actively participate in its ongoing development with responsibility and ethics at the forefront.

The objective of this chapter is to offer readers a firm foundation in the basics of machine learning, enabling a more profound comprehension of its fundamental principles and methodologies.

1.2 WHAT IS MACHINE LEARNING?

At its core, machine learning involves the development of algorithms capable of learning from data without being explicitly programmed [3]. Unlike traditional rule-based programming, where instructions are explicitly defined, machine learning algorithms iteratively improve their performance by discerning patterns and relationships within datasets.

Machine learning represents a paradigm shift in the way computers are programmed and operated. In contrast to conventional programming methods, which rely on providing explicit instructions to carry out tasks, machine learning algorithms are crafted to learn from data and enhance their performance gradually. Essentially, machine learning strives to empower computers to autonomously uncover patterns, connections, and insights from data without the need for explicit programming tailored to specific tasks.

The underlying principle of machine learning revolves around the concept of learning from experience. This learning process involves algorithms analyzing large datasets, identifying patterns, and using these patterns to make predictions or decisions. By iteratively refining their models based on new data, machine learning algorithms can adapt and improve their performance continuously.

Machine learning can be generally classified into three primary categories: supervised learning, unsupervised learning, and reinforcement learning.

Supervised learning: in supervised learning, models are trained on labeled datasets, where inputs are paired with corresponding outputs [4]. Through this process, algorithms learn to predict output values for unseen inputs.

In supervised learning, the algorithm is trained using a labeled dataset, where each example is associated with a corresponding output or target variable.The objective is to grasp the relationship between inputs and outputs, empowering the algorithm to forecast outcomes for unseen data. This type of learning is analogous to a teacher providing labeled examples to a student, guiding them to learn the correct patterns and relationships in the data.

Unsupervised learning: unsupervised learning entails training algorithms on unlabeled data, aiming to reveal concealed patterns or structures within the dataset [4]. Unlike supervised learning, there are no provided explicit labels or target variables during training. Instead, the algorithm autonomously discerns significant patterns or groupings within the data. This type of learning is akin to exploring a dataset without prior knowledge or guidance, allowing the algorithm to uncover novel insights and relationships.

Techniques such as clustering, dimensionality reduction, and density estimation are commonly employed in unsupervised learning.

Reinforcement learning: reinforcement learning centers on agents learning to interact with environments by taking actions and receiving feedback in the form of rewards or penalties [5]. The aim is to develop a policy that maximizes the total rewards obtained over time.

Reinforcement learning entails an agent learning to interact with an environment by executing actions and obtaining feedback through rewards or penalties. The objective is to acquire a policy that maximizes the total rewards accumulated over time. In reinforcement learning, the algorithm learns through trial and error, exploring different actions and observing their consequences in the environment. This type of learning is analogous to how humans learn from experience, adapting their behavior based on the outcomes of their actions.

Overall, machine learning represents a powerful approach to solving complex problems and extracting value from data. By enabling computers to learn from data and make data-driven decisions, machine learning has the potential to revolutionize industries, drive innovation, and enhance our understanding of the world around us.

1.3 KEY CONCEPTS AND TECHNIQUES

In machine learning, understanding key concepts and mastering fundamental techniques is essential for developing effective models and extracting meaningful insights from data. This section delves into some of the core concepts and techniques that underpin machine learning workflows.

Feature engineering: feature engineering refers to the practice of choosing, modifying, or generating new features from raw data to enhance the efficacy of machine learning models [6]. Features are the individual measurable properties or characteristics of the data that serve as input variables to the model. Competent feature engineering can have a substantial impact on the predictive accuracy and generalization capability of a model. Techniques in feature engineering include scaling, normalization, encoding categorical variables, and creating derived features through mathematical transformations or domain-specific knowledge.

Model selection and evaluation: model selection entails the process of determining the optimal algorithm and its corresponding hyperparameters for a specific machine learning task. This decision relies on various factors including the characteristics of the data, the complexity of the problem, and the desired performance metrics. After selecting a set of candidate models, model evaluation techniques are employed to assess their performance and generalization capability. Common evaluation metrics include accuracy, precision, recall, F1-score, and area under the ROC curve (AUC). Methods like cross-validation and holdout validation are employed to assess a model's performance on unseen data and address concerns like overfitting.

Cross-validation: cross-validation is a commonly utilized method to gauge a model's performance on unseen data by dividing the dataset into multiple subsets for training and validation purposes [7]. It helps mitigate issues related to overfitting and provides more reliable performance estimates.

It also serves as a means of evaluating the performance of machine learning models by dividing the dataset into several folds or subsets. Each fold is used as a validation set while the others are utilized for training. This process is iterated multiple times, with each fold taking turns as the validation set.

Cross-validation offers a more reliable estimate of model performance compared to a single train-test split, as it utilizes multiple data subsets for both training and validation.

Hyperparameter tuning: hyperparameters are predetermined parameters that control different aspects of how a learning algorithm behaves before the actual learning process begins. Examples of hyperparameters include the learning rate in gradient descent methods, the depth of decision trees, or the number of hidden layers in a neural network.

Hyperparameter tuning, also known as hyperparameter optimization, involves systematically searching for the optimal set of hyperparameters that yield the best performance for a given machine learning task. Techniques for hyperparameter tuning include grid search, random search, and more advanced optimization algorithms such as Bayesian optimization.

Model interpretability and explainability: as machine learning models become increasingly complex and are deployed in critical applications, there is a growing need for model interpretability and explainability. Interpretability refers to the ability to understand and interpret how a model makes predictions, while explainability refers to the ability to provide explanations or justifications for individual predictions. Techniques for enhancing model interpretability include feature importance analysis, partial dependence plots, SHAP (SHapley Additive exPlanations) values, LIME (Local Interpretable Model-agnostic Explanations), and model-specific interpretation methods tailored to specific algorithms such as decision trees and linear models.

Mastering these key concepts and techniques is crucial for developing robust and interpretable machine learning models, optimizing their performance, and gaining actionable insights from data. Effective feature engineering, model selection, evaluation, and interpretability are essential components of the machine learning workflow, enabling practitioners to tackle real-world problems and drive innovation across various domains.

1.4 ETHICAL CONSIDERATION IN MACHINE LEARNING

The widespread adoption of machine learning technologies raises ethical concerns related to privacy, fairness, transparency, and accountability [8]. Adhering to ethical principles is essential to ensure the responsible development and deployment of machine learning systems.

As machine learning technologies become increasingly pervasive in society, it is crucial to address the ethical implications and considerations associated with their development, deployment, and usage. Ethical considerations in machine learning encompass a broad range of issues, including privacy, fairness, transparency, accountability, and societal impact. This section delves into some of the key ethical considerations in machine learning.

Privacy: one of the foremost ethical concerns in machine learning is the protection of individuals' privacy and the responsible handling of sensitive data. Machine learning algorithms often rely on vast amounts of personal data to make predictions or derive insights, raising concerns about data privacy and the potential for unauthorized access or misuse of personal information. It is imperative for organizations and practitioners to implement robust data privacy measures, such as data anonymization, encryption, and access controls, to safeguard individuals' privacy rights.

Fairness and bias: machine learning algorithms have the potential to perpetuate or exacerbate biases present in the data on which they are trained. Biases in training data can lead to unfair or discriminatory outcomes, particularly in high-stakes domains such as hiring, lending, and criminal justice. Addressing algorithmic bias requires careful attention to dataset composition, feature selection, and algorithm design to mitigate biases and ensure fairness and equity in decision-making processes. Techniques such as fairness-aware machine learning and algorithmic auditing can help identify and mitigate biases in machine learning systems.

Transparency and expalinability: as machine learning models become increasingly complex, there is a growing need for transparency and explainability in algorithmic decision-making. Individuals affected by algorithmic decisions have the right to understand how those decisions are made and the factors that influence them. Providing explanations or justifications for algorithmic decisions enhances transparency, fosters trust, and enables stakeholders to assess the reliability and fairness of machine learning systems. Techniques for enhancing transparency and explainability include model interpretability methods, such as feature importance analysis, local explanations, and model-specific interpretation techniques.

Accountability and oversight: accountability is essential in ensuring that machine learning practitioners and organizations are held responsible for the outcomes of their algorithms. Establishing clear lines of accountability and oversight mechanisms is crucial for addressing issues of algorithmic accountability and ensuring that machine learning systems operate ethically and responsibly. Regulatory frameworks, industry standards, and ethical guidelines can help establish norms and best practices for responsible AI development and deployment.

Societal impact: machine learning technologies have the potential to have profound societal impacts, influencing everything from employment patterns and economic inequality to healthcare outcomes and social interactions. It is essential to consider the broader societal implications of machine learning systems and to ensure that their development and deployment align with societal values and goals. Ethical considerations should encompass the broader societal impacts of machine learning, including considerations of social justice, environmental sustainability, and democratic governance.

Addressing these ethical considerations requires a multi-stakeholder approach involving collaboration between technologists, policymakers, ethicists, and civil society organizations. By prioritizing ethics and responsible AI development, we can harness the transformative potential of machine learning while safeguarding individual rights, promoting fairness and equity, and advancing the collective well-being of society.

1.5 APPLICATIONS OF MACHINE LEARNING

Machine learning finds applications across diverse domains, including healthcare, finance, transportation, and entertainment. In healthcare, it facilitates disease diagnosis, personalized treatment planning, and drug discovery [9]. In finance, machine learning powers fraud detection, risk assessment, and algorithmic trading systems [9]. Other notable applications include autonomous vehicles, recommendation systems, natural language processing, and computer vision.

This section explores some of the key applications of machine learning.

Healthcare: in healthcare, machine learning is employed for a wide range of tasks, including disease diagnosis, treatment planning, drug discovery, and personalized medicine. Machine learning algorithms analyze medical imaging data such as X-rays, MRIs, and CT scans to detect abnormalities and assist radiologists in diagnosing diseases like cancer, cardiovascular conditions, and neurological disorders. Additionally, machine learning models analyze electronic health records (EHRs) and genomic data to identify patterns and risk factors for diseases, enabling personalized treatment plans tailored to individual patients.

Finance: the finance sector heavily depends on machine learning for various functions like fraud detection, risk evaluation, algorithmic trading, and managing customer relationships. Machine learning algorithms scrutinize extensive financial data, such as transaction logs, market patterns, and client actions, to detect fraudulent behavior, evaluate credit risk, and enhance trading tactics. Additionally, machine learning-powered chatbots and virtual assistants provide personalized financial advice and support to customers, enhancing customer engagement and satisfaction.

Transportation: machine learning holds a vital position in the transportation industry, driving autonomous vehicles, traffic management systems, and predictive maintenance solutions. Through the analysis of sensor data derived from vehicles, road infrastructure, and traffic cameras, machine learning algorithms enable autonomous navigation, forecast traffic congestion, and enhance route optimization. Additionally, machine learning models predict equipment failures and maintenance needs in transportation fleets, reducing downtime and improving operational efficiency.

E-commerce and recommendation syatems: in e-commerce, machine learning algorithms power recommendation systems that analyze user behavior and preferences to provide personalized product recommendations. These algorithms leverage techniques such as collaborative filtering, content-based filtering, and matrix factorization to suggest products, movies, music, and articles tailored to individual users' tastes and preferences. Recommendation systems enhance user engagement, increase sales, and improve customer satisfaction in online retail platforms and streaming services.

Natural language processing (NLP) and language translation: natural language processing (NLP) technologies enable computers to understand, interpret, and generate human language. Machine learning algorithms analyze text data to perform tasks such as sentiment analysis, named entity recognition, and machine translation. NLP applications include virtual assistants, chatbots, language translation services, and text summarization tools. Machine learning models trained on large text corpora learn to understand and generate human-like language, facilitating communication and information retrieval across different languages and cultures.

Computer vision: computer vision technologies empower computers to interpret and scrutinize visual data obtained from images and videos. Through machine learning algorithms, pixel data is analyzed to execute tasks like object detection, image classification, and facial recognition. Applications of computer vision encompass autonomous vehicles, surveillance systems, medical imaging, and augmented reality. By training machine learning models on extensive image datasets, computers acquire the capability to recognize and comprehend visual patterns, thereby enabling them to perceive and interpret the visual world akin to humans.

1.6 FUTURE DIRECTIONS

The future of machine learning holds promising avenues for innovation, driven by advancements in deep learning, reinforcement learning, and interdisciplinary collaborations [10]. Key areas of focus include the development of interpretable and robust models, addressing challenges related to data privacy and security, and exploring synergies with emerging technologies such as robotics and quantum computing.

As we look ahead, several key trends and areas of focus are poised to shape the future of machine learning:

Advancement in deep learning: deep learning, a subset of machine learning that uses artificial neural networks with multiple layers of abstraction, continues to drive significant progress in various domains. Future advancements in deep learning are expected to focus on improving model efficiency, scalability, and interpretability. Techniques such as self-attention mechanisms, capsule networks, and neural architecture search are poised to enhance the capabilities of deep learning models and enable them to tackle increasingly complex tasks.

Interdisciplinary collaboration: machine learning intersects with a wide range of disciplines, including computer science, statistics, mathematics, neuroscience, and domain-specific fields such as healthcare, finance, and climate science. Future advancements in machine learning are likely to be driven by interdisciplinary collaborations that leverage insights and methodologies from diverse domains. Collaborative efforts between researchers, practitioners, policymakers, and stakeholders will facilitate the development of innovative solutions to complex problems and address pressing societal challenges.

Interpretability and trustworthy AI: as machine learning systems become increasingly complex and are deployed in high-stakes applications, there is a growing need for model interpretability and trustworthy AI. Future research efforts will focus on developing techniques and methodologies for enhancing the interpretability, transparency, and accountability of machine learning models. Explainable AI (XAI) approaches, model debugging tools, and techniques for quantifying uncertainty and robustness will play a crucial role in building trust and confidence in machine learning systems.

Ethical and responsible AI: ethical considerations are paramount in the development and deployment of machine learning technologies. Future directions in machine learning will prioritize ethical and responsible AI practices, including fairness, transparency, privacy, and accountability. Regulatory frameworks, industry standards, and ethical guidelines will play a crucial role in promoting ethical AI development and ensuring that machine learning systems operate in a manner that respects individual rights, promotes equity, and aligns with societal values.

Continual learning and lifelong learning: traditional machine learning approaches often assume static datasets and fixed environments. However, real-world applications often involve dynamic, evolving scenarios where data distributions and tasks change over time. Future research in machine learning will focus on continual learning and lifelong learning approaches that enable models to adapt and learn from new data, retain knowledge from previous tasks, and generalize across diverse environments. Techniques such as meta-learning, transfer learning, and online learning will facilitate the development of more adaptive and resilient machine learning systems.

Human-centred AI: human-centered AI approaches prioritize the design and development of machine learning systems that augment human capabilities, enhance user experiences, and promote human well-being. Future directions in machine learning will emphasize human-centered design principles, user-centric methodologies, and interdisciplinary collaborations that integrate human factors and values into the development process. Human-AI collaboration tools, interactive machine learning interfaces, and user-centered evaluation metrics will enable the co-creation of AI systems that are intuitive, inclusive, and aligned with human needs and preferences.

In conclusion, the future of machine learning is characterized by continuous innovation, interdisciplinary collaboration, and a commitment to ethical and responsible

AI development. By embracing these future directions and addressing emerging challenges, machine learning has the potential to drive transformative advancements across various domains, empower individuals and communities, and contribute to the collective well-being of society.

1.7 CONCLUSION

In conclusion, machine learning stands at the forefront of technological innovation, driving transformative changes across industries, society, and the way we interact with the world. Throughout this chapter, we have explored the foundational concepts, key techniques, ethical considerations, applications, and future directions of machine learning.

Machine learning, a branch of artificial intelligence, enables computers to acquire knowledge from data and make informed decisions without the need for explicit programming. Through supervised, unsupervised, and reinforcement learning paradigms, machine learning algorithms can tackle a wide range of tasks, from predictive modeling and pattern recognition to decision-making and optimization.

Key concepts such as feature engineering, model selection, cross-validation, and hyperparameter tuning play pivotal roles in developing robust and effective machine learning models. Ethical considerations, including privacy, fairness, transparency, accountability, and societal impact, are paramount in ensuring that machine learning technologies are deployed responsibly and ethically.

Machine learning finds applications across diverse domains, including healthcare, finance, transportation, e-commerce, natural language processing, and computer vision. From disease diagnosis and fraud detection to personalized recommendations and autonomous vehicles, machine learning technologies have revolutionized industries, enhanced productivity, and improved quality of life for individuals worldwide.

Looking ahead, the future of machine learning holds immense promise, with advancements in deep learning, interdisciplinary collaborations, interpretability, ethical AI, continual learning, and human-centered design driving innovation and progress. By embracing these future directions and addressing emerging challenges, machine learning will continue to push the boundaries of what is possible, empower human creativity and ingenuity, and shape the future of technology and society.

In this ever-evolving landscape of machine learning, it is crucial to remain vigilant, ethical, and mindful of the broader societal implications of our technological advancements. By fostering a culture of responsible innovation, collaboration, and inclusivity, we can harness the full potential of machine learning to address pressing challenges, promote human well-being, and create a better future for all.

REFERENCES

1 C. M. Bishop, "Pattern recognition and machine learning," Springer, 2006.
2 I. Goodfellow, Y. Bengio, and A. Courville, "Deep learning," MIT Press, 2016.
3 T. M. Mitchell, "Machine learning," McGraw-Hill, 1997.

4 R. S. Sutton and A. G. Barto, "Reinforcement learning: An introduction," MIT press, 2018.
5 I. Guyon and A. Elisseeff, "An introduction to variable and feature selection," *Journal of Machine Learning Research*, vol. 3, pp. 1157–1182, 2006.
6 M. Sokolova and G. Lapalme, "A systematic analysis of performance measures for classification tasks," *Information Processing & Management*, vol. 45, no. 4, pp. 427–437, 2009.
7 R. Kohavi, "A study of cross-validation and bootstrap for accuracy estimation and model selection," in Proceedings of the 14th international joint conference on Artificial intelligence-Volume 2, pp. 1137–1143, Morgan Kaufmann Publishers Inc., 1995.
8 S. Barocas, M. Hardt, and A. Narayanan, "Fairness and machine learning," fairmlbook.org, 2019.
9 M. I. Jordan and T. M. Mitchell, "Machine learning: Trends, perspectives, and prospects," *Science*, vol. 349, no. 6245, pp. 255–260, 2015.
10 Y. LeCun, Y. Bengio, and G. Hinton, "Deep learning," *Nature*, vol. 521, no. 7553, pp. 436–444, 2015.

2 AI Applications in Production

Nalayini C.M., V. Sathya, Shruthi Arunkumar, and M. Dinesh Babu

2.1 INTRODUCTION TO AI IN SMART MANUFACTURING

2.1.1 Smart Manufacturing

Smart manufacturing stands as the vanguard of a transformative wave sweeping through industrial landscapes. At its essence, it is a new way of making things by adding smart technologies to make everything connected. The main feature of smart manufacturing is its ability to use live data, automation and connections throughout the whole process of making things. The aim is to make things work better, cut down on expenses, and improve overall output. Smart manufacturing is different from normal making because it uses ideas from Industry 4.0. This includes cyber-physical systems, internet of things (IoT), and cloud computing together to create a highly adaptive and responsive production environment.

The heart of smart manufacturing is data driven decision making. In order to generate large amounts of data we can use IoT and sensors. All of the data that is collected and extracted from it can be used after it is thoroughly examined in order to gain valuable insights. By doing so, anyone can make wise and informed decisions at every stage of the manufacturing process. The main intent of smart manufacturing is to create an environment or space that is capable of adapting, learning and evolving on its own through various insights rather than just producing goods. Once smart manufacturing becomes more mainstream, the benefits such as increased productivity, better quality of products, improved decision making, etc. become much clearer. Smart manufacturing is also very flexible as it can be modified to meet any change in the market demand. Thus it is an ingenious approach that allows people to maximize their profit and also lay a flexible yet solid foundation for the future.

2.1.2 Overview of Digital Twin Technology

Digital twin is the basis for smart manufacturing. It is a concept that helps us to visualize things that we present in the human world inside a virtual world. This technology helps us to share our real-time product in the virtual world for worldwide sharing. This helps manufacturers to produce a dynamic product which allows them to have a greater insight into these products.

DOI: 10.1201/9781003480860-2

Digital twins work hand in hand with their physical products, thus always exchanging information. This close connection helps watch, study, and improve things in real-time. Sensors and IoT are connected to the internet which helps to make sure that the digital twin of the product shows what is happening right now within the system. This allows us to create a link between the physical and virtual world. It ensures that manufacturers are able to inspect various characteristics of the product without having to compromise on the actual production process. Testing also becomes an easier process which enables us to predict possible future outcomes which could help in avoiding potential failures that could disrupt the process.

In manufacturing, digital twins are a key component for creating a virtual thread that draws out throughout the entire life cycle of the product. This helps manufacturers to have a complete picture, beginning from the design phase up till the production phase. This accelerates growth and increases our understanding of every component. This helps to improve the products during production since there is constant optimization with the production plans.

This study looks at artificial intelligence's (AI) potential in cities and provides a framework that highlights important aspects for incorporating AI into cities. Subculture, metabolism, and governance are essential to ensuring that smart cities are implemented in line with Sustainable Development Goal 11 and the New Urban Agenda [1].

Digital twins can be made mainstream throughout the entire manufacturing sector or even to complex systems. By allowing the people involved in the manufacturing sector to be able to create digital copies or replicas of their product, each and every component can be analyzed more thoroughly. This can increase scalability, thus improving production and improving the quality of manufacturing. It is important for us to be aware of the growing importance of smart manufacturing in this fast paced tech driven world as it allows us to bring forth more innovative ideas and achieve the impossible.

2.2 FOUNDATIONS OF AI IN PRODUCTION

2.2.1 AI

Artificial intelligence is reshaping the capabilities of the manufacturing industry day by day. As more advancements are made in AI, there is an increase in the possibilities of more AI integration with manufacturing. This applies into various manufacturing application cases that improved adaptability, increased efficiency and made more data driven decisions.

Fundamentally, AI in smart manufacturing utilizes data to its maximum capacity. Huge amounts of data are used to its fullest. Vast quantities of real-time data from the sensors and IoT that are connected to the networked systems become a valuable resource for training these AI models. One of the subsets of AI is machine learning, which uses algorithms to analyze the data, find similar looking patterns, and to make meaning with them. By doing so, AI systems gain the ability to predict outcomes, improve workflow, and streamline the decision making process.

Predictive analysis is the core factor that powers AI in smart manufacturing. With the help of this, AI models are able to predict possible mishaps that could occur during the production process. By being proactive, we can save bottleneck situations that may arise in the future, thus saving money and effort. For industries that depend on a constant flow of manufacturing, such as water treatment plants, the ability to predict issues that may happen before they occur is groundbreaking.

This paper explores the role of AI in Industry 4.0. It has a particular focus on the digital twins as a model for the cyber physical systems (CPS). It covers tech such as 5G,edge computing, and mainly concentrating on how AI can improve automation procedures. It also focuses on real-world applications of robots in motion prediction [2]. This looks at how AI can be applied in the automobile industry. It also considers factors such as dynamism and organizational flexibility. The use of big data to draw conclusions and to draw attention to the constraints present in production [3]. AI producers are based on sellers that manage continuous online games and also new business models that are analyzed through studies conducted for recreational AI [4].

Autonomous decision making is the other aspect of AI integration that enables smart manufacturing. Systems that have been fed real-time data are able to make much faster and quick decisions. This helps increase productivity and improves the results. In a fast paced environment with an imprecise workplace environment, the adoption of this helps to simplify the decision making process.

AI's importance in clever manufacturing extends to the field of exceptional warranty. A form of artificial intelligence referred to as computer imaginative and prescient can perceive defects by studying visible facts from manufacturing methods. This ensures that only items that meet strict necessities make it to market. This now improves the first rate of the product however additionally lowers waste and related prices.

The integration of AI in intelligent design marks a seismic shift from reactive to proactive, manual to autonomous. It envisions a future where intelligence is not just a product but the very cornerstone of design success, intertwining efficiency, innovation, and sustainable development.

2.2.2 Machine Learning Algorithms in Manufacturing

Machine learning (ML) algorithms form a foundation of sorts for combining AI with the manufacturing industry. These algorithms help to minimize the amount of human interaction that is usually needed by systems in order to train them using pre-processed data collections, help in identifying patterns and making informed decisions. These algorithms are crucial as they help manufacturers to predict future outcomes and scenarios.

As mentioned briefly in Section 2.3.1, predictive maintenance is one of the popular methods that are used in smart manufacturing. Continuously evaluating historical data, real-time data helps make better decisions on when certain unforeseen and unfortunate events may occur. This helps to minimize damage and downtime which may disrupt the production process. By identifying these patterns that are linked to cause errors, ML algorithms can help to enrich the quality of the products by allowing real-time modifications to the process.

Machine learning algorithms optimize industrial processes such as supply chain management, demand forecasting, and resource allocation due to their inherent flexibility. The more these algorithms continue learning and developing, the greater their proficiency becomes in managing intricate and dynamic production situations. This study methodically examines the literature on AI and machine learning applications within the manufacturing sector. It specifically focuses on the changes in this field from 1999 until the present, both before as well as after introducing Industry 4.0 [5]. Through the gathering, interpretation, and assessment of trained data from beyond, this work not only creates simulation logs but also proposes a systematic approach to integrating predictive security into business processes. The method by integrating actual, predicted, and simulated data enhances the estimation of remaining useful life (RUL) using concepts such as the digital twin [6]. This paper investigates the combination of predictive maintenance techniques, with a focus on the effect of flexibility on renovation efficiency in manufacturing systems, using virtual twins, a replace carrier, and a labelling interface in the context of a bendy hybrid assembly machine [7].

Fundamentally, production systems have the ability to automatically enhance and improve the production process due to machine learning algorithms. As the manufacturing sector moves into a period of smart automation, these algorithms can be considered as the driving factor for these models that are made which helps manufacturers to upgrade their production landscape and increase competition.

2.2.3 Role of Robotics and Automation

We have discussed how AI and ML is the driving force behind the major operational factor transformation that may be fixed via the integration of robotics and automation. By using robotics in manufacturing, it increases the standards with regards to efficiency and flexibility.

Smart AI enabled robots are able to perform a wide variety of tasks in the manufacturing sector, ranging from simple tasks that require repetition to complex assembling tasks. Robotics is a technology that can work along with humans to increase productivity. Their ability to quickly adapt to their ever so changing surroundings, pick up on cues, and to be able to work harmoniously amongst humans allows them to be much preferred in a manufacturing environment. By combining humans and robots we can solve labor issues and also ensure the safety of workers.

Automation that is enabled by AI helps eliminate the need for manual intervention during production. This method simplifies the process and ensures a common agreement amongst the people, lowers the possibility of errors from occurring, and also increases the speed of the production cycle. AI driven automation has also spread its roots into supply chain management, logistics, as well as decision making support that ensures an overall well organized production architecture.

All three of these, automation, robotics, and AI, work together hand in hand to produce smart factories that offer higher efficiency and an adaptable manufacturing facility. Quick shifts in supply chain dynamics, demand, and manufacturing requirements are all possible due to the real-time data that is obtained from these linked systems in the network. This results in a manufacturing landscape that is capable of reacting to the complexities of today's market.

2.3 DIGITAL TWIN TECHNOLOGY

The use of digital twin technology, especially in manufacturing, provides a virtual counterpart of the real system and procedures which is revolutionizing the industry. Digital twins have several applications and possible integrations that can be done to enhance their capabilities which are elaborated in the following sections.

2.3.1 Applications of Digital Twin in Manufacturing

There are several applications for the digital twin technology that are altering both the design and manufacturing process along with the product maintenance process. At its core, a digital twin is a virtual replica of the behavior and physical characteristics of a product that is connected to any system. In this scenario, the replica provides manufacturers with unimaginable opportunities and valuable insights throughout the product lifecycle.

- **Product design:**
  - ***Virtual prototyping:*** by using this technology, engineers can test and simulate the developed prototypes in the virtual reality (VR) before the actual manufacturing process which helps us to reduce the need for expensive physical prototypes.
  - ***Accelerated design iterations:*** this technology also helps manufacturers to ensure that the final product is precise and satisfies all of the standards that need to be met.
- **Production:**
  - ***Real-time monitoring:*** real-time production line monitoring is possible by using digital twins. This helps to provide information on how well each component performs in a simulated environment.
  - ***Optimization capabilities:*** these virtual replicas that are created can be used by manufacturers to modify and improve the quality of the product and improve the overall efficiency of the manufacturing process.
- **Predictive maintenance:**
  - ***Continuous equipment monitoring:*** digital twins constantly monitor the condition of the product in real time allowing manufacturers to take action immediately before potential issues can occur.
  - ***Enhanced reliability:*** this method increases the lifetime and durability of the products being manufactured and also the manufacturing equipment by making predictive maintenance a reality.
- **Product lifecycle management:**
  - ***Continuous evolution:*** digital twins also help during the product life cycle. By giving continuous, real-time data to the manufacturers, it helps to enhance the production process and makes it easier.

TABLE 2.1
Key Aspects and Benefits of AI Integration

ASPECT	DESCRIPTION
Integration Purpose	Increases the intelligence and adaptability of digital twins through artificial intelligence.
Data Analysis	To find patterns, anomalies, and trends, AI systems examine large datasets gathered from sensors and internet of things devices.
Predictive Capabilities	Converts digital twins into forecasting instruments for instantaneous decision-making.
Quality Optimization	Enables quick remedial action and process optimization, ensuring high-quality output.
Predictive Maintenance	AI and digital twins work together to predict maintenance needs through continuous monitoring, which minimizes downtime and lowers operating costs.
Autonomous Decision-Making	In dynamic industrial contexts, AI-driven digital twins autonomously modify production parameters, distribute resources, and streamline workflows.
Adaptability	Makes it easier to respond to unforeseen obstacles and adjust to changing demands without the need for human intervention.

- ***Usage data capture:*** by collecting usage data from users, the system generates a feedback loop that helps with future design revision processes and also helps to improve overall product enhancement.

2.3.2 Integration with AI for Enhanced Capabilities

With the use of digital twin technology in conjunction with artificial intelligence (AI), its capacities are raised to a new level that provides for synergizing opportunities and, consequently, effective adaptability capabilities. Better decision-making through application in real-time production environment increases overall efficiency regarding operating processes. Table 2.1 shows the key aspects and benefits of AI integration.

In this regard, the introduction of digital twin technology with AI is a giant step in manufacturing capabilities. Digital twins create a detailed virtual model of physical assets, while the integration of AI increases intelligence and capability driven by environmental factors or stresses. This dynamic pair is shaping the ludic paradigm of manufacturing, showing a picture into tomorrow's world with products and processes rather copied in the digital reality and made better, faster, and intelligent by way of AI.

2.4 LITERATURE SURVEY

In the cutthroat realm of modern business, businesses face various challenges. They propose an intelligent production scheduling system that will seamlessly integrate with the enterprise resource planning (ERP) systems large corporations employ. However, certain issues with current ERP systems necessitate resolution [8]. This paper primarily focuses on the development of the DADO machine, a parallel computer featuring tree-like architecture. It also introduces the TREAT match algorithm commonly employed in running production systems. The study reveals that this approach consistently outshines its counterpart, the conventional RETE match algorithm, not only in terms of effectiveness but also when operated on single threaded computers [9]. This study investigates the relation between Industry 4.0 and back-sourcing by returning production to high cost overseas locations. It concludes that reshoring companies usually continue to restrict their adoption of advanced technologies only as an approach where price is still a consideration. Companies reshore after giving priority to quality over value and investing in new technologies, particularly for the latest innovations of their products [10]. This paper concentrates on the links between Industry 4.0, sustainable manufacturing, and circular economy systematic model. It provides a research framework that includes these concepts in an approach of supply chain management so as to understand the relation between institutional forces, concrete resources, and human capital for adoption within Industry 4.0 result in capacities advantages on sustainable manufacturing capacity and circular economy capacity among others [11]. In this paper, analysis of various approaches and strategies used in formulating virtual twins with particular emphasis on perspective modeling are provided. It involves the revolutionary aspect on how digital twins contribute to cyber-bodily sensate systems development and operations, address real challenging scenarios, and undermine enablers supporting technologies. It also looks at current trends in this domain [12]. This study conducts a comprehensive analysis of the development and perception level from DT in business including critical elements, current reputation, and major packages as well as describing challenging situations alongside future perspectives for sector [13].To enhance the management of predictive monitoring and independence in production methods, a statistics directed approach based on digital twin skills is proposed for smart production systems from this paper. This is achieved by applying deep reinforcement learning for the dynamic scheduling and its use of a virtual engine having interfaces between both physical backgrounds [14]. The study discusses the usefulness of digital twin-as-a-service (DTaaS) models for modeling and forecasting commercial approaches within the context of Industry 4.0 and cloud computing [15].Using the 5G connectivity, this study uses an edge-based digital twin for robotics application, enough automation offload control, and computer enhance Industry 4.0 deployment [16]. Considering that its components are compute, manipulation and communication it is presented as the reference version of C2PS in this article for cloud-based cyber physical systems (CPS) virtual dual structure. It also presents the use of cars applying a supporting application [17].The subject considers the impact of aspect, fog, and cloud computing on boosting intelligent falls by using combination sign (DT) with cyber-physical plant equipment via factor plane, device level, and domain degree items stages [18]. It offers a virtual twin-driven collaborative data management approach for metal additive manufacturing, as this framework will allow for significant control, process management, and sophisticated analytics during the product life cycle [19]. This gives a two-stage DTL (deep transfer learning) based on the digital twin assisted fault diagnosis, which allows real-time manufacturing phase monitoring and specific problem pinpointing later in development [20]. The challenge has numerous aspects, including the subject of utility and hierarchy, fieldwork type, field size, universality functionality; conducts an extensive review of recent works related to permission technology and techniques for digital twin modeling [21]. In addition to the historical development of big data in production statistics as well as presenting a theoretical framework for its use (underlining practical strategies using value added by big data applications towards more effective competition), the paper explores changes resulting from creating intelligent manufacturing associated with revolutionary functions of advanced technologies. The case studies, with illustrations, are used for evaluation of massive records in support of various production elements including monitoring, modeling, and optimization [22]. The research considers various neural network methods and evolutionary computing to make semiconductor manufacturing more rational. It focuses on high-speed long-term record keeping, clever automation, and displays an advanced feature choice algorithm to enhance the predictive technology offered at industrial procedures [23]. This paper proposes a facts driven resource structure that makes use of tactix methods as well as cyber-physical systems in the smart production area and also to enhance decal field by providing an easy accessibility of information for previous planning decision calling. The aim behind establishing Industry 4.0 is to allow the processes of production within a decentralized format and autonomy involving the combination between physicalities and logic systems responsible for various segments needed in the production process [24]. By using a case study of the mechatronic robot, the research provides a virtual dual model for systems that are very difficult. This model includes both a typical form representing domain-specific partial designs to allow for the handling of interactions and an increase in fee at any stage of life [25]. This information can then be used to enhance methodologies and inform advanced modeling techniques. It examines ways in which it will issue out these methods that include identifying structural responses, preventative maintenance, and strengthening initiatives using a case study of the Milan Cathedral [26].This paper considers that the physics-based-simulation models and digital twin will enable predictive preservation to manufacturing assets by determining the remaining useful life (RUL) for mechanical devices. The approach is general, cautious, and observes all the records of real-time devices accompanied by physic based models grafted with simulation adjustments so as to accurately estimate RUL. It addresses problems with predicting the initial issues, adaptability and considering the physics aspect in general [27]. In this paper, many authors made a comprehensive survey of existing literature concerned with the implementation of predictive remodeling using digital twins as a solution for one problem – there are insufficient failure records in traditional structures. The main contribution of the review is to the software engineering method for predictive maintenance enabled by digital twins, which synthesizes 42 articles examining different targets, utility domains, digital twin structures, illustration kinds, tactics, abstraction tiers, design styles, communication protocols, twinning parameters, and challenges [28]. This paper discusses the use of DT and AR technology in a collaborative approach used for development of business solutions where detection is achieved using predictive methods. It promotes Industry 4.0 and focuses on how the solution is used for the purpose of business [29] .

2.5 AI IN SMART MANUFACTURING

The concept of smart manufacturing has been transformed through the introduction and widespread usage of Artificial Intelligence (AI) which not only provides uncomparable levels productivity, resilience, and creativity but also dramatically enhances industrial durability. The main applications of AI in smart manufacturing along with a focus on computer vision are robotics and automation, natural language processing (NLP), and machine learning (ML).

By applying an extended TOE framework, as well as of the integrated TAM-TOE version, this study investigates how environmental factors influence people to adopt digital manufacturing technologies originating from Industry 4.0 along with socio-technicalities that affect such adoption. The avenue taken in a bid to ascertain relationships and moderation outcomes is reading organizational readiness, compatibility, and partner guide [30].

2.5.1 Machine Learning Applications

The basis of artificial intelligence (AI) in smart manufacturing is machine learning which provides a diversity of packages that once and for all retune methods to fabricate optimally. Through ML algorithms, learning patterns in the data are done and predictions made to assist the human users to decide after taking informed steps.

Addressing the predictive maintenance domain, machine learning algorithms go through several sets of historical data from sensor devices in predicting when the equipment will collapse. This preventive maintenance saves a lot in terms of cost, reduces downtime, and preserves precious manufacturing time. Understandably, ML is critical for demand forecasting due to the fact that it ensures the production satisfies market demands in a timely manner, optimizes resource utilization rate and has little spare inventory. Detail goes beyond the product quality enhancement, reducing waste and ensuring that only products of high standard become marketable. With machine learning on the rise, intelligent packages in smart manufacturing as a part of the future are becoming a reality where production techniques are automated but at the same time increasingly intelligent and self-optimizing [31–36].

2.5.2 Robotics and Automation in Manufacturing

AI powered automation and robotics are modifying the manufacturing landscape by improving flexibility and performance. Robots can perform a wide range of jobs ranging from intricate assembly procedures to work amongst humans when instructed by AI algorithms.

Robots that have AI capabilities work independently since they do not need an external force to prompt them to do work. They are also able to learn from their environment. This method helps the supply chain industry, including delivery chain control and logistics. This leads to a more cohesive and unified device that helps to streamline processes, reduces errors, and fastens the production life cycle.

This isn't about replacing human work with robots. It is more about optimizing the production process by enhancing human capabilities and also creating a more adaptable and productive environment in production. Smart manufacturing can be positioned as a hub for innovations.

2.5.3 NLP for Smart Factories

By emulating human understanding and perception of language, natural language processing (NLP) makes intelligent production communicate between man as well as machine in a humanlike manner. NLP is used in a smart factory environment to substitute human decision-making, accelerate process, and to facilitate intuitive interfaces.

It is possible to say commands out loud or write them using an input device. They are understood and responded to by the NLP frameworks, which appear more efficient and natural. Such automatization is especially useful in the complicated production scenarios and it requires absolute precision of changes and errorless communication. Utilization of natural language instructions empowers workers to interact with computers, develop queries on systems, and understand the outputs due to NLP-aided interfaces.

Moreover, NLP allows for the extraction of information from informal information sources among other unstructured materials like manuals and reviews maintenance logs. This information can be then used to enhance methodologies and inform advanced modeling techniques. Including NLP in the intelligent production will create a friendly environment requiring less or no machinery, integrating human skills and AI architecture without interruptions.

2.5.4 Computer Vision in Production Environments

The term "computer vision" refers to the technologies utilized in robotics and automation that allow machines to function by commands such as seeing a path as part of navigating its environment. Human-robot teams can complement each other's strengths because robots that are armed with computer vision systems can work together with individuals; these technologies also cope well with the changing environments and feature an incomparably high degree of accuracy. The use of this software promotes flexibility and safety in the intelligent manufacturing processes.

Creativity is dependent upon machines understanding and interpreting visual data, and this knowledge base of a subset of artificial intelligence named computer vision gives them that ability. Production environments refer to the use of computer vision systems that utilize video imaging through cameras and sensors in order to capture, analyze, and quantify visual statistics. This is an opportunity for the development of different applications that increase effectiveness and help with aesthetic development.

Quality management, of which structures are made aware of said flaws or inconsistent patterns or any deflections from set standards, is implemented almost wholly by computer vision. This minimizes waste and increases the overall level of quality in terms of products that are good by ensuring that easiest assembling tasks are applied and the highest standards maintained under all conditions as they move through the production line.

Additionally, thanks to its ability to deliver up-to-date information on production activities, computer vision contributes to the method of optimization. It can follow the movement of material, highlight the performance of equipment, and find bottlenecks; that means it allows it to orientate on record-driven decisions and continuous improvement.

2.6 KEY APPLICATIONS OF AI IN PRODUCTION

2.6.1 Predictive Maintenance Using AI

Production system management is being revolutionized by AI enabled predictive maintenance. Conventional renovation procedures are reactive in nature and can result in delays, increased costs, and inefficiencies. This dynamic is changed by artificial intelligence which introduces predictive preservation models that can evaluate sensors and machine data to predict when a system is going to malfunction. Artificial intelligence algorithms look at both historical and modern data to find patterns that can point to cognitive issues. Because of their ability to plan ahead, maintenance teams can minimize downtime and the likelihood of catastrophic errors by precisely scheduling interventions when needed. In AI driven production environments predictive remodeling is critical because it maximizes maintenance costs and prolongs equipment life.

2.6.2 Quality Control and Assurance

AI plays a vital part in the improvement of outstanding fine control and validation approaches used during the manufacturing process. Traditional fine control strategies are effective but are also time consuming and may lead to human errors. AI in computer vision and machine learning help to enhance the precision and efficiency of management techniques. Computer vision systems that are equipped with AI are able to visually inspect products in order to identify flaws, irregularities, or deviations from fixed standards. This helps to guarantee the production of high quality goods that fulfill strict standards, minimizing waste production, etc. The ability of AI models to continuously evolve over time enables flexibility, thus enhancing precision over time and adapting to the constant changes in production requirements.

2.6.3 Supply Chain Optimization

The modern supply chain is flexible and complex, which means that intelligent solutions should be created to optimize the efficiency of organizational action and quick responsiveness. Delivery chain optimization followed by AI applies contemporary real life records, processing algorithm process predictive modeling, and state of the art information analytics to improve decision making faster speeds. Using artificial intelligence, the system analyzes and reviews historical information related to customer demands and current trends in the industry. Thus, this allows them to have accurate expectations of what is needed and by when it will be required. This leads to an increase in production foresight that enables them to employ resources, regulate production timelines and effectively manage the levels of inventory. The ultimate outcome is a supply chain that is responsive to fluctuations in demand from consumers, cost-effective, and yet retains a high level of customer satisfaction.

2.6.4 Process Automation and Efficiency

An essential function in manufacturing settings is product automation, which can be made through AI innovation. That involves the industrial processes being driven by AI based automated units in adding more than just simple rule guided functions to these processes.

The machine learning algorithms allow structures to process data from records and streamline the ways of manufacturing. This may entail increasing basic operating efficiency, robotizing of monotonous tasks, and real-time adjustments in most cases of production parameters. The adoption of AI-based automated cycle also enables real-time changes to accommodate demand preferences, keeping the production sequences somewhat adapted and eco-friendly.

2.7 LAYERED COMPONENTS OF THE DIGITAL TWIN ECOSYSTEM

Figure 2.1 shows the layers of the digital twin ecosystem.

- **Digital twin environment (core layer)**: the base layer generally provides a solid start, mostly in 3D, with a simulation engine producing a virtual model that could offer an excellent kick off. This ensures that the underlying model, and hence what the underlying instance reflects, is a well-represented simulation of the real system. The connection interface provides for efficient communication with other parts of this unit.
- **Data sources layer**: the entry layer receives real-time data from sensors, edge devices, IoT devices, and data streams. It is the interface between the physical and digital worlds.
- **Data processing layer**: this layer uses high-tech tools such as machine learning algorithms, data analysis software, interpretation systems, and components that pull out certain aspects. Finding significant trends and insights from the incoming data is its responsibility.
- **Control layer**: the control layer interfaces with the physical world through actuators, automated systems, engines, and robotics. It performs operations by utilizing processed data to modify the behavior of real-world systems.
- **Intelligent layer**: at the very top, artificial intelligence, or AI, makes all the decisions. This area includes sophisticated decision-making algorithms, intelligent cognitive computing components, and future-predictive technologies. This level builds an extremely sophisticated and adaptable network by utilizing data from the levels below to handle the big-picture decisions.

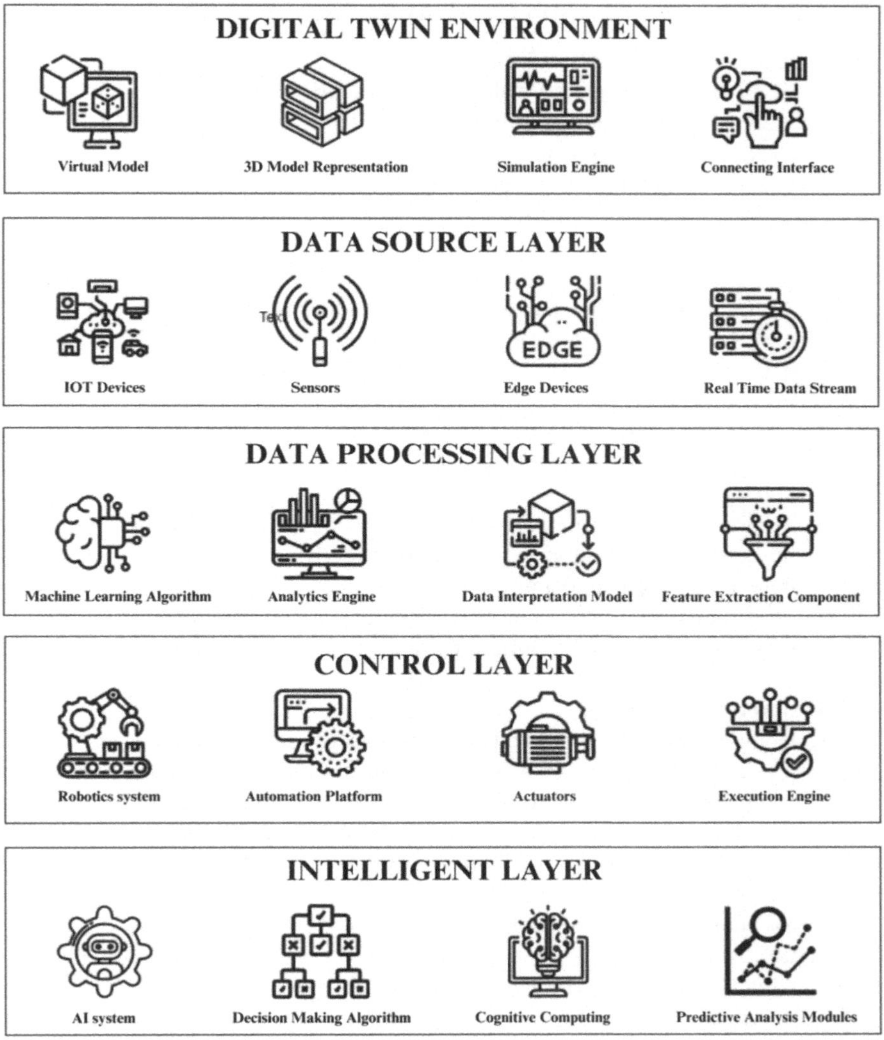

FIGURE 2.1 Digital Twin Ecosystem Architecture.

By layering these different structures on top of one another it becomes possible to represent the transition between them as an integrated and interdependent ecosystem wherein data streams continuously from the physical environment into more refined processing and analytical treatment, ultimately leading to intelligent decision making and control action. The digital twin technology acts as the backbone that gives perfect consistency and precision for an accurate physical counterpart at all times.

2.8 REAL-WORLD EXAMPLES OF AI IN PRODUCTION

- **Healthcare: IBM Watson Health**
  - To enhance their operational performance, IBM has integrated AI technologies into their manufacturing systems. Their AI-driven solutions encompass computer vision for optimal process control, predictive maintenance to ensure machinery reliability, and strategies to boost efficiency and profitability in intelligent factories.
- **Manufacturing: Siemens**
  - In order to achieve the best results in their operations, Siemens has adopted AI technologies that have been incorporated in manufacturing. Their AI powered solution contains computer image based Optimal Control, the machine Predictive maintenance, and smart factories' maximum efficiency profitability.
- **Finance: JPMorgan Chase**
  - Leveraging AI extensively, JPMorgan Chase employs advanced algorithms for fraud detection, risk management, and customer support. Their AI systems meticulously analyze vast financial data – a process that enhances transaction security by identifying potential fraud through detection of unusual patterns. This proactive approach guarantees not only robust protection but also efficient customer service. It is a crucial strategy in an industry where the landscape of financial transactions constantly evolves.
- **Technology: Google**
  - Google extensively integrates AI across its entire suite of services and products. In the search engine realm specifically, Google leverages sophisticated AI algorithms to heighten search result quality. Furthermore, they harness artificial intelligence in applications like Google Photos for impeccable photo recognition and Google Translate for seamless language translation – even within the multifaceted capacities of their virtual assistant: Google Assistant.
- **Retail: Amazon**
  - Amazon applies AI in several aspects of its operation, majorly as a way of providing customers with recommendations when making purchases. The AI algorithms are built using data on consumer behavior and preferences, with which products to make suggestions. AI also plays a role in improving logistics and plans to manage the supply chain.
- **Automotive: Tesla**
  - AI in more advanced intervention such as Tesla's Autopilot provides an impressive example of its application in the automotive industry. Combining a variety of sensors, cameras, and other AI algorithms in Tesla cars indicates the first signs of the potential activity of artificial intelligence that leads to semi-autonomous driving and navigation features.
- **Social media: Facebook**
  - Machine learning is applied by Facebook to predict users' recommendations, filter and control content as well as deliver customized

contents. AI algorithms make decisions regarding what data is acceptable and unacceptable, they also base some of the decision-making process on user reaction or preference; content selected to appear on users' timeline depends automatically upon not only relevance but also viewer acceptance.

2.9 EMERGING TRENDS IN AI FOR MANUFACTURING

- **Edge computing integration:**
  - ***Trend:*** the integration of AI and edge computing is developing at a much faster pace. By doing so it reduces the latency as it specializes in processing data on the network's surface, thus enhancing real-time decision making process in smart manufacturing.
  - ***Impact:*** edge computing is preferred in a place where brief judgments are crucial. It is perfect for places where decisions must be taken immediately, like predictive renovation.
- **Explainable AI (XAI):**
  - ***Trend:*** XAI is a bunch of tools and frameworks that help us to have an idea on what kind of outcomes can be expected from the predictions made by the machine learning model that was developed by the manufacturer. The more complex a structure becomes, the greater the demand for XAI.
  - ***Impact:*** explainable AI is very important in areas where a small wrong decision can impact the entire production process severely. Since AI are able to detect and rely on a judgment at a higher level, this will increase the adaptability of this amongst manufacturers.
- **AI-driven supply chain resilience:**
  - ***Trend:*** supply chain consists of the entire process from collecting raw materials up to the entire product being developed. By bringing AI into this process, manufacturers can improve their supply chain resilience. AI algorithms can be used to predict and avoid any disruptions that may occur. This also helps in improving the traditional supply chain system.
  - ***Impact:*** through the use of AI to give real-time visibility and a flexible selection making of their supply chains, manufacturers may also respond in a more positive way to unforeseen events such as a pandemic that may happen.
- **AI for sustainable manufacturing:**
  - ***Trend:*** sustainable manufacturing refers to the process of creating products in a much more eco-friendly way so that the impact on the environment is minimal. The integration of AI in this process is gaining global popularity. One of the few trends that are seen are waste reduction, energy efficient manufacturing, resource optimization, etc.
  - ***Impact:*** AI-driven sustainability initiatives help support the eco-friendly production methods and lead global initiatives into advanced sustainability and an eco-friendly production process.
- **Robotic process automation (RPA) advancements:**
  - ***Trend:*** robots are now able to handle difficult tasks on their own, leading to improvement in production and performance across industries.

  - ***Impact:*** due to RPA's qualities, robotic systems can perform tasks with a much higher accuracy in production. This also makes it more flexible and adaptable.
- **Generative design using AI:**
  - ***Trend:*** by using AI and machine learning algorithms, generative layout is gaining more popularity in the manufacturing sector. An AI driven generative layout techniques help to improve product design by offering innovative approaches and testing a variety of layout options.
  - ***Impact:*** AI powered generative layout increases the productivity and creativeness of the product design methodologies, thus leading to the development of advanced, flawless and ideal designs.
- **Digital twin enhancements:**
  - ***Trend:*** evolving landscapes in manufacturing are a result of lasting traits of the digital twin era that are made possible via AI integration. AI fastens the analytical capacities of the digital twin by considering intricate simulations and enhancements.
  - ***Impact:*** better understanding of the production tactics and strategies that are applied by using AI's integration in digital twin to allow for much better choices and better performances.

2.10 CONCLUSION

When AI and digital twins are combined, they have the potential to revolutionize and revamp the manufacturing sector by optimizing the production process and by also making more informed decisions. This also ensures real-time adaptability in case of a sudden change.

Online tracking, analyzations, and decision making is possible due to this very integration which helps us to get a more in depth perspective of the production environment. This dynamic and broad view of manufacturing is possible by the integration of digital twin technology, which encourages digital replicas, predictive analysis, and also promotes operational visibility. All of the algorithms involved help in recognising patterns that help in making predictions. As we saw in Section 2.9, only a few sectors have successfully started to adopt the use of smart manufacturing in their daily process. Thus, the combination of AI and digital twins provide exhaustive feedback in each of these situations.

REFERENCES

1. Allam, Zaheer, and Zaynah A. Dhunny. "On big data, artificial intelligence and smart cities." *Cities* 89 (2019): 80–91.
2. Groshev, Milan, et al. "Toward intelligent cyber-physical systems: Digital twin meets artificial intelligence." *IEEE Communications Magazine* 59.8 (2021): 14–20.
3. Metaxiotis, Kostas S., John E. Psarras, and Kostas A. Ergazakis. "Production scheduling in ERP systems: An AI-based approach to face the gap." *Business Process Management Journal* 9.2 (2003): 221–247.

4. Miranker, Daniel Paul. *TREAT: A new and efficient match algorithm for AI production systems*. Columbia University, 1987.
5. Cioffi, Raffaele, et al. "Artificial intelligence and machine learning applications in smart production: Progress, trends, and directions." *Sustainability* 12.2 (2020): 492.
6. Werner, Andreas, Nikolas Zimmermann, and Joachim Lentes. "Approach for a holistic predictive maintenance strategy by incorporating a digital twin." *Procedia Manufacturing* 39 (2019): 1743–1751.
7. Barthelmey, André, et al. "Dynamic digital twin for predictive maintenance in flexible production systems." IECON 2019-45th Annual Conference of the IEEE Industrial Electronics Society. Vol. 1. IEEE, 2019.
8. Bag, Surajit, et al. "Role of institutional pressures and resources in the adoption of big data analytics powered artificial intelligence, sustainable manufacturing practices and circular economy capabilities." *Technological Forecasting and Social Change* 163 (2021): 120420.
9. Riedl, Mark Owen, and Alexander Zook. "AI for game production." 2013 IEEE Conference on Computational Inteligence in Games (CIG). IEEE, 2013.
10. Ancarani, Alessandro, and Carmela Di Mauro. "Reshoring and Industry 4.0: How often do they go together?." *IEEE Engineering Management Review* 46.2 (2018): 87–96.
11. Bag, Surajit, and Jan Harm Christiaan Pretorius. "Relationships between industry 4.0, sustainable manufacturing and circular economy: Proposal of a research framework." *International Journal of Organizational Analysis* 30.4 (2022): 864–898.
12. Rasheed, Adil, Omer San, and Trond Kvamsdal. "Digital twin: Values, challenges and enablers from a modeling perspective." *IEEE Access* 8 (2020): 21980–22012.
13. Tao, Fei, et al. "Digital twin in industry: State-of-the-art." *IEEE Transactions on Industrial Informatics* 15.4 (2018): 2405–2415.
14. Xia, Kaishu, et al. "A digital twin to train deep reinforcement learning agent for smart manufacturing plants: Environment, interfaces and intelligence." *Journal of Manufacturing Systems* 58 (2021): 210–230.
15. Borodulin, Kirill, et al. "Towards digital twins cloud platform: Microservices and computational workflows to rule a smart factory." Proceedings of the10th International Conference on Utility and Cloud Computing. 2017.
16. Girletti, Luigi, et al. "An intelligent edge-based digital twin for robotics." 2020 IEEE Globecom Workshops (GC Wkshps. IEEE, 2020).
17. Alam, Kazi Masudul, and Abdulmotaleb El Saddik. "C2PS: A digital twin architecture reference model for the cloud-based cyber-physical systems." *IEEE Access* 5 (2017): 2050–2062.
18. Qi, Qinglin, et al. "Modeling of cyber-physical systems and digital twin based on edge computing, fog computing and cloud computing towards smart manufacturing." *International Manufacturing Science and Engineering Conference*. Vol. 51357. American Society of Mechanical Engineers, 2018.
19. Liu, Chao, et al. "Digital twin-enabled collaborative data management for metal additive manufacturing systems." *Journal of Manufacturing Systems* 62 (2022): 857–874.
20. Xu, Yan, et al. "A digital-twin-assisted fault diagnosis using deep transfer learning." *IEEE Access* 7 (2019): 19990–19999.
21. Tao, Fei, et al. "Digital twin modeling." *Journal of Manufacturing Systems* 64 (2022): 372–389.
22. Tao, Fei, et al. "Data-driven smart manufacturing." *Journal of Manufacturing Systems* 48 (2018): 157–169.

23. Ghahramani, Mohammadhossein, et al. "AI-based modeling and data-driven evaluation for smart manufacturing processes." *IEEE/CAA Journal of Automatica Sinica* 7.4 (2020): 1026–1037.
24. Rossit, Daniel Alejandro, Fernando Tohmé, and Mariano Frutos. "A data-driven scheduling approach to smart manufacturing." *Journal of Industrial Information Integration* 15 (2019): 69–79.
25. Erkoyuncu, John Ahmet, Peter Butala, and Rajkumar Roy. "Digital twins: Understanding the added value of integrated models for through-life engineering services." *Procedia Manufacturing* 16 (2018): 139–146.
26. Angjeliu, Grigor, Dario Coronelli, and Giuliana Cardani. "Development of the simulation model for Digital Twin applications in historical masonry buildings: The integration between numerical and experimental reality." *Computers & Structures* 238 (2020): 106282.
27. Aivaliotis, Panagiotis, Konstantinos Georgoulias, and George Chryssolouris. "The use of digital twin for predictive maintenance in manufacturing." *International Journal of Computer Integrated Manufacturing* 32.11 (2019): 1067–1080.
28. van Dinter, Raymon, Bedir Tekinerdogan, and Cagatay Catal. "Predictive maintenance using digital twins: A systematic literature review." *Information and Software Technology* 151 (2022): 107008.
29. Rabah, Souad, et al. "Towards improving the future of manufacturing through digital twin and augmented reality technologies." *Procedia Manufacturing* 17 (2018): 460–467.
30. Chatterjee, Sheshadri, et al. "Understanding AI adoption in manufacturing and production firms using an integrated TAM-TOE model." *Technological Forecasting and Social Change* 170 (2021): 120880.
31. Nalayini, C. M., Katiravan, J., Geetha, S., Eunaicy, J. I. C. "A novel dual optimized IDS to detect DDoS attack in SDN using using Hyper Tuned RFE and deep grid network." *Cyber Security and Applications*, Elsevier Vol. 2, 2024. https://doi.org/10.1016/j.csa.2024.100042.'
32. Nalayini, C. M., Katiravan, J., Sathya, V. "Intrusion detection in cyber physical systems using multichain, Malware analysis and intrusion detection in cyber-physical systems." *IGI Global Platform.* https://doi.org/10.4018/978-1-6684-8666-5.ch009, June 2023.
33. Nalayini, C. M., Katiravan Jeevaa. "A new IDS for detecting DDoS attacks in wireless networks using spotted Hyena optimization and fuzzy temporal CNN." *Journal of Internet Technology* 24.1 (2023).
34. Nalayini, C. M., Jeevaa Katiravan. "Detection of DDoS attack using machine learning algorithms." *Journal of Emerging Technologies and Innovative Research* 9.7 (2022): f223–f232.
35. Nalayini, C. M., Katiravan, J., Sathyabama, A. R., Rajasuganya, P. V., Abirami, K. (2023). Identification and Detection of Credit Card Frauds Using CNN. In: Mishra, M., Kesswani, N., Brigui, I. (eds.) *Applications of Computational Intelligence in Management & Mathematics. ICCM 2022. Springer Proceedings in Mathematics & Statistics*, Vol. 417. Springer, Cham. https://doi.org/10.1007/978-3-031-25194-8_22.
36. Nalayini C. M., Gayathri, T. (2022). A Comparative Analysis of Standard Classifiers with CHDTC to Detect Credit Card Fraudulent Transactions. In: Sivasubramanian, A., Shastry (eds.) *Electrical Engineering*, Vol. 792. Springer, Singapore. https://doi.org/10.1007/978-981-16-4625-6_99.

3 Data Analytics and Artificial Intelligence for Predictive Maintenance in Manufacturing

M. Sivakumar, M. Maranco, and N. Krishnaraj

3.1 INTRODUCTION

Data analytics is a methodical process that involves analyzing, refining, converting, and modeling data in order to discover important insights, draw educated conclusions, and support decision-making. This methodology utilizes several methodologies and instruments to examine and decipher complex datasets. Conversely, artificial intelligence (AI) refers to the progress of computer systems that can do tasks typically linked to human intelligence. These tasks include acquiring knowledge, logical thinking, finding solutions to problems, understanding sensory information, recognizing speech, and comprehending language. AI systems are created to imitate or reproduce human cognitive abilities, and they can be categorized into two primary types: narrow or weak AI and general or strong AI. Data analytics and artificial intelligence (AI) are interconnected disciplines that frequently enhance one another, as data analytics serves as the fundamental basis for numerous AI applications. Predictive maintenance is an anticipatory maintenance approach that employs data analytics and machine learning to forecast the probable occurrence of equipment or machinery failure. This enables maintenance personnel to carry out timely repair procedures, preventing failures from happening. This technique differs from traditional or reactive maintenance, which involves doing maintenance activities based on a predetermined timetable or in response to a failure. Predictive maintenance is essential in the industrial sector as it enables firms to enhance their maintenance strategy, save downtime, and enhance overall equipment performance. Predictive maintenance is implemented in the manufacturing industry through many processes including condition monitoring, data collection and analysis, alerts and notifications, maintenance planning and optimization, root cause analysis, and integration with enterprise systems. Condition monitoring involves the installation of sensors on production equipment to monitor a range of characteristics, including vibration, temperature,

DOI: 10.1201/9781003480860-3

pressure, and other pertinent metrics. Furthermore, with the utilization of the internet of things (IoT), sensors and devices can be interconnected, enabling the instantaneous gathering of data and facilitating communication between equipment and centralized systems. Data collection and analysis involve the continual acquisition and storage of data from sensors and several other sources. The data encompasses details regarding the performance of the equipment, its operating conditions, and the historical records of maintenance. In addition, we employ sophisticated analytics techniques to examine extensive datasets and detect patterns, anomalies, and trends that could signify possible problems or deterioration in the condition of equipment. Within the context of alerts and notifications, automated alerts and notifications are dispatched to maintenance crews when the predictive algorithms identify potential faults. These alerts provide details regarding the anticipated malfunction, its level of seriousness, and suggested courses of action. Furthermore, by classifying alarms according to the immediacy and gravity of the anticipated malfunction, maintenance teams can effectively prioritize their interventions.

Maintenance planning and optimization use predictive methods to schedule maintenance tasks during scheduled periods of inactivity, hence reducing any negative effects on production schedules and by effectively distributing resources, such as manpower and spare components, according to the anticipated maintenance requirements of various equipment. Predictive maintenance systems often include features for doing root cause analysis, which aids maintenance teams in comprehending the fundamental factors that contribute to equipment deterioration or malfunctions. Integration with enterprise systems involves the integration of predictive maintenance systems with enterprise resource planning (ERP) systems to provide smooth communication between maintenance, production, and other business operations. Through the use of predictive maintenance in the manufacturing sector, organizations can achieve substantial advantages such as enhanced equipment dependability, decreased periods of inactivity, optimized maintenance expenses, and improved overall operational effectiveness. Adopting this proactive strategy enables firms to maintain their competitiveness in a rapidly changing and challenging industry. Data analytics and artificial intelligence (AI) have become powerful tools in the realm of predictive maintenance in manufacturing. This cutting-edge strategy utilizes state-of-the-art technologies to forecast equipment malfunctions, enabling proactive interventions and reducing unexpected periods of inactivity. Manufacturers can shift from conventional reactive maintenance tactics by utilizing real-time data from sensors and IoT devices, along with advanced machine learning algorithms. The amalgamation of data analytics and AI not only amplifies the dependability of equipment but also leads to substantial cost reductions and enhanced operational efficiency. Within this particular framework, we investigate the fundamental elements and advantages of utilizing data analytics and artificial intelligence for the purpose of predictive maintenance in the manufacturing industry.

3.2 RELATED WORKS

Lee et al. [1] emphasized the importance of maintenance in reducing disruptions during production and highlighted the necessity of implementing a cost-effective maintenance program. It is advisable to use predictive maintenance for important parts, and using big data in this context improves the visibility of the system and the speed of decision-making. The chapter introduces a framework for managing maintenance policies on a big data platform. It demonstrates the implementation of this framework in a semiconductor manufacturing plant that utilizes sensor monitoring. The utilization of artificial intelligence involves the classification of failure patterns and the anticipation of machine conditions. This approach emphasizes the current strategy that combines the power of AI and big data to improve maintenance optimization in semiconductor production. In their study, Lee et al. [2] focused on the prevalent problem of unexpected periods of inactivity in manufacturing machinery caused by the lack of a systematic maintenance strategy. It emphasizes the disadvantages of regular maintenance, including longer periods of non-operation and higher maintenance expenses. As Industry 4.0 emerges, there is an increasing emphasis on employing predictive maintenance (PdM) solutions to reduce downtime expenses and improve equipment accessibility. Predictive maintenance (PdM) is widely recognized as a crucial facilitator for sustainable manufacturing practices as it effectively extends the operational lives of components. The research presents AI-driven algorithms designed for predictive maintenance, with a specific focus on monitoring vital components such as cutting tools and spindle motors in machine tool systems. The paper introduces a modeling approach that relies on data, with a particular focus on its application in the analysis of tool wear and bearing failures. This study makes a valuable contribution to the development of predictive maintenance methods, utilizing artificial intelligence to enhance the efficiency of equipment and promote sustainability in the manufacturing industry.

Lawrence et al. [3] examined the scarcity of empirical data about the combination of artificial intelligence-powered big data analytics, predictive maintenance systems, and internet of things (IoT) in sustainable Industry 4.0 wireless networks. Utilizing data from reputable sources such as BDO, Capgemini, The Economist Intelligence Unit, EEF, McKinsey, PAC, PwC, and Vodafone, the report conducts thorough analyses and estimations to investigate the impact of sustainable manufacturing internet of things (IoT) on industrial facilities. The report highlights the use of smart devices to monitor data flows, integrate disruptive technologies in cyber-physical system-based smart factories, and employ artificial intelligence and deep learning for decision-making and process planning. Descriptive statistics are computed using the collected survey data, providing valuable insights into the capacity of these technologies to drive sustainability in Industry 4.0. Samatas et al. [4] investigated patterns in predictive maintenance by utilizing machine learning in the context of the fourth industrial revolution and improvements in the internet of things (IoT). Industries are progressively embracing artificial intelligence technologies to enhance the efficiency of industrial processes. The research analyzes predictive maintenance trends, taking into account the industries involved, the dominant artificial intelligence models (mostly machine learning), and the types of IoT sensors used. The production sector is the most prominent, accounting for 54.55% of all publications. Prominent artificial intelligence models comprise artificial neural networks (28.95%), support vector machine (18.42%), and random forest (14.47%). Furthermore, there are 12 distinct categories of sensors, with temperature and vibration sensors being the most prevalent, constituting 60.71% and 46.42% of the total usage, respectively. This study offers significant insights regarding the utilization of machine learning in the field of predictive maintenance across several sectors.

Çınar et al. [5] explored the incorporation of machine learning (ML) into artificial intelligence (AI) for predictive maintenance (PdM) in the context of Industry 4.0 (I4.0) and smart manufacturing. Due to the emergence of Industry 4.0 and the increased availability of data resulting from digital transformation, machine learning techniques are widely used to monitor the condition of industrial equipment. The primary focus is on automating the process of identifying and diagnosing faults in order to minimize periods of inactivity, enhance the efficiency of components, and extend their remaining lifespan. Predictive maintenance (PdM) is regarded as essential for the long-term viability of intelligent manufacturing within the framework of Industry 4.0 (I4.0), with machine learning (ML) methods attracting significant attention from researchers. The study provides a thorough examination, organizing research according to machine learning algorithms, machinery and equipment, data gathering devices, data classification, and significant contributions. The 2022 research undertaken by Cheng et al. [6] is a valuable resource that offers substantial insights and recommendations for future investigations in the emerging subject of predictive maintenance (PdM). Their systematic review focuses on the application of PdM (predictive maintenance) with visual assistance in the context of Industry 4.0. The assessment synthesizes existing research and emphasizes key areas of inadequate comprehension, particularly in sectors such as utilities, power generation, industry, and energy consumption. It underscores the need of doing comprehensive analysis on data acquired from sensors put on machines in manufacturing plants. The evaluation includes essential components such as anomaly detection, planning/scheduling, exploratory data analysis (EDA), and the incorporation of explainable artificial intelligence (XAI) into the visual analytics of PdM. The findings emphasize the extensive utilization of anomaly detection in research linked to predictive maintenance (PdM). However, the current body of literature does not have a complete framework that successfully integrates data-driven and knowledge-driven techniques in the predictive maintenance (PdM) field of the manufacturing business. Additional research is required to examine the integration of maintenance personnel's input in the subsequent phases of PdM architecture, as suggested by the evaluation. Before implementing PdM with minimal human intervention, it is necessary to thoroughly investigate and address the significant problems and limitations that exist.

In 2020, Rousopoulou et al. [7] introduced a cognitive analytics system that has the ability to learn independently and autonomously. This system was designed to tackle the difficulties associated with predictive maintenance in the context of Industry 4.0. Their main focus was to quickly detect anomalies in industrial data, specifically related to injection molding machinery. The researchers implemented a methodical strategy and employed a blend of supervised and unsupervised learning models to construct an ensemble prediction model. This model combines the results of multiple algorithms to get a consensus. The system's main benefit is its cognitive process, which involves a real-time self-retraining feature that is based on a unique double-oriented evaluation objective. This objective includes both components that rely on data analysis and components that are based on mathematical models. The program was implemented on a real-time monitoring system, providing the capacity to identify errors in incoming data streams. The system aims to assist operators of injection molding machines with maintenance tasks, showcasing the advancements that may be achieved by implementing artificial intelligence in the context of Industry 4.0. A study conducted in 2021 by Ayvaz et al. [8] aimed to create a data-driven predictive maintenance system specifically designed for manufacturing production lines. This system utilized real-time data from internet of things (IoT) sensors. The main objective of this system is to detect signals that suggest possible faults before they occur, utilizing a range of machine learning techniques. The study demonstrated the efficacy of the system in identifying signs of possible breakdowns and effectively avoiding specific production interruptions, using real-world industrial IoT data. Comparative evaluations of machine learning algorithms have shown that ensemble strategies such as random forest and boosting methods like XGBoost fared better than individual algorithms. The evaluations identified the most effective models, which were then included into the factory's manufacturing system.

Keleko et al. [9] explored the important field of industrial maintenance within the framework of Industry 4.0 (I4.0) and the progress made in Predictive Maintenance 4.0 (PdM4.0). The study employs bibliometric analysis to provide significant insights into the utilization of artificial intelligence (AI) in PdM4.0. The research uncovers the potential benefits of PdM4.0, including enhanced productivity, cost reduction, early identification of malfunctions, and prediction of equipment lifespan. The bibliometric study examines concepts, application areas, methods, and trends in the utilization of artificial intelligence for real-time predictive maintenance, utilizing technologies such as Biblioshiny, VOSviewer, and Power BI. The findings emphasize the primary factors, institutions, nations, and cooperative networks that play a significant role. There has been a notable increase in papers focusing on data-driven models, hybrid methods, and digital twin frameworks for prognosis, diagnostics, and anomaly detection in the context of Industry 4.0 and Predictive Maintenance 4.0. The analysis also examines potential obstacles such as data collecting, ethical issues, socio-economic effects, and transparency. It suggests a precise definition for reliable artificial intelligence in the context of the fourth industrial revolution. The authors Yu et al. [10] presented a thorough big data ecosystem specifically developed for identifying and diagnosing faults in predictive maintenance. This ecosystem was built to operate within the context of industrial internet of things (IoT)-based smart manufacturing in an Industry 4.0 system. The ecosystem leverages authentic industrial big data from extensive worldwide manufacturing plants and tackles issues pertaining to data acquisition, integration, conversion, retention, analysis, and display. The proposed architecture utilizes advanced technologies such as data lakes, NoSQL databases, Apache Spark, Apache Drill, Apache Hive, OPC Collector, and various techniques to achieve real-time processing. It also addresses data and network security concerns through transformation protocols, authentication, and data encryption methods. The implementation of fault identification and diagnosis utilizes a distributed principal component analysis (PCA) model based on MapReduce. The framework has been effectively incorporated into an operational industrial production system, offering advanced notifications of defects several days in advance. An in-depth examination of a specific instance where there were disruptions in 2014 showcases the efficacy of the suggested system in improving anticipatory maintenance for intelligent manufacturing in the Industry 4.0 era.

Bekar et al. [11] examined the changing field of predictive maintenance in manufacturing, with a specific emphasis on incorporating artificial intelligence methods for monitoring machine health and making predictions about future performance. The paper highlights the increasing reliance on data-driven machine learning algorithms and emphasizes the crucial importance of data pre-processing in achieving optimal generalization performance. The paper presents a sophisticated method that utilizes unsupervised machine learning techniques to preprocess and analyze data in predictive maintenance. The goal is to provide data that is of excellent quality and well-organized. The suggested approach's suitability is showcased through an industry case study in the manufacturing sector, where datasets are carefully examined to detect data quality problems and uncover concealed information in intriguing subsets. The devised methodology allows for the methodical retrieval of valuable and discerning data, serving as a basis for decision-making assistance and the creation of predictive maintenance prognostic models. Unal et al. [12] introduced the digital twin pipeline framework in the COGNITWIN project. This framework facilitates the creation of hybrid and cognitive digital twins. This framework comprises a sequence of pipeline procedures that incorporate advanced big data and artificial intelligence (AI) approaches, specifically tailored for digital twins. The pipeline comprises four primary components: data acquisition, data representation, AI/machine learning, and visualization and control. Following the BDV reference model, the technical selections for big data and AI give preference to a hybrid digital twin approach, which entails merging data-driven digital twins with first-order physical models. This study investigates the actual use of maintenance for spiral welded steel industrial machinery, with a particular focus on the crucial role played by the digital twin in enabling predictive maintenance. The ongoing architectural growth aims to incorporate cognitive digital twins, which include learning, understanding, and strategic abilities, as well as domain and human expertise. The pilot project seeks to reduce energy consumption and limit machine downtime through the implementation of data-driven artificial intelligence techniques and predictive analytics models. The digital twin pipeline depicted is considered appropriate for similar situations in the process industry.

Daniyan et al. [13] conducted a study that concentrated on creating training modules in learning factories. These modules were specifically developed to improve human capacity development by incorporating artificial intelligence (AI) systems. The training involves the utilization of an artificial neural network (ANN) integrated with a dynamic time series model. Its purpose is to educate maintenance staff on the process of monitoring and analyzing data from the internet of things (IoT) and other sources. The goal is to predict the condition and potential breakdown of railcar wheel bearings. The training modules encompass data gathering, pre-processing, network training, feature extraction, and predictive model modules. The demonstrations utilize historical wheel-bearing temperature data that has been pre-processed and trained repeatedly using the Levenberg Marquardt algorithm in

MATLAB 2018a. The results illustrate the AI's capability to diagnose the states of railcar wheel bearings, forecast the remaining useful life (RUL), and identify the most suitable maintenance intervals. Rosati et al. [14] developed and evaluated a decision support system (DSS) designed for predictive maintenance (PdM) in the specific domain of sophisticated processing and measuring devices. The DSS consists of data gathering, feature extraction, a predictive model, cloud storage, and data analysis. This approach differentiates itself from current machine learning models by placing a strong emphasis on feature extraction and machine learning prediction models that are fueled by specific topics collected from both the lower and upper levels of the production system. The experimental findings shown that the suggested method achieves a harmonious balance between prediction accuracy, computational complexity, and interpretability. By incorporating a cloud-based architecture, it becomes possible to receive immediate alerts regarding operational concerns, optimize maintenance schedules, and save service expenses by maximizing the amount of time a system is operating and improving efficiency.

Drakaki, et al. [15] conducted a comprehensive analysis of current developments in predictive maintenance (PdM) for induction motors (IM). The study specifically examined the use of multi-agent system (MAS) and deep learning (DL) techniques for the purpose of fault detection and diagnosis (FD/D). Predictive maintenance (PdM) utilizes artificial intelligence (AI), cyber-physical systems (CPS), and big data analytics as essential components of smart manufacturing and Industry 4.0. Instant messaging (IMs) plays a vital role in industrial environments, and neural network-based fault detection and diagnosis (FD/D) has been widely used. In recent times, DL approaches have surfaced, providing effective analysis of vast amounts of sensor data. This study investigates the utilization of MAS (multi-agent systems) and NNs (neural networks) as a decision support tool for incident management fault detection and diagnosis. The report examines current advancements, patterns, and delineates prospective avenues for further research in this field. In their study, Pech et al. [16] conducted a thorough examination of previous academic literature on predictive maintenance (PdM) and intelligent sensors in smart factories. Their analysis specifically concentrated on recent advancements in this field and identified future research challenges. The study utilized burst analysis, systematic review technique, and keyword co-occurrence analysis to investigate the growing number of publications that explore the topic of predictive maintenance (PdM) and Industry 4.0 technologies. The authors proposed the idea of smart and intelligent predictive maintenance (SIPM) after conducting a comprehensive analysis of pertinent literature. They emphasized the growing significance of PdM within the framework of Industry 4.0. The paper's contribution is its succinct synopsis and thorough examination of the most recent advancements in utilizing intelligent sensors for predictive maintenance in smart factories. In their study, Hosseinzadeh et al. [17] addressed the problem of reducing production interruptions caused by unexpected tool deterioration or inadequate workpiece quality in the manufacturing sector. They accomplished this by making precise predictions of machine failures and identifying potential troublesome scenarios. The study evaluates different techniques for detecting defects, including machine learning (ML), deep learning (DL), and deep hybrid learning (DHL). It utilizes a synthetic predictive maintenance dataset generated by the School of Engineering at the University of Applied Sciences in Berlin, Germany. The results indicate that the deep forest and gradient boosting algorithms achieved a significantly high degree of accuracy, above 90%.

Table 3.1 summarizes the focus, methodology and key findings/contributions for the related works above.

3.3 DATA ANALYTICS IN PREDICTIVE MAINTENANCE

Data analytics, under the framework of predictive maintenance, pertains to the utilization of statistical analysis, machine learning, and other data-driven methodologies to derive valuable information from the data obtained from industrial equipment, as seen in Figure 3.1. The objective is to forecast the occurrence of equipment failure in order to carry out preventative maintenance actions, hence minimizing downtime and optimizing maintenance expenses. Utilizing data analytics in predictive maintenance allows firms to transition from reactive or scheduled maintenance methods to a proactive and data-centric strategy. This leads to enhanced equipment reliability, decreased downtime, and minimized maintenance expenses.

3.3.1 Data Collection

Data analytics, under the realm of predictive maintenance, pertains to the utilization of statistical analysis, machine learning, and other data-driven methodologies to derive valuable information from the data gathered from industrial equipment, as depicted in Figure 3.1. The objective is to anticipate the occurrence of equipment failure in order to carry out proactive maintenance actions, hence minimizing downtime and optimizing maintenance expenses. By employing data analytics in predictive maintenance, companies can shift from reactive or scheduled maintenance approaches to a proactive and data-driven strategy. As a result, this leads to improved reliability of the equipment, reduced periods of inactivity, and decreased costs associated with maintenance. Social media and web scraping entail the retrieval of data from online platforms for the purpose of analyzing user behavior, sentiment, and trends. Transaction records, encompassing bank transactions and point-of-sale systems, offer useful data for study. Surveillance and imaging techniques, such as video surveillance and medical imaging devices like MRI or CT scans, provide valuable information for security, traffic monitoring, and healthcare analysis. user comments and reviews obtained from platforms such as Yelp or Amazon play a significant role in sentiment analysis and provide valuable information into user behavior. Organizations can utilize their existing databases and data warehouses, together with appropriate external databases, as important sources of information for analysis purposes. Government and public datasets, which are accessible through open data sources, offer publicly available datasets for doing specialized studies. Smart devices and wearables, such as fitness trackers, provide health and wellness information. Each technique addresses distinct data requirements, and the selection depends on the characteristics of the analysis and the desired insights pursued by the data analyst or researcher.

TABLE 3.1
Comparison of Related Works

Paper Title & Year	Focus	Methodology	Key Findings/Contributions
Lee et al. [1]	Maintenance policy, Predictive maintenance, Big Data, AI	Maintenance Policies Management framework, Application in semiconductor manufacturing	Integration of big data enhances system transparency and decision-making speed for optimized maintenance in semiconductor manufacturing.
Lee et al. [2]	Unplanned downtime, Predictive maintenance, Industry 4.0	AI-based algorithms for predictive maintenance, Focus on critical elements like cutting tools and spindle motors	AI-driven predictive maintenance strategies mitigate downtime costs and enhance equipment availability.
Lawrence et al. [3]	AI-driven big data analytics, IoT, Sustainable Industry 4.0 wireless networks	Descriptive statistics from compiled survey data	Sustainable manufacturing IoT in industrial plants through smart devices, AI, and deep learning.
Samatas et al. [4]	Predictive maintenance trends, Machine Learning, IoT	Analysis of predictive maintenance trends across industries	Insights into prevalent ML models, IoT sensors, and industries adopting predictive maintenance.
Çınar et al. [5]	Predictive maintenance, Machine Learning, Industry 4.0	Comprehensive review categorizing research based on ML algorithms, machinery, data acquisition, and contributions	Insights into recent trends and developments in ML-based predictive maintenance for Industry 4.0.
Cheng et al. [6]	Predictive maintenance with visual aids, Industry 4.0	Synthesis of evidence on PdM, identification of knowledge gaps	Lack of a comprehensive framework integrating data-driven and knowledge-driven techniques in PdM.
Rousopoulou et al. [7]	Cognitive analytics, Predictive maintenance, Industry 4.0	Cognitive mechanism with real-time self-retraining, Application on injection molding machines	Real-time self-retraining capability utilizing a dual-focused evaluation aim.
Ayvaz et al. [8]	Data-driven predictive maintenance, IoT sensors, Machine Learning	Successful identification of indicators of potential failures, Integration of best-performing ML models	Effectiveness of data-driven predictive maintenance in identifying potential failures and preventing production stops.

(*continued*)

TABLE 3.1 (Continued)
Comparison of Related Works

Paper Title & Year	Focus	Methodology	Key Findings/Contributions
Keleko et al. [9]	Predictive Maintenance 4.0, AI, Bibliometric analysis	Comprehensive bibliometric study on AI applied to PdM4.0	Insights into key contributors, institutes, countries, and collaborative networks in AI-based PdM4.0.
Yu et al. [10]	Fault detection and diagnosis, Predictive maintenance, Big Data, IoT	Implementation of a real-time industrial production system, MapReduce-based distributed PCA model	Successful integration of a fault detection and diagnosis system into a real-time industrial production system.
Bekar et al. [11]	Data pre-processing, Machine Learning, Predictive maintenance	Intelligent approach using unsupervised ML techniques for data pre-processing	Formulated approach for systematic extraction of useful information for decision support in predictive maintenance.
Unal et al. [12]	Digital Twin Pipeline Framework, Industry 4.0, Predictive maintenance	Framework for Hybrid and Cognitive Digital Twins, Application in industrial machinery maintenance	Integration of Digital Twin pipeline into the industry for predictive maintenance and downtime reduction.
Daniyan et al. [13]	Learning factories, Artificial Intelligence, IoT	Development of training modules for railcar wheel bearing prediction using AI	Feasibility of AI in predicting railcar wheel bearing conditions and optimal maintenance times.
Rosati et al. [14]	Decision Support System, Predictive maintenance, Cloud-based architecture	Development of a DSS for predictive maintenance, Integration into a cloud-based architecture	Real-time warnings about operational risks, optimization of maintenance schedules, and reduced service costs.
Drakaki, et al. [15]	Predictive Maintenance for Induction Motors, Multi-Agent System, Deep Learning	Review of recent trends in PdM for Induction Motors, Focus on MAS and DL	Insights into recent developments and trends in PdM for Induction Motors using MAS and DL.
Pech et al. [16]	Smart factories, Predictive maintenance, Intelligent sensors	Proposed Smart and Intelligent Predictive Maintenance	An examination of current patterns in intelligent sensors for the purpose of predictive maintenance in smart factories.
Hosseinzadeh et al. [17]	Machine failure prediction, Defect detection models, Machine Learning	Evaluation of defect detection models using synthetic predictive maintenance dataset	Deep Forest and Gradient Boosting algorithms achieved relatively high accuracy.

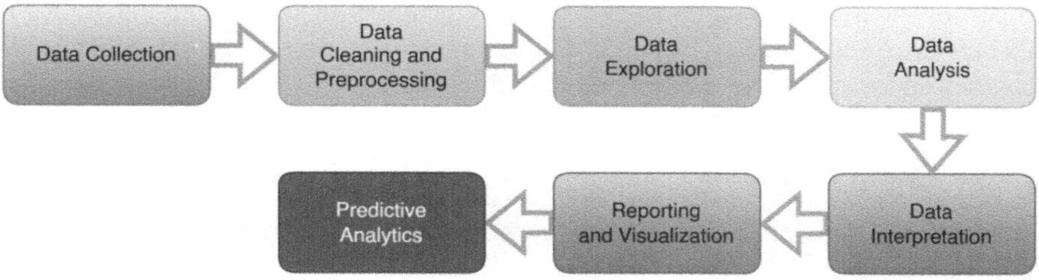

FIGURE 3.1 Data Analytics Process.

3.3.2 Data Cleaning and Preprocessing

Ensuring data quality is an essential step in the data preprocessing phase to provide precise and dependable analysis. Dealing with missing values is the process of detecting and locating them within the dataset, and subsequently employing imputation techniques such as mean, median, or mode replacement, depending on the characteristics of the data. Addressing duplicate entries necessitates the identification of such instances and subsequent determination of whether to eliminate or combine them, taking into account the accuracy and reliability of the data. To address outliers, one must first identify them using statistical methods or visualizations. Then, a decision must be made on whether to eliminate or change the outliers in order to reduce their impact on the analysis. Data transformation techniques like as normalization, log transformation, and Box-Cox transformation are used to standardize numerical data and address skewed distributions. Techniques such as one-hot encoding, label encoding, and the generation of dummy variables are used to handle categorical data.

Dealing with inconsistent data entails the process of standardizing formats, such as date and time, and cleansing text data by eliminating extraneous characters and rectifying typographical errors. Feature engineering encompasses methods such as generating novel features, dividing data into bins or discrete categories, and aggregating information to offer further insights and summarize data. Imbalanced datasets are mitigated by employing resampling strategies to equalize class distributions. In order to mitigate the influence of skewed distributions, log transformation is employed on skewed data. Data integration refers to the process of merging datasets from several sources, with the aim of ensuring that variables are aligned and consistent. Data splitting involves dividing the dataset into distinct training and testing sets to evaluate the model. In the case of time series data, it is necessary to ensure adequate time alignment in order to handle irregular time intervals. Data scaling is accomplished by employing methods such as Min-Max scaling and Z-score standardization to standardize numerical characteristics. To eliminate unnecessary or unrelated features, feature selection or dimensionality reduction approaches, such as principal component analysis (PCA), are employed to preserve the most pertinent information. Data validation is conducted by the utilization of cross-validation methodologies in order to guarantee the integrity of the data and evaluate the performance of the model. The approach of data cleaning and preprocessing is contingent upon the dataset's properties, analytical objectives, and the specifications of the selected analytics or machine learning models. Efficient data cleaning and preprocessing enhance the accuracy and resilience of studies and model results.

3.3.3 Data Exploration

Data exploration is an essential stage in the data analysis process, which employs many methodologies to acquire a deep understanding of the structure and attributes of a dataset. Descriptive statistics are a first stage in data analysis that calculate measures such as mean, median, mode, and standard deviation to identify numerical characteristics. Counts and frequencies are used to provide an overview of the distribution of categorical variables. Data visualization is crucial in the process of exploration, as it utilizes various tools such as histograms, box plots, bar charts, pie charts, and scatter plots to facilitate the comprehension of numerical and categorical data distributions. Correlation analysis, conducted using correlation matrices and scatter plots, uncovers the connections between numerical variables. Cross-tabulations, specifically contingency tables, provide valuable insights into the relationships between categorical variables. Distribution analysis utilizes methods such as kernel density plots and QQ plots to visually represent the probability density and evaluate conformity to theoretical distributions. Time series analysis, employing time plots and seasonal decomposition, enables the examination of temporal trends and patterns. Dimensionality reduction techniques, such as principal component analysis (PCA) and t-distributed stochastic neighbor embedding (t-SNE), assist in identifying significant features and visualizing data that have a large number of dimensions. Outlier detection techniques, such as Z-score and box plots, facilitate the discovery of possible anomalies in numerical data. Evaluating the significance of features using machine learning models facilitates the identification of influential features. Data profiling, which encompasses quality assessments for missing values, duplicates, and inconsistent data, guarantees the dependability of a dataset.

Tableau or Power BI tools enable the creation of dashboards that encourage interactive exploration, offering dynamic exploration experiences. Statistical tests, such as hypothesis testing, are used to confirm hypotheses or investigate disparities between groups. Segmentation analysis enables the examination of subsets by applying particular criteria to divide data into distinct segments. Geospatial analysis, employing maps, facilitates the visualization of geographical patterns and distributions. Furthermore, the application of natural language processing (NLP) methods, such as text analysis, enables the extraction of significant insights from written content, so improving the whole process of exploration and comprehension. Data exploration approaches facilitate a comprehensive understanding of the dataset's structure and intricacies, establishing the foundation for later analysis and decision-making. Data exploration is a repetitive procedure that entails integrating many methodologies in order to acquire a thorough comprehension of the dataset. It is a necessary preliminary step to more complex analyses and modeling endeavors, aiding analysts in making well-informed decisions regarding the upcoming stages of the data analysis workflow.

3.3.4 Data Analysis

Data analysis utilizes statistical approaches, machine learning algorithms, and other analytical techniques to extract significant insights from the data. The purpose of this stage is to address precise inquiries and reveal recurring patterns or trends. Predictive maintenance utilizes diverse data analysis techniques to extract valuable information from data and forecast the probable occurrence of equipment failure. These strategies utilize statistical methodologies, machine learning algorithms, and data-driven approaches to proactively detect possible problems. Predictive maintenance utilizes diverse data analysis techniques to improve machine dependability and minimize unexpected periods of inactivity. Descriptive analytics, such as summary statistics and time series analysis, are used to comprehend the average values and past patterns in equipment performance. The failure mode and effects analysis (FMEA) method employs root cause analysis to discover the underlying variables that contribute to equipment failures, hence offering valuable insights for the development of predictive models. The risk priority number (RPN), a crucial parameter in failure mode and effects analysis (FMEA), can be calculated using equation (3.1) as stated below:

$$\text{RPN} = \text{Severity (S)} \times \text{Occurrence (O)} \times \text{Detection (D)} \tag{3.1}$$

Here, S quantifies the seriousness of the impact of a failure mode, O represents the likelihood or probability of a failure mode occurring and D signifies the likelihood of detecting a failure mode before it reaches the end-user

Correlation and regression analyses evaluate the connections between variables, utilizing methods such as correlation matrices and regression models. Machine learning is crucial in leveraging both supervised and unsupervised learning approaches, such as decision trees, random forests, support vector machines, and clustering algorithms, to perform predictive modeling and anomaly detection. Survival analysis, encompassing reliability analysis and prognostics and health management (PHM), facilitates the evaluation of equipment dependability throughout its lifespan and the anticipation of its remaining usable life (RUL). Digital twins, which are part of the prognostics and health management (PHM) field, entail the creation of digital clones of physical assets in order to replicate and analyze their real-time behavior. Bayesian approaches, such as Bayesian networks, employ probabilistic modeling to capture relationships while accounting for uncertainty and iteratively refining predictions with the incorporation of new data. Feature engineering entails the process of carefully choosing and crafting meaningful features that can be used to accurately anticipate instances of equipment failures. Optimization techniques, such as maintenance scheduling optimization, employ algorithms to ascertain maintenance schedules that are cost-efficient.

Ensemble methods, such as bagging or boosting, improve the effectiveness of predictive models, whereas deep learning techniques like neural networks are capable of handling intricate data patterns and vast datasets. Probabilistic approaches for failure prediction provide both predictions and probabilities of equipment failures. Threshold-based monitoring utilizes alarm systems that are programmed with predetermined thresholds derived from previous data. These thresholds are used to detect and notify deviations in equipment behavior that are significant. To summarize, the wide range of data analysis techniques used in predictive maintenance helps with proactive equipment management and improves operating efficiency. The selection of data analysis techniques is contingent upon the characteristics of the data, the particular maintenance goals, and the resources at hand. Utilizing a combination of various techniques in a cohesive manner frequently results in more precise and resilient predictive maintenance solutions.

3.3.5 Data Interpretation

Converting the analytical findings into practical and effective insights. This requires comprehending the ramifications of the discoveries and making well-informed choices based on the examination. Data interpretation encompasses the process of comprehending the outcomes derived from data analysis and formulating significant inferences. This is a crucial stage in the data analysis process, where the analyst or decision-maker converts the statistical discoveries or trends into practical and implementable insights. Data interpretation is a methodical and complex procedure that is essential for obtaining significant insights from studied data. The analysis starts by ensuring a comprehensive understanding of the context and goals, highlighting the significance of aligning with business or research inquiries, and acknowledging the consequences of the discoveries. By examining descriptive statistics and conducting visual exploration, the analyst acquires valuable insights into the central tendencies, spreads, and interactions among variables. Comparing groups and evaluating statistical significance, using measures like as p-values and confidence intervals, enhances the rigor of the interpretation process. The ensuing procedures entail establishing connections between the data and the previously formed hypotheses, differentiating between correlation and causation, and taking into account the practical meaning that goes beyond statistical metrics. When utilizing predictive models, the assessment of confidence and accuracy involves the evaluation of metrics such as precision, recall, and the ROC curve, as outlined in equations (3.2) and (3.3).

$$\text{Precision} = \frac{\text{True Positives}}{\text{True Positives} + \text{False Positives}} \tag{3.2}$$

$$\text{Recall} = \frac{\text{True Positives}}{\text{True Positives} + \text{False Negatives}} \tag{3.3}$$

The procedure also involves verifying assumptions and exploring alternate explanations to ensure the strength and reliability of the results in different settings. Furthermore, data interpretation involves connecting the obtained insights to the overall business or research goals, evaluating their impact on decision-making, and predicting future consequences. Effective communication is crucial, requiring the delivery of explanations in a clear and actionable way to stakeholders, while avoiding superfluous technical language. The iterative nature of the process enables the confirmation and refinement of interpretations through the incorporation of additional insights or input, so ensuring a continual cycle of progress. In summary, this all-encompassing method of analyzing data intends to support well-informed decision-making and strategic planning. Effective data interpretation necessitates the integration of statistical acumen, specialized knowledge in a particular field, and the ability to think critically. The process entails combining the numerical results with qualitative understanding to obtain significant and practical insights.

3.3.6 Reporting and Visualization

The importance of reporting and visualization in predictive maintenance is vital for effectively communicating insights and discoveries to stakeholders. Concise and efficient reporting enables decision-makers to comprehend the predictive maintenance outcomes, make well-informed choices, and implement suitable measures. Efficient reporting and visualization in predictive maintenance necessitate the establishment of crucial indicators and KPIs that are in line with maintenance objectives, the selection of suitable visualization methods, and the development of interactive dashboards. Real-time monitoring screens provide immediate updates on the condition of equipment, notifications of any issues, and information about ongoing maintenance tasks, allowing for timely responses. The visualization of predictive maintenance forecasts and warnings can highlight the likelihood of failure, suggested actions, and levels of severity. Time series plots depict past equipment performance, facilitating the discovery of patterns and trends. Visualizations of failure analysis illustrate the underlying reasons and circumstances that lead to equipment breakdowns, providing valuable information for implementing preventive actions. Heatmaps facilitate the detection of anomalies by swiftly spotting atypical patterns that demand attention. Geospatial visualizations depict the distribution of equipment and maintenance, specifically for enterprises that have many sites. By visualizing the allocation of resources such as labor, spare parts, and maintenance schedules, it is possible to optimize resource consumption according to recommendations. Presenting performance measures of predictive models, such as accuracy and precision, guarantees that stakeholders comprehend the dependability of forecasts. Integrating interactive components enables users to delve into specific aspects, while explicit annotations and labels improve the comprehensibility. Ensuring uniform design and color schemes throughout visualizations establishes a coherent appearance, facilitating understanding and preventing ambiguity. By incorporating narrative context into visualizations, one can effectively elucidate results, emphasize important insights, and assist stakeholders in interpreting the information. The implementation of a feedback mechanism allows for the collection of input from stakeholders in order to continuously enhance the reporting and visualization processes to meet changing requirements. Efficient reporting and visualization enable stakeholders to make well-informed decisions, optimize maintenance plans, and improve overall operational efficiency in the context of predictive maintenance.

3.3.7 Predictive Analytics

Predictive analysis in the context of predictive maintenance is the utilization of statistical models and machine learning algorithms to anticipate equipment breakdowns or maintenance requirements by leveraging historical and real-time data. The objective is to preemptively detect problems before to their occurrence, therefore preventing unexpected periods of inactivity or expensive malfunctions. Predictive analysis in the context of predictive maintenance adheres to a systematic and sequential approach aimed at improving the accuracy of equipment health prediction and mitigating potential breakdowns. Commencing with the process of feature selection, the act of identifying relevant variables enhances the effectiveness and comprehensibility of the predictive model. Subsequently, data splitting entails dividing datasets into training and testing sets, enabling effective model training and evaluation on unseen data. The range of predictive models available includes regression models, time series models, machine learning models, and anomaly detection models, each designed to address unique problems. The training phase entails facilitating the model to acquire knowledge of patterns and connections within the data, subsequently fine-tuning the hyperparameters to achieve optimal performance. Ultimately, the model's accuracy, precision, recall, and other metrics are assessed through validation and evaluation using a testing dataset, which confirms its usefulness in real-world scenarios. This methodical approach guarantees a strong predictive maintenance model that is capable of providing precise equipment health projections and actively preventing failures. Predictive analysis in the context of predictive maintenance is a continuous procedure that encompasses the integration of subject understanding, data proficiency, and machine learning methodologies. The efficacy of the prediction models relies on the data's quality, the selected algorithms, and the ongoing refining informed by real-world performance feedback. Data analytics is extensively utilized across diverse sectors such as finance, healthcare, marketing, and technology to acquire a competitive advantage, enhance operational efficiency, and make well-informed judgments.

3.4 DATA ANALYTICS APPROACHES TO PREDICTIVE MAINTENANCE IN MANUFACTURING

Several data analytics approaches are employed for predictive maintenance in manufacturing, each with its strengths and applications.

3.4.1 Descriptive Analytics

Descriptive analytics plays a crucial role in predictive maintenance in manufacturing by providing a basis for well-informed decision-making through the analysis of historical data, thereby generating valuable insights. By employing techniques such as descriptive statistics, visualization approaches, and failure analysis, it assists in understanding the attributes of maintenance data, recognizing typical failure patterns, and determining the most important problems. Downtime analysis and equipment health monitoring entail the continuous evaluation of machinery conditions and the detection of patterns in downtime incidents. Root cause analysis is a method used to identify the fundamental causes of failures, whereas asset utilization analysis and performance benchmarking are used to assess efficiency and compare the performance of assets. Work order analysis examines past maintenance tasks in order to optimize maintenance schedules and improve overall operations. Descriptive analytics is an important initial step before implementing advanced predictive maintenance solutions. It allows firms to extract valuable insights from historical data in order to improve manufacturing operations.

3.4.2 Diagnostic Analytics

Diagnostic analytics is essential for predictive maintenance in the manufacturing industry. It entails a thorough analysis of historical and real-time data to identify the underlying reasons behind equipment breakdowns and performance problems. Organizations can identify the precise causes of mechanical breakdowns by utilizing different diagnostic approaches, including fault tree analysis, failure mode and effects analysis (FMEA), and pattern recognition. Diagnostic analytics aids in the detection of abnormalities and deviations from typical operational states, enabling maintenance crews to gain a comprehensive understanding of equipment malfunctions. It surpasses descriptive analytics by investigating the root causes of these errors, enabling the creation of more precise predictive models. Conducting this level of research is crucial for creating focused maintenance plans, minimizing downtime, and enhancing the overall dependability and effectiveness of industrial operations.

3.4.3 Predictive Analytics

Predictive analytics plays a crucial role in predictive maintenance within the industrial sector. It utilizes sophisticated statistical algorithms and machine learning approaches to anticipate the occurrence of equipment breakdowns. Predictive analytics use historical and real-time data to detect patterns and trends that occur before machinery failures. By adopting this proactive strategy, organizations can anticipate probable breakdowns and schedule maintenance tasks appropriately, so avoiding unexpected periods of inactivity and decreasing overall maintenance expenses. Predictive maintenance models employ regression analysis, time series forecasting, or advanced machine learning algorithms to utilize sensor data, equipment conditions, and performance metrics in order to accurately predict the timing of necessary attention for specific components. By utilizing data, manufacturers can transition from reactive to proactive maintenance procedures, resulting in improved operational efficiency and increased longevity of essential apparatus.

3.4.4 Prescriptive Analytics

Prescriptive analytics is essential in predictive maintenance in the manufacturing industry since it provides practical insights and recommendations based on the predictions made by predictive models. Predictive analytics anticipates possible equipment malfunctions, whereas prescriptive analytics takes an additional step by suggesting the most effective measures to address these problems. The system offers decision-makers valuable insights into the optimal maintenance strategies, advising on whether to implement preventive maintenance, replace components, or schedule repairs based on the projected probability of failure. Prescriptive analytics seeks to enhance decision-making processes by optimizing criteria such as cost-effectiveness, resource usage, and overall operational efficiency. By incorporating prescriptive analytics into predictive maintenance processes, manufacturers may proactively identify issues and promptly apply cost-effective remedies, reducing interruptions and optimizing the efficiency and dependability of their equipment.

3.4.5 Prognostic Analytics

Prognostic analytics is a crucial element of predictive maintenance in the manufacturing industry, with a specific focus on predicting the remaining usable life (RUL) of equipment and components. Predictive analytics identifies possible problems, while prognostic analytics goes farther by evaluating the remaining operational life of a certain component or system. Prognostic analytics models utilize both historical and real-time data to evaluate the present state of machinery and predict the probable timing of a failure. This data facilitates proactive scheduling for maintenance operations, guaranteeing that interventions are executed precisely at the necessary times, reducing downtime, and maximizing the longevity of assets. Prognostic analytics enables firms to shift from reactive or preventative maintenance procedures to more strategic and resource-efficient methods, leading to enhanced overall equipment effectiveness and operational excellence in the production setting.

3.4.6 Cognitive Analytics

Cognitive analytics is essential for enhancing predictive maintenance in the industrial sector through the integration of intelligent technologies that imitate human cognitive processes. This technique surpasses conventional analytics by incorporating artificial intelligence (AI) technology such as machine learning, natural language processing, and pattern identification. Cognitive analytics improves the ability to comprehend, acquire knowledge, and adjust to intricate and ever-changing manufacturing environments within the framework of predictive maintenance. These systems have the capability to evaluate extensive quantities of data from diverse sources, such as sensors and IoT devices, in order to detect patterns, anomalies, and potential failure modes. The cognitive part of the system allows for ongoing learning from new data, adjusting its models, and enhancing accuracy as time progresses. Cognitive analytics enhances manufacturers' decision-making process, maintenance plans, and overall manufacturing efficiency by offering in-depth insights about equipment health and prospective difficulties.

3.4.7 Digital Twin Technology

The utilization of digital twin technology is essential for the advancement of predictive maintenance in the manufacturing industry, as it enables the creation of virtual replicas of physical assets. These digital twins provide the continuous monitoring, analysis, and simulation of equipment performance, thereby connecting the physical and digital realms. Manufacturers can obtain valuable insights into the health of their physical assets, discover anomalies, and predict probable breakdowns by consistently collecting data from sensors and inputting it into the digital twin. This enables proactive decision-making, optimizing maintenance plans and minimizing downtime. Predictive modeling is improved by the capability to simulate different operational situations, which in turn promotes data-driven and intelligent manufacturing methods.

3.4.8 Deep Learning

The reviewed studies encompass different facets of predictive maintenance (PdM) in the manufacturing sector, with a focus on integrating artificial intelligence (AI), machine learning (ML), and data analytics to maximize equipment performance, minimize downtime, and improve overall operational efficiency. Notable topics encompass the utilization of predictive maintenance (PdM) in Industry 4.0, the significance of big data, internet of things (IoT), and cognitive analytics in identifying and diagnosing faults. Scientists investigate the difficulties, patterns, and future paths in predictive maintenance (PdM), emphasizing the importance of trustworthy data, the requirement for efficient data pre-processing, and the rise of cognitive and digital twin technologies. Furthermore, the research explores particular implementations of PdM, such as maintenance strategies in semiconductor manufacturing, the creation of intelligent training modules, and the utilization of AI-powered decision support systems. Collectively, these articles offer valuable perspectives on the changing nature of predictive maintenance (PdM), highlighting its essential significance in contemporary intelligent manufacturing settings.

3.4.9 Time Series Analysis

Time series analysis is used in predictive maintenance in manufacturing to examine sequential data and detect patterns, trends, and abnormalities that occur over time. Time series analysis is essential for forecasting equipment failures, estimating the remaining usable life, and optimizing maintenance schedules. The process utilizes methodologies including autoregressive integrated moving average (ARIMA), exponential smoothing techniques, and Fourier analysis to construct models and predict future values using past data. Manufacturing firms can boost overall equipment reliability by proactively addressing anticipated faults and reducing downtime through the utilization of time-dependent patterns. Time series analysis is crucial in the creation of predictive maintenance models as it offers a temporal viewpoint for decision-making and enables early interventions to avoid unexpected disruptions.

3.4.10 Fuzzy Logic

Predictive maintenance in manufacturing utilizes fuzzy logic to address the presence of uncertainty and imprecision in the data related to equipment health and performance.

Fuzzy logic enables the representation of imprecise or subjective data, making it especially valuable in scenarios where obtaining precise numerical values is difficult. Fuzzy logic systems can be utilized in predictive maintenance to analyze sensor data, equipment states, and other pertinent variables, thereby offering a sophisticated and contextually-aware framework for decision-making. Fuzzy logic facilitates the depiction of linguistic terms, enabling maintenance models to integrate human expertise and subjective evaluations. This technique improves the flexibility of predictive maintenance systems in handling real-world intricacies and variations, hence enhancing the efficacy and resilience of decision support for maintenance activities in manufacturing environments.

3.5 ROLE OF ARTIFICIAL INTELLIGENCE IN PREDICTIVE MAINTENANCE

The role of artificial intelligence (AI) in predictive maintenance within manufacturing is pivotal, bringing about significant improvements in operational efficiency, cost savings, and overall equipment reliability. Here are key aspects of AI's role in predictive maintenance in manufacturing.

3.5.1 Data Analysis and Pattern Recognition

Advanced algorithms: AI employs sophisticated machine learning algorithms to analyze vast amounts of historical and real-time data. This allows for the identification of subtle patterns, trends, and anomalies that might be indicative of impending equipment failures.

3.5.2 Predictive Modeling

Machine learning models: artificial intelligence enables the creation of prediction algorithms that can accurately anticipate equipment malfunctions. These models utilize historical data, sensor readings, and operational characteristics to forecast the timing of required maintenance, facilitating proactive interventions.

3.5.3 Anomaly Detection

Identification of abnormal behavior: AI-powered anomaly detection systems excel in recognizing deviations from normal operating conditions. This capability is crucial for early detection of potential issues before they escalate into major failures.

3.5.4 Real-time Monitoring

Continuous surveillance: AI enables real-time monitoring of equipment health by processing and analyzing data as it is generated. This immediate response allows for timely decision-making and intervention to prevent downtime.

3.5.5 Remaining Useful Life (RUL) Estimation

Prognostic analytics: AI plays a key role in estimating the remaining useful life of machinery. By considering various factors and trends, AI models can provide insights into how much longer equipment is expected to operate efficiently.

3.5.6 Adaptive Learning

Continuous improvement: AI models can adapt and evolve over time by incorporating new data. This adaptive learning ensures that predictive maintenance systems become more accurate and effective as they encounter new scenarios and information.

3.5.7 Integration with IoT and Sensors

Data fusion: AI seamlessly integrates with internet of things (IoT) devices and sensors, utilizing the wealth of data they generate. This integration enhances the accuracy of predictions by considering a diverse range of parameters.

3.5.8 Automated Decision-making

Automated responses: AI allows for automated decision-making processes based on predictive insights. Maintenance tasks, scheduling, and resource allocation can be automatically triggered, optimizing the response to potential issues.

3.5.9 Reduced False Positives

Refinement of algorithms: AI algorithms can be refined to minimize false positives, ensuring that maintenance interventions are genuinely needed. This helps in avoiding unnecessary downtime and resource allocation.

3.5.10 Cognitive Computing

Natural language processing (NLP): AI's cognitive computing capabilities, including NLP, can be employed for analyzing unstructured data sources, such as maintenance reports and documentation. This aids in gaining a comprehensive understanding of equipment health.

3.5.11 Cost Optimization

Efficient resource allocation: AI-driven predictive maintenance optimizes the allocation of resources, ensuring that maintenance efforts are focused on critical areas, thereby reducing overall maintenance costs.

Artificial intelligence revolutionizes predictive maintenance in manufacturing by harnessing the power of data analytics, machine learning, and real-time monitoring. This proactive approach helps manufacturers move from reactive to predictive strategies, ensuring enhanced equipment reliability and operational excellence.

3.6 EXISTING CHALLENGES IN PREDICTIVE MAINTENANCE

While the integration of data analytics and artificial intelligence (AI) for predictive maintenance in manufacturing offers substantial benefits, it also comes with its set of challenges. Some of the existing challenges include the following.

3.6.1 Data Quality and Availability

Insufficient data and poor data quality pose significant challenges in the realm of predictive maintenance. In certain instances, the availability of historical data for accurate predictive modeling may be limited, especially for newer equipment or within industries with constrained data collection practices. The effectiveness of predictive maintenance models is contingent on the richness and completeness of historical datasets. Additionally, poor data quality, characterized by inaccuracies, missing values, or inconsistencies in data formats, can further impede the reliability of predictive maintenance systems. Addressing these issues requires a concerted effort to enhance data collection practices, ensure data accuracy, and implement strategies for handling missing or inconsistent information, ultimately fortifying the foundation for robust predictive maintenance analyses.

3.6.2 Complexity of Manufacturing Processes

Navigating the landscape of predictive maintenance in manufacturing encounters hurdles due to multifaceted systems and dynamic environments. The intricacies of manufacturing processes often entail complex, interconnected systems, introducing challenges in accurately modeling and predicting failures. The dynamic nature of manufacturing environments, marked by rapid changes in production processes, equipment configurations, or operating conditions, adds another layer of complexity. Static predictive models may struggle to adapt swiftly to these dynamic shifts, highlighting the need for agile and adaptable approaches to effectively predict and mitigate potential failures in manufacturing systems. Addressing these challenges involves leveraging advanced modeling techniques that can account for the intricate interdependencies within multifaceted systems and accommodate the rapid changes inherent in dynamic manufacturing environments.

3.6.3 Integration with Legacy Systems

The integration of modern predictive maintenance solutions with legacy systems in manufacturing poses challenges, primarily related to compatibility and data silos. Compatibility issues arise due to the intricate nature of merging new technologies with existing legacy systems, necessitating a thoughtful assessment of interoperability. Legacy systems often function in isolated data silos, hindering the seamless aggregation and analysis of data across the entire manufacturing environment. This isolation can impede the flow of information and compromise the effectiveness of predictive maintenance strategies, highlighting the importance of strategic planning and technology alignment when implementing advanced maintenance solutions in legacy manufacturing infrastructures.

3.6.4 Model Interpretability and Trust

The challenges associated with interpretability and trust in predictive maintenance systems revolve around the use of black-box models, especially prevalent in advanced machine learning, such as deep learning models. These models are often perceived as complex and difficult to interpret, creating challenges for operators who need to understand and trust the predictions they generate. The lack of explainability in these models can be a significant concern, particularly in industries where regulatory compliance and transparency are crucial. Achieving a balance between the predictive power of advanced models and the interpretability required for user trust becomes essential in developing effective and reliable predictive maintenance solutions. Addressing these challenges is critical for widespread adoption and successful implementation of predictive maintenance strategies in various industrial settings.

3.6.5 Scalability and Resource Constraints

Scalability and resource constraints present challenges in the implementation of predictive maintenance, particularly in smaller manufacturing facilities. The demand for substantial computational resources for the deployment and maintenance of AI models can strain the capabilities of facilities with limited resources. Additionally, there may be a shortage of skilled personnel proficient in implementing and managing AI systems, creating a skill gap that further impedes widespread adoption. Overcoming these challenges involves finding solutions that are resource-efficient and accessible to facilities with varying levels of technical expertise.

3.6.6 Costs and Return on Investment (ROI)

The implementation of a predictive maintenance system requires an initial investment encompassing costs for technology adoption, training, and infrastructure upgrades. However, manufacturers often encounter challenges in promptly quantifying and demonstrating the return on investment (ROI), particularly in the short term. This difficulty arises from the need to establish clear metrics and assess tangible benefits, emphasizing the importance for companies to develop robust strategies for measuring the effectiveness of their predictive maintenance initiatives over time.

3.6.7 Security and Privacy Concerns

Security and privacy concerns are paramount in predictive maintenance implementation. Storing and transmitting sensitive equipment and operational data pose risks related to cybersecurity, with the potential for unauthorized access being a prominent worry. Moreover, privacy compliance becomes a critical consideration, particularly when dealing with information related to equipment health and maintenance schedules. Adhering to stringent data privacy regulations is essential to ensure the protection of sensitive data and maintain compliance with legal frameworks governing privacy in industrial contexts. Addressing these concerns is vital for fostering trust among stakeholders and safeguarding the integrity of predictive maintenance systems in manufacturing environments.

3.6.8 Cultural Resistance and Change Management

Cultural resistance and change management present significant challenges in the deployment of predictive maintenance systems. Resistance within the organizational culture can hinder the seamless integration of AI-driven solutions, emphasizing the need for proactive change management strategies. Adequate training and awareness programs are crucial to overcoming resistance, ensuring that personnel at all levels understand the benefits and processes associated with predictive maintenance. Effectively managing cultural shifts and fostering a culture of openness to technological advancements are key factors in successful implementation, requiring a collaborative effort to align organizational values with the transformative potential of AI in maintenance practices.

3.7 DATASETS FOR PREDICTIVE MAINTENANCE IN MANUFACTURING

Several benchmark datasets are available as presented in Table 3.2 for predictive maintenance in manufacturing, providing a standardized way to evaluate and compare different predictive maintenance models.

3.8 CONCLUSION

Data analytics and artificial intelligence (AI) have become integral components in revolutionizing the field of predictive maintenance, particularly in manufacturing. The synergy between data analytics and AI empowers organizations to transition from reactive to proactive maintenance strategies. Predictive maintenance, driven by advanced technologies such as machine learning and IoT, enables the prediction of equipment failures, facilitating timely interventions and minimizing unplanned downtime. This innovative approach not only enhances equipment reliability but also leads to substantial cost savings and improved operational efficiency. By leveraging real-time data analytics and AI capabilities, manufacturers can optimize maintenance strategies, reduce downtime, and ultimately stay competitive in dynamic industries. The seamless integration of these technologies with enterprise systems further streamlines communication between different business functions, contributing to a more synchronized and efficient manufacturing ecosystem. As industries continue to embrace the transformative potential of data analytics and AI in predictive maintenance, the outlook for increased reliability, cost-effectiveness, and operational excellence remains promising.

TABLE 3.2
Datasets for Predictive Maintenance in Manufacturing

Dataset Name	Description
NASA Prognostics Center of Excellence (PCoE) Datasets [18]	Diverse datasets, including turbofan engine degradation simulations with different fault modes.
PHM Data Challenge Datasets [19]	Real-world failure data from various industrial systems, provided through Prognostics and Health Management (PHM) Data Challenges.
C-MAPSS Datasets [20]	Simulation datasets for benchmarking predictive maintenance algorithms for aircraft engines. Includes sensor measurements with various fault modes.
IMS Datasets [21]	Bearing fault diagnosis datasets from the Institute of Machine Tools and Industrial Management (IMS) at the Technical University of Munich.
SECOM Manufacturing Dataset [22]	Derived from semiconductor manufacturing processes, this dataset includes sensor data to predict failures in the manufacturing process.
Turbofan Engine Degradation Simulation Data [23]	Simulated degradation data generated using C-MAPSS for turbofan engines, commonly used for benchmarking predictive maintenance models.
Condition Monitoring of Hydraulic Systems Data [24]	Sensor data from hydraulic test rigs aiming to predict component failures in hydraulic systems, suitable for benchmarking predictive maintenance models.

REFERENCES

1. Lee, C. K. M., Cao, Y., & Ng, K. H. (2017). Big data analytics for predictive maintenance strategies. In *Supply Chain Management in the Big Data Era*, pp. 50–74. Hershey, PA: IGI Global.
2. Lee, W. J., Wu, H., Yun, H., Kim, H., Jun, M. B., & Sutherland, J. W. (2019). Predictive maintenance of machine tool systems using artificial intelligence techniques applied to machine condition data. *Procedia Cirp*, *80*, 506–511.
3. Lawrence, J., & Durana, P. (2021). Artificial intelligence-driven big data analytics, predictive maintenance systems, and internet of thingsbased real-time production logistics in sustainable industry 4.0 wireless networks. *Journal of Self-Governance & Management Economics*, *9*(4), 62–75.
4. Samatas, G. G., Moumgiakmas, S. S., & Papakostas, G. A. (2021, May). Predictive maintenance-bridging artificial intelligence and IoT. In *2021 IEEE World AI IoT Congress (AIIoT)*, Seattle, WA, pp. 0413–0419. IEEE.
5. Çınar, Z. M., Abdussalam Nuhu, A., Zeeshan, Q., Korhan, O., Asmael, M., & Safaei, B. (2020). Machine learning in predictive maintenance towards sustainable smart manufacturing in industry 4.0. *Sustainability*, *12*(19), 8211.
6. Cheng, X., Chaw, J. K., Goh, K. M., Ting, T. T., Sahrani, S., Ahmad, M. N., ... & Ang, M. C. (2022). Systematic literature review on visual analytics of predictive maintenance in the manufacturing industry. *Sensors*, *22*(17), 6321.

7. Rousopoulou, V., Nizamis, A., Vafeiadis, T., Ioannidis, D., & Tzovaras, D. (2020). Predictive maintenance for injection molding machines enabled by cognitive analytics for industry 4.0. *Frontiers in Artificial Intelligence*, *3*, 578152.
8. Ayvaz, S., & Alpay, K. (2021). Predictive maintenance system for production lines in manufacturing: A machine learning approach using IoT data in real-time. *Expert Systems with Applications*, *173*, 114598.
9. Keleko, A. T., Kamsu-Foguem, B., Ngouna, R. H., & Tongne, A. (2022). Artificial intelligence and real-time predictive maintenance in industry 4.0: A bibliometric analysis. *AI and Ethics*, *2*(4), 553–577.
10. Yu, W., Dillon, T., Mostafa, F., Rahayu, W., & Liu, Y. (2019). A global manufacturing big data ecosystem for fault detection in predictive maintenance. *IEEE Transactions on Industrial Informatics*, *16*(1), 183–192.
11. Bekar, E. T., Nyqvist, P., & Skoogh, A. (2020). An intelligent approach for data pre-processing and analysis in predictive maintenance with an industrial case study. *Advances in Mechanical Engineering*, *12*(5), 1687814020919207.
12. Unal, P., Albayrak, Ö., Jomâa, M., & Berre, A. J. (2022). Data-driven artificial intelligence and predictive analytics for the maintenance of industrial machinery with hybrid and cognitive digital twins. In *Technologies and Applications for Big Data Value*, pp. 299–319. Cham: Springer International Publishing.
13. Daniyan, I., Mpofu, K., Oyesola, M., Ramatsetse, B., & Adeodu, A. (2020). Artificial intelligence for predictive maintenance in the railcar learning factories. *Procedia Manufacturing*, *45*, 13–18.
14. Rosati, R., Romeo, L., Cecchini, G., Tonetto, F., Viti, P., Mancini, A., & Frontoni, E. (2023). From knowledge-based to big data analytic model: a novel IoT and machine learning based decision support system for predictive maintenance in Industry 4.0. *Journal of Intelligent Manufacturing*, *34*(1), 107–121.
15. Drakaki, M., Karnavas, Y. L., Tzionas, P., & Chasiotis, I. D. (2021). Recent developments towards industry 4.0 oriented predictive maintenance in induction motors. *Procedia Computer Science*, *180*, 943–949.
16. Pech, M., Vrchota, J., & Bednář, J. (2021). Predictive maintenance and intelligent sensors in smart factory. *Sensors*, *21*(4), 1470.
17. Hosseinzadeh, A., Chen, F. F., Shahin, M., & Bouzary, H. (2023). A predictive maintenance approach in manufacturing systems via AI-based early failure detection. *Manufacturing Letters*, *35*, 1179–1186.
18. Ramasso, E., & Saxena, A. (2014). Performance benchmarking and analysis of prognostic methods for CMAPSS datasets. *International Journal of Prognostics and Health Management*, 5(2), 1–15.
19. Jia, X., Huang, B., Feng, J., Cai, H., & Lee, J. (2018, November). A review of PHM data competitions from 2008 to 2017: Methodologies and analytics. In *Proceedings of the Annual Conference of the Prognostics and Health Management Society*, Philadelphia, PA, pp. 1–10. PHM Society.
20. Asif, O., Haider, S. A., Naqvi, S. R., Zaki, J. F., Kwak, K. S., & Islam, S. R. (2022). A deep learning model for remaining useful life prediction of aircraft turbofan engine on C-MAPSS dataset. *IEEE Access*, *10*, 95425–95440.
21. Hayasaka, T., Goto-Inoue, N., Ushijima, M., Yao, I., Yuba-Kubo, A., Wakui, M., ... & Setou, M. (2011). Development of imaging mass spectrometry (IMS) dataset extractor software, IMS convolution. *Analytical and Bioanalytical Chemistry*, *401*, 183–193.
22. Moldovan, D., Cioara, T., Anghel, I., & Salomie, I. (2017, September). Machine learning for sensor-based manufacturing processes. In *2017 13th IEEE International*

Conference on Intelligent Computer Communication and Processing (ICCP), Cluj-Napoca, Romania, pp. 147–154. IEEE.

23. Saxena, A., & Goebel, K. (2008). Turbofan engine degradation simulation data set. *NASA Ames Prognostics Data Repository*, *18*, 878–887.
24 S. S. Chawathe (2019). Condition Monitoring of Hydraulic Systems by Classifying Sensor Data Streams. In *IEEE 9th Annual Computing and Communication Workshop and Conference (CCWC)*, Las Vegas, NV, pp. 0898–0904. IEEE.

4 Scalability and Deployment of Emerging Technologies in Predictive Maintenance

Ritesh Kumar Dewangan and Vinita Dewangan

4.1 INTRODUCTION

Predictive maintenance has gained prominence in various industries due to its ability to anticipate equipment failures and minimize downtime (Khan et al., 2020). As organizations increasingly adopt emerging technologies such as artificial intelligence (AI), machine learning (ML), and the internet of things (IoT) for predictive maintenance, ensuring scalability and effective deployment becomes paramount (Li et al., 2020).

In today's rapidly evolving industrial landscape, the effective management of equipment and machinery is crucial for ensuring optimal performance and productivity. Traditional approaches to maintenance, such as reactive and preventive maintenance, often fall short in addressing the complexities and uncertainties inherent in modern industrial environments. In response to these challenges, predictive maintenance has emerged as a proactive and data-driven approach to equipment maintenance, enabling organizations to anticipate and prevent failures before they occur.

Predictive maintenance leverages advanced data analytics, machine learning algorithms, and the internet of things (IoT) to analyze equipment performance data in real-time, identify patterns, and predict potential failures. By monitoring key indicators of equipment health and performance, organizations can proactively schedule maintenance activities, optimize asset lifecycles, and minimize downtime. The adoption of predictive maintenance has become increasingly prevalent across industries, ranging from manufacturing and energy to transportation and healthcare, as organizations recognize its potential to enhance operational efficiency and reduce maintenance costs.

However, the successful implementation of predictive maintenance requires careful consideration of scalability and deployment strategies. As organizations accumulate vast amounts of data from sensors, equipment logs, and other sources, scalability becomes a critical concern. The ability to scale predictive maintenance systems to handle large volumes of data and accommodate increasing computational demands is essential for ensuring reliability and performance. Furthermore, efficient deployment

 DOI: 10.1201/9781003480860-4

practices are necessary to seamlessly integrate predictive maintenance solutions into existing workflows, maximize the value derived from emerging technologies, and drive tangible business outcomes.

In this chapter, we explore the challenges and opportunities associated with the scalability and deployment of emerging technologies in predictive maintenance. We begin by providing an overview of predictive maintenance, highlighting its proactive approach to equipment maintenance and the role of emerging technologies. We then delve into the scalability challenges encountered in predictive maintenance systems, addressing issues such as handling big data, computational complexity, real-time processing, and resource constraints. Subsequently, we explore scalable architectures for predictive maintenance systems, including cloud-based solutions, edge computing, hybrid architectures, and scalable data management practices.

Following discussions on scalability, we shift focus to deployment methodologies for predictive maintenance systems. We examine agile deployment practices, DevOps principles, continuous integration/continuous deployment (CI/CD), and containerization and orchestration technologies. Additionally, we discuss best practices for successful deployment, covering various aspects such as data management, model versioning and monitoring, security and privacy considerations, and collaboration and communication strategies.

To illustrate concepts and strategies discussed, we provide case studies and real-world examples of successful deployments of predictive maintenance systems across different industries. Finally, we explore future directions and emerging trends in predictive maintenance, highlighting advancements in AI and machine learning, the evolution of edge computing, integration with Industry 4.0 technologies, and considerations for sustainability and green technologies.

By examining the challenges, strategies, and innovations in the scalability and deployment of emerging technologies in predictive maintenance, this chapter aims to provide valuable insights and guidance for organizations seeking to optimize their maintenance practices and unlock the full potential of their assets.

4.1.1 Overview of Predictive Maintenance

Predictive maintenance is a proactive maintenance strategy that leverages data analytics, machine learning algorithms, and sensor technology to predict equipment failures before they occur. By analyzing historical data and real-time sensor data, predictive maintenance systems can identify patterns and anomalies indicative of potential issues. This approach allows organizations to schedule maintenance activities more efficiently, minimize downtime, and reduce overall maintenance costs compared to traditional reactive or scheduled maintenance approaches.

4.1.2 Importance of Scalability and Deployment

Scalability and deployment are critical aspects of predictive maintenance systems, especially as industries increasingly adopt emerging technologies. Scalability refers to the system's ability to handle growing volumes of data, increasing computational demands, and expanding user bases without sacrificing performance. Effective deployment ensures that predictive maintenance solutions are implemented smoothly, integrated into existing workflows, and accessible to relevant stakeholders. Without scalability and efficient deployment strategies, organizations may struggle to fully realize the benefits of predictive maintenance or encounter challenges in managing and utilizing the technology effectively.

4.1.3 Emerging Technologies in Predictive Maintenance

Emerging technologies such as artificial intelligence (AI), machine learning (ML), and the internet of things (IoT) are revolutionizing predictive maintenance practices. AI and ML algorithms can analyze complex data sets to identify patterns and predict equipment failures with high accuracy. IoT devices equipped with sensors collect real-time data on equipment health and performance, enabling proactive maintenance actions. Additionally, advancements in edge computing enable data processing and analysis to occur closer to the data source, reducing latency and improving responsiveness in predictive maintenance systems. By harnessing these emerging technologies, organizations can enhance the scalability, accuracy, and efficiency of their predictive maintenance efforts, driving improvements in asset reliability and operational performance.

4.2 SCALABILITY CHALLENGES

Handling large volumes of data is one of the primary challenges in scaling predictive maintenance systems. This data includes sensor readings, equipment logs, maintenance records, and environmental factors, among others (Dubey et al., 2019). As the complexity of machine learning algorithms increases or the size of the datasets grows, computational resources become a bottleneck (Kusiak & Verma, 2018).

4.2.1 Handling Big Data

One of the primary challenges in scaling predictive maintenance systems is managing the vast volumes of data generated by industrial equipment and sensors. This data includes sensor readings, equipment logs, maintenance records, and environmental factors, among others. As the number of monitored assets increases and the frequency of data collection rises, organizations must have robust data storage, processing, and analytics infrastructure in place. Traditional databases and processing systems may struggle to handle such large volumes of data efficiently, necessitating the adoption of scalable big data technologies like distributed file systems, NoSQL databases, and data lakes.

4.2.2 Computational Complexity

Predictive maintenance often involves complex machine learning algorithms trained on large datasets to predict equipment failures or maintenance needs. As the complexity of these algorithms increases or the size of the datasets grows, computational resources become a bottleneck. Scaling predictive maintenance systems requires adequate computational resources, including CPUs, GPUs, or specialized hardware accelerators, to train and deploy machine learning models efficiently. Additionally, organizations may need to implement parallel computing techniques or distributed processing frameworks to distribute workloads across multiple computing nodes and improve performance.

4.2.3 Real-time Processing

Real-time processing is crucial for timely detection and response to equipment failures or anomalies in predictive maintenance systems. However, scaling real-time processing capabilities presents challenges, particularly in high-volume environments with strict latency requirements. Processing streaming data from IoT sensors or equipment telemetry streams in real-time requires low-latency processing frameworks and efficient event processing pipelines. Organizations may leverage technologies like Apache Kafka, Apache Flink, or stream processing platforms to ingest, process, and analyze streaming data in real-time while ensuring scalability and fault tolerance.

4.2.4 Resource Constraints

Resource constraints, including hardware limitations, network bandwidth constraints, and budgetary constraints, can hinder the scalability of predictive maintenance systems. Organizations may face challenges in acquiring sufficient computational resources, storage capacity, or network bandwidth to support the growing demands of predictive maintenance initiatives. Furthermore, constraints on IT infrastructure, such as data center space, power, and cooling, may limit the organization's ability to scale hardware resources vertically. To overcome resource constraints, organizations may explore cloud computing solutions, edge computing architectures, or hybrid deployments to leverage external resources and optimize resource utilization while managing costs effectively.

4.3 SCALABLE ARCHITECTURES

Cloud-based solutions, which provide scalable and flexible resources for predictive maintenance systems, allow organizations to utilize platforms such as Amazon Web Services (AWS), Microsoft Azure, or Google Cloud Platform (GCP) to provision compute, storage, and analytics services on-demand (Kumar et al., 2020; Choudhury et al., 2021). This scalability empowers organizations to effectively manage large volumes of data and meet increasing computational demands, thereby ensuring reliability and performance in predictive maintenance operations. Edge computing brings computational capabilities closer to the data source, enabling real-time processing and analysis of data generated by IoT sensors or equipment (Sharma & Chen, 2019).

4.3.1 Cloud-Based Solutions

Cloud computing offers scalable and flexible resources for predictive maintenance systems. Organizations can leverage cloud platforms such as Amazon Web Services

(AWS), Microsoft Azure, or Google Cloud Platform (GCP) to provision compute, storage, and analytics services on-demand. Cloud-based solutions provide scalability by allowing organizations to dynamically scale resources up or down based on workload demands. Additionally, cloud platforms offer managed services for big data processing, machine learning, and IoT, enabling organizations to focus on building predictive maintenance applications rather than managing infrastructure. By deploying predictive maintenance solutions in the cloud, organizations can achieve scalability, reliability, and cost-effectiveness.

4.3.2 Edge Computing

Edge computing brings computational capabilities closer to the data source, enabling real-time processing and analysis of data generated by IoT sensors or equipment. Edge computing architectures distribute computing tasks across edge devices deployed at or near the source of data generation, reducing latency and bandwidth usage. In predictive maintenance, edge computing allows organizations to perform local data preprocessing, anomaly detection, or predictive modeling at the edge before transmitting relevant insights to the cloud or centralized data center. By offloading processing tasks to edge devices, organizations can achieve scalability while maintaining real-time responsiveness and preserving bandwidth for critical operations.

4.3.3 Hybrid Architectures

Hybrid architectures combine cloud-based and edge computing components to leverage the strengths of both approaches. In a hybrid architecture for predictive maintenance, edge devices collect and preprocess data locally, while cloud-based resources handle complex analytics, machine learning model training, and centralized management. Hybrid architectures provide scalability by distributing workloads across edge and cloud environments based on workload characteristics, latency requirements, and resource availability. Organizations can dynamically adjust the distribution of computing tasks between edge and cloud components to optimize performance, scalability, and cost-efficiency according to changing requirements.

4.3.4 Scalable Data Management

Scalable data management is essential for handling the large volumes of data generated by predictive maintenance systems. Organizations need robust data storage, processing, and analytics capabilities to manage data efficiently and extract actionable insights. Scalable data management solutions include distributed file systems, NoSQL databases, and data lakes that can scale horizontally to accommodate growing data volumes. Additionally, organizations may implement data governance policies, data quality controls, and data lifecycle management practices to ensure data consistency, integrity, and accessibility. By adopting scalable data management practices, organizations can effectively manage and analyze data for predictive maintenance while accommodating future growth and scalability requirements.

4.4 DEPLOYMENT METHODOLOGIES

DevOps principles emphasize collaboration, automation, and continuous delivery across development and operations teams (Lee et al., 2015). In the context of predictive maintenance, DevOps practices facilitate the seamless integration of development, deployment, and operations activities (International Organization for Standardization [ISO], 2018). Continuous integration/continuous deployment (CI/CD) pipelines automate the process of building, testing, and deploying code changes, enabling organizations to deliver updates more frequently and reliably (Khan et al., 2020).

4.4.1 Agile Deployment Practices

Agile deployment practices emphasize iterative and incremental development, allowing organizations to deliver value to users more quickly and adapt to changing requirements. In the context of predictive maintenance, agile methodologies enable teams to prioritize and implement features based on user feedback, business needs, and technological advancements. By breaking down the deployment process into smaller, manageable tasks or sprints, organizations can accelerate the deployment of emerging technologies while maintaining flexibility and responsiveness to evolving demands. Agile deployment practices also promote collaboration between cross-functional teams, including data scientists, engineers, domain experts, and end-users, fostering a culture of continuous improvement and innovation.

4.4.2 DevOps Principles

DevOps principles emphasize collaboration, automation, and continuous delivery across development and operations teams. In the context of predictive maintenance, DevOps practices facilitate the seamless integration of development, deployment, and operations activities, streamlining the deployment process and improving overall efficiency. By automating repetitive tasks, such as code deployment, testing, and infrastructure provisioning, organizations can reduce manual errors, increase deployment frequency, and enhance reliability. DevOps also encourages the adoption of infrastructure as code (IaC) and configuration management tools, enabling organizations to manage and provision infrastructure resources programmatically and consistently. Through DevOps practices, organizations can accelerate the deployment of emerging technologies in predictive maintenance while maintaining high levels of quality and reliability.

4.4.3 Continuous Integration/Continuous Deployment (CI/CD)

Continuous integration/continuous deployment (CI/CD) is a software development practice that involves automating the integration, testing, and deployment of code changes. In the context of predictive maintenance, CI/CD pipelines automate the process of building, testing, and deploying machine learning models, enabling organizations to deliver updates more frequently and reliably. CI/CD pipelines integrate with version control systems, such as Git, to automatically trigger builds and tests whenever changes are made to the codebase. By automating the deployment of machine learning models, organizations can reduce the time and effort required to deploy new predictive maintenance features or improvements. CI/CD pipelines also facilitate rapid experimentation and iteration, allowing organizations to quickly validate hypotheses and incorporate feedback into future iterations of the predictive maintenance system.

4.4.4 Containerization and Orchestration

Containerization and orchestration technologies, such as Docker and Kubernetes, enable organizations to package, deploy, and manage applications and services in scalable, portable, and efficient environments. In the context of predictive maintenance, containerization allows organizations to encapsulate machine learning models, data preprocessing pipelines, and other components into lightweight, self-contained units called containers. These containers can be deployed consistently across different environments, including development, testing, and production, without the need for manual configuration or dependencies. Container orchestration platforms, such as Kubernetes, automate the deployment, scaling, and management of containerized applications, ensuring high availability, scalability, and resilience. By leveraging containerization and orchestration technologies, organizations can accelerate the deployment of emerging technologies in predictive maintenance while maximizing resource utilization and minimizing operational overhead.

4.5 BEST PRACTICES FOR SUCCESSFUL DEPLOYMENT

Effective data management is crucial for the success of predictive maintenance deployments (Kumar et al., 2020). This involves ensuring the quality, integrity, and accessibility of data used for training machine learning models and making maintenance predictions (Kusiak & Verma, 2018). Model versioning and monitoring are essential for tracking the performance and behavior of machine learning models deployed in predictive maintenance systems (Choudhury et al., 2021).

4.5.1 Data Management

Effective data management is crucial for the success of predictive maintenance deployments. This involves ensuring the quality, integrity, and accessibility of data used for training machine learning models and making maintenance predictions. Best practices for data management include establishing data governance policies to define data ownership, privacy, and security requirements. Organizations should implement data quality controls to identify and address issues such as missing values, outliers, and inconsistencies in the data. Additionally, data lineage and provenance tracking mechanisms should be in place to trace the origins and transformations of data throughout its lifecycle. By implementing robust data management practices, organizations can improve the accuracy and reliability of predictive maintenance models and ensure compliance with regulatory requirements.

4.5.2 Model Versioning and Monitoring

Model versioning and monitoring are essential for tracking the performance and behavior of machine learning models deployed in predictive maintenance systems. Organizations should establish version control mechanisms to manage different versions of machine learning models, including model training data, hyperparameters, and evaluation metrics. Continuous monitoring of model performance metrics, such as accuracy, precision, recall, and F1-score, helps detect drifts or degradation in model performance over time. Organizations should also monitor the input data distribution to detect changes in data patterns or characteristics that may affect model performance. By actively monitoring model versions and performance metrics, organizations can identify and address issues promptly, ensuring the reliability and effectiveness of predictive maintenance models in production environments.

4.5.3 Security and Privacy Considerations

Security and privacy considerations are paramount when deploying predictive maintenance systems, especially in industrial environments where sensitive data and critical infrastructure are involved. Organizations should implement measures to protect data confidentiality, integrity, and availability throughout the predictive maintenance lifecycle. This includes encrypting data at rest and in transit, implementing access controls and authentication mechanisms to restrict unauthorized access to sensitive information, and establishing auditing and logging mechanisms to monitor and track data access and usage. Additionally, organizations should adhere to privacy regulations and standards, such as the general data protection regulation (GDPR) and the California consumer privacy act (CCPA), to ensure compliance with legal and regulatory requirements. By prioritizing security and privacy considerations, organizations can build trust and confidence in their predictive maintenance systems and mitigate the risk of data breaches or unauthorized access.

4.5.4 Collaboration and Communication

Collaboration and communication are essential for successful deployment and adoption of predictive maintenance systems across organizations. Cross-functional collaboration between data scientists, engineers, domain experts, and end-users facilitates the alignment of technical requirements with business objectives and user needs. Regular communication and feedback loops enable stakeholders to provide input, share insights, and address concerns throughout the deployment process. Organizations should establish clear channels of communication and collaboration, such as regular meetings, workshops, and documentation repositories, to facilitate knowledge sharing and decision-making. By fostering a culture of collaboration and communication, organizations can ensure that predictive maintenance deployments meet the needs and expectations of all stakeholders and drive tangible business value.

4.6 CASE STUDIES

Case studies provide real-world examples of how organizations have implemented and benefited from scalable deployment of emerging technologies in predictive maintenance. These case studies offer valuable insights into the challenges faced, strategies employed, and outcomes achieved in various industrial settings. Here are examples of potential case studies. Cloud-based predictive maintenance has been successfully implemented in the manufacturing industry, leading to significant reductions in unplanned downtime and increased operational efficiency (Li et al., 2020). Edge computing technologies have enabled energy companies to enhance predictive maintenance capabilities for critical infrastructure, such as power plants and distribution networks (Sharma & Chen, 2019). Hybrid cloud-edge architectures have provided transportation companies with scalability, flexibility, and real-time responsiveness in their predictive maintenance operations (Dubey et al., 2019).

4.6.1 Cloud-Based Predictive Maintenance in Manufacturing

This case study focuses on a manufacturing company that implemented a cloud-based predictive maintenance solution to optimize equipment uptime and reduce maintenance costs. The company integrated IoT sensors into their production machinery to collect real-time data on equipment health and performance. Leveraging cloud-based analytics and machine learning algorithms, the company developed predictive maintenance models to forecast equipment failures and schedule proactive maintenance activities. The deployment of the predictive maintenance solution resulted in significant reductions in unplanned downtime, increased operational efficiency, and improved asset reliability. The case study highlights the scalability and cost-effectiveness of cloud-based predictive maintenance solutions in the manufacturing industry.

4.6.2 Edge Computing Deployment in Energy Sector

This case study examines how an energy company deployed edge computing technologies to enhance predictive maintenance capabilities for its critical infrastructure, such as power plants and distribution networks. By deploying edge devices equipped with sensors and edge computing capabilities at remote sites, the company was able to perform real-time data processing and analysis locally, without relying on centralized data centers or cloud resources. This approach reduced latency, improved responsiveness, and enabled timely detection of equipment anomalies or performance deviations. The case study demonstrates the scalability and reliability of edge computing architectures for predictive maintenance applications in the energy sector.

4.6.3 Hybrid Cloud-Edge Architecture in Transportation

This case study explores how a transportation company implemented a hybrid cloud-edge architecture to support predictive maintenance for its fleet of vehicles and infrastructure assets. The company deployed edge devices onboard vehicles and at transportation hubs to collect telemetry data and perform initial data processing and analysis. The processed data was then transmitted to the cloud for further analysis, model training, and centralized management. By combining edge computing with cloud-based analytics, the company achieved scalability, flexibility, and real-time responsiveness in its predictive maintenance operations. The case study illustrates the benefits of hybrid architectures for predictive maintenance applications in the transportation industry.

These case studies provide practical examples of how organizations have successfully deployed emerging technologies in predictive maintenance, showcasing the scalability, effectiveness, and business impact of these solutions in diverse industrial contexts. By analyzing real-world examples, readers can gain valuable insights and best practices for implementing scalable deployment strategies in their own predictive maintenance initiatives.

4.7 FUTURE DIRECTIONS AND EMERGING TRENDS

Advancements in AI and machine learning are expected to drive improvements in prediction accuracy, reliability, and scalability in predictive maintenance systems (Khan et al., 2020). Edge computing is poised to play a crucial role in the future of predictive maintenance, particularly in industries with stringent latency and bandwidth requirements (Sharma & Chen, 2019). The integration of predictive maintenance with other Industry 4.0 technologies, such as digital twins and augmented reality, holds immense potential for enhancing asset management and optimization (Lee et al., 2015).

4.7.1 Advancements in AI and Machine Learning

As AI and machine learning technologies continue to mature, we anticipate significant advancements in predictive maintenance models and algorithms. This includes the development of more sophisticated machine learning techniques, such as deep learning and reinforcement learning, capable of handling complex data patterns and dynamics. Additionally, we expect to see increased automation and autonomy in predictive maintenance systems, enabling machines to learn and adapt to changing operating conditions without human intervention. Advancements in AI and machine learning will drive improvements in prediction accuracy, reliability, and scalability, paving the way for more intelligent and efficient predictive maintenance solutions.

4.7.2 Evolution of Edge Computing

Edge computing is poised to play a crucial role in the future of predictive maintenance, particularly in industries with stringent latency and bandwidth requirements. We anticipate continued growth and innovation in edge computing architectures, enabling organizations to perform more advanced data processing and analysis at the edge. This includes the integration of edge AI and machine learning capabilities, enabling real-time inference and decision-making directly on edge devices. As edge computing technologies evolve, we expect to see greater adoption of edge-native predictive maintenance solutions, offering scalability, resilience, and responsiveness in distributed environments.

4.7.3 Integration with Industry 4.0 Technologies

The integration of predictive maintenance with other Industry 4.0 technologies, such as digital twins, augmented reality, and cyber-physical systems, holds immense potential for enhancing asset management and optimization. Digital twins, virtual replicas of physical assets or systems, enable organizations to simulate and analyze the behavior of assets in real-time, facilitating predictive maintenance planning and optimization. Augmented reality technologies provide maintenance technicians with immersive and interactive experiences, enabling them to visualize equipment data and instructions in the field. Cyber-physical systems combine physical components with digital technologies, enabling autonomous and adaptive maintenance operations. Integration with Industry 4.0 technologies will enable organizations to unlock new insights, efficiencies, and capabilities in predictive maintenance.

4.7.4 Sustainability and Green Technologies

Sustainability and green technologies are becoming increasingly important considerations in predictive maintenance, driven by growing environmental concerns and regulatory pressures. Organizations are exploring eco-friendly approaches to maintenance, such as condition-based maintenance and energy-efficient operation strategies, to minimize resource consumption and environmental impact. Additionally, the adoption of renewable energy sources and energy-efficient technologies, such as smart grids and electric vehicles, presents new opportunities and challenges for predictive maintenance. By incorporating sustainability principles into predictive maintenance strategies, organizations can achieve cost savings, regulatory compliance, and environmental stewardship goals simultaneously.

These future directions and emerging trends underscore the dynamic nature of predictive maintenance and the ongoing innovation in technology and industry practices. By staying informed and proactive, organizations can prepare for future challenges and opportunities, driving continuous improvement and innovation in predictive maintenance initiatives.

4.8 CONCLUSION

In conclusion, this chapter has provided a comprehensive examination of the scalability and deployment of emerging technologies in predictive maintenance. We began by discussing the proactive nature of predictive maintenance and its reliance on advanced data analytics, machine learning algorithms, and IoT sensors. Throughout the chapter, we addressed the scalability challenges encountered in predictive maintenance systems, including handling big data, managing computational complexity, ensuring real-time processing, and overcoming resource constraints.

Furthermore, we explored scalable architectures such as cloud-based solutions, edge computing, and hybrid architectures, highlighting their role in accommodating growing data volumes and computational demands. Additionally, we discussed deployment methodologies and best practices for successful deployment, emphasizing the importance of efficient data management, model versioning, and collaboration among stakeholders.

Through case studies and real-world examples, we illustrated the practical implementation of predictive maintenance solutions in various industries, showcasing their impact on operational efficiency and asset reliability. Looking ahead, we identified future directions and emerging trends in predictive maintenance, including advancements in AI and machine learning, the evolution of edge computing, and the integration with Industry 4.0 technologies.

In summary, the scalability and deployment of emerging technologies in predictive maintenance are essential for organizations seeking to optimize their maintenance practices, minimize downtime, and maximize the value of their assets. By leveraging scalable architectures, deploying robust methodologies, and embracing innovation, organizations can unlock new opportunities for predictive maintenance optimization and drive sustainable business outcomes in the digital age.

REFERENCES

Choudhury, A., Kumar, M., Goh, M., & Agarwal, A. (2021). Predictive Maintenance and Edge Computing in the Industrial Internet of Things. In *Proceedings of the International Conference on Industrial Engineering and Operations Management (IEOM)*, Dubai, UAE, 10–13 March 2021. DOI: 10.1109/IEOM48322.2021.9453893

Dubey, R., Gunasekaran, A., Childe, S. J., Wamba, S. F., Papadopoulos, T., & Hazen, B. (2019). Exploring the relationship between big data analytics capability and competitive performance: The mediating roles of dynamic and operational capabilities. *Journal of Business Research*, 98, 276–287. DOI: 10.1016/j.jbusres.2018.11.031

International Organization for Standardization (ISO). (2018). *ISO 55000:2018 Asset management – Overview, principles and terminology*. Geneva, Switzerland: ISO. Retrieved from www.iso.org/standard/55088.html.

Khan, M. Z. U., Karim, A., & Rana, M. A. (2020). A review of predictive maintenance of industrial systems using machine learning. *Journal of Industrial Information Integration*, 17, 100116. DOI: 10.1016/j.jii.2019.100116.

Kumar, A., Patel, R. B., & Venugopal, D. (2020). Cyber-physical systems and digital twin for predictive maintenance of manufacturing processes: A comprehensive review. *Robotics and Computer-Integrated Manufacturing*, 63, 101897. DOI: 10.1016/j.rcim.2019.101897

Kusiak, A., & Verma, A. (2018). Predictive maintenance of multi-stage manufacturing processes. *Journal of Manufacturing Systems*, 47, 88–94. DOI: 10.1016/j.jmsy.2018.05.005

Lee, J., Bagheri, B., & Kao, H. A. (2015). A Cyber-Physical Systems architecture for Industry 4.0-based manufacturing systems. *Manufacturing Letters*, 3, 18–23. DOI: 10.1016/j.mfglet.2014.12.00.

Li, X., Wang, X., & Cheng, L. (2020). Predictive Maintenance in Manufacturing: Challenges and Opportunities. *IEEE Transactions on Industrial Informatics*, 16(3), 1890–1899. DOI: 10.1109/TII.2019.293658.

Sharma, A., & Chen, J. (2019). Edge Computing in Industrial Internet of Things: A Survey. *IEEE Access*, 7, 103270–103287. DOI: 10.1109/ACCESS.2019.292801.

5 AI Models for Predictive Maintenance

Roopa Devi E M, R Shanthakumari, S Dhanushya, and G Kiruthika

5.1 INTRODUCTION

With increasing advancements in technology and the necessity to meet the growing demand for goods and services by people, the industrial operations and the level of automation has increased rapidly in the past decade. There is an increasing pressure on the industries and other sectors to use machinery which has maximum lifespan while minimizing operational failures and downtime. The main objective of industrial operations is optimizing asset performance and minimizing downtime. The increasing demand is due to the rising population and development in technology. In Industry 4.0, the paradigm of maintenance strategies has witnessed a transformative shift towards proactive and data driven approaches. Over time, predictive maintenance has become a cornerstone for seeking to optimize their asset management process. Preemptive maintenance interventions are made possible by the integration of DL and ML models, which has become crucial in predicting equipment breakdowns. [1] In-depth reviews of the various ML and DL models used in predictive maintenance are conducted in this study, along with a detailed analysis of their uses, advantages, and potential future prospects.

In contrast to conventional reactive methods, predictive maintenance offers a preventative strategy that predicts likely equipment failures using historical data analytics. Predictive maintenance techniques combined with ML and DL models have shown to be a powerful force in bringing in a new era of effectiveness and economy. Predictive maintenance models have a wide range of applications, much like industrial systems are complex. Organizations can utilize regression models, which vary from basic linear regression to advanced time series forecasting methods, to forecast the equipment's remaining usable life (RUL) by utilizing sensor data from the past. Classification approaches that offer useful insights into the probability of equipment breakdown within predetermined time intervals include logistic regression and ensemble techniques like random forests. Support vector machines and isolation forests are two examples of anomaly detection algorithms that greatly aid in the early detection of anomalous patterns suggestive of possible equipment failures. [2] Predictive maintenance is now more sophisticated because of deep learning models' ability to identify complex relationships and patterns. The comprehension of equipment behavior over time has been enhanced by the remarkable performance of convolutional neural networks (CNNs), recurrent neural networks

DOI: 10.1201/9781003480860-5

(RNNs) and long short-term memory (LSTM) networks in interpreting sequential and sensor data.

The objective is to present a comprehensive overview of these models, clarifying their distinctive features and investigating possible collaborations for improved predictive maintenance results. The analysis aims to give organizations the information they need to make informed decisions when implementing, upgrading, or refining predictive maintenance strategies. It emphasizes the critical role of incorporating domain knowledge, continuous model retraining, and adaptability to evolving industrial ecosystems. A more complex understanding will emerge as we go deeply into each model, helping to shape the course of predictive maintenance in the industrial environment going forward.

5.2 MACHINE LEARNING MODELS

Using machine learning models, predictive maintenance forecasts and prevents equipment breakdowns by evaluating past data, tracking current events, and determining when maintenance is necessary. Predictive maintenance offers an active approach to equipment management in the context of industrial operations, where downtime can be expensive and perhaps even deadly. In predictive maintenance, a machine learning model is a computational system that uses real-time and historical data to recognize patterns, trends, and deviations. This allows the system to forecast the equipment's future state of health. The basic objective is to forecast the likelihood of a machine or component failing so that maintenance tasks can be scheduled and carried out precisely when needed, minimizing downtime, cutting expenses, and enhancing overall operational efficiency. Predictive maintenance's essential element is the choice of algorithms. Various machine learning algorithms can be applied according to the kind of data and the particular needs of the maintenance task. Commonly used algorithms include regression models, time series models, survival analysis models, and ensemble methods.

Figure 5.1 shows a machine learning process for predictive maintenance.

5.2.1 K-Nearest Neighbor (KNN)

One machine learning algorithm is KNN that can be applied to tasks involving regression and classification. In predictive maintenance, KNN can be applied to forecast equipment failures and determine the likelihood of maintenance requirements based on historical data. KNN classifies or predicts determined by averaging the k-nearest data points or the majority class. The term "nearest" is defined by a distance metric, commonly the Euclidean distance or other measures like Manhattan or Minkowski. During the training phase, the algorithm stores the historical data points and their corresponding labels or outcomes. In predictive maintenance, these data points include features such as sensor readings, time, and maintenance records. [3] Whenever a new data point is predicted, KNN identifies the k-nearest neighbors within the feature area. The dominant class in terms of classification or mean for regression, of these neighbors determines the prediction for the new data point.

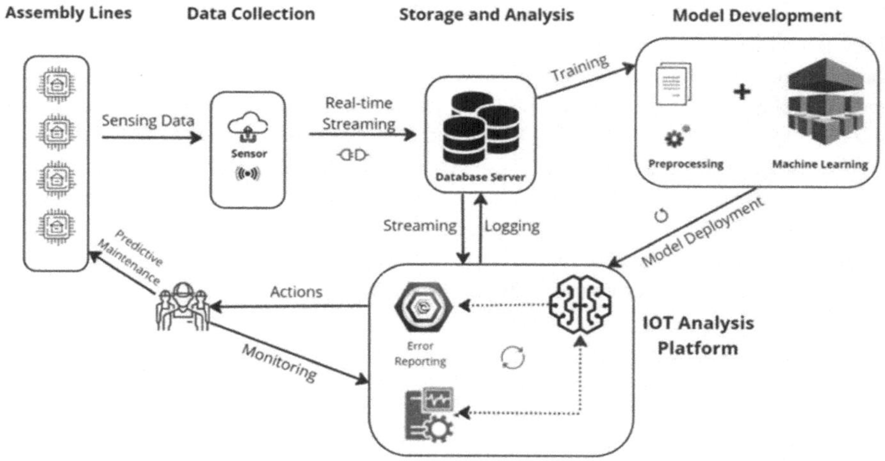

FIGURE 5.1 Predictive Maintenance using Machine Learning Models.

Application in predictive maintenance:

- **Feature selection**: KNN considers the entire feature space for predictions. In predictive maintenance, relevant features such as temperature, vibration, or usage hours are used to identify patterns indicative of potential failures.
- **Scalability**: KNN's simplicity and scalability make it applicable to various equipment types and sizes. It can adapt to different maintenance scenarios without extensive model tuning.
- **Dynamic systems**: KNN is effective for predictive maintenance in dynamic systems where equipment behavior may change over time. It captures non-linear relationships and adapts to evolving conditions.

Challenges and considerations:

- **Computationally intensive**: when working with large datasets, KNN can be computationally demanding. Optimization techniques and efficient data structures like KD-trees are often employed to address this challenge.
- **Determining optimal k**: the choice of the parameter k (number of neighbors) is crucial. Cross-validation and performance metrics help in selecting the optimal k for a given predictive maintenance task.

5.2.2 Support Vector Machine

Support vector machines is a powerful machine learning algorithm which can be effectively utilized for predictive maintenance tasks. SVM is primarily used for classification, and its capability to manage relationships that are both linear and non-linear in data makes it suitable for identifying patterns related to equipment failures and maintenance needs. A supervised learning method called SVM was created for classification applications. It seeks to identify the optimal hyperplane in a high-dimensional space for dividing data points belonging to various classes. SVM seeks to optimize the margin, which measures the separation between the nearest support vectors (data points) from each class and the hyperplane. This ensures better generalization to new, unseen data. [4] SVM can handle non-linear relationships through the use of kernel functions. Kernels establish a higher-dimensional space from the input features, allowing SVM to find non-linear decision boundaries. In predictive maintenance, SVM can categorize equipment health based on features such as sensor readings, operational parameters, and historical maintenance records.

Application in predictive maintenance:

- **Feature space mapping**: SVM excels at mapping high-dimensional feature spaces, making it suitable for predictive maintenance where multiple features contribute to equipment health assessments. It can handle complex relationships between these features.
- **Handling non-linearity**: SVM's kernel trick enables it to capture non-linear dependencies in the data. This is crucial in predictive maintenance scenarios where failure patterns may not follow linear trends.
- **Binary classification**: SVM is often used for binary classification tasks in predictive maintenance, where the goal is to determine whether an equipment failure is likely within a specific timeframe.
- **Outlier detection**: SVM can be used for outlier detection, identifying data points that deviate significantly from normal behavior. This is valuable in predictive maintenance for identifying anomalies or early signs of potential failures.

Challenges and considerations:

- **Parameter tuning**: the performance of SVM depends on parameter tuning, including the choice of the kernel function and other hyperparameters. Cross-validation is often used to find optimal parameter values.
- **Interpretability**: SVM models may be less interpretable compared to simpler models like decision trees. Understanding the decision boundaries and the impact of individual features can be challenging.

Handling imbalanced data:

- In scenarios where maintenance events are rare (imbalanced data), proper techniques, such as adjusting class weights, should be applied to prevent bias towards the majority class.

5.2.3 Regression Models

5.2.3.1 Linear Regression

Linear regression is a common method used for predictive analysis. It involves using the linear relationships between a target variable and independent variables to forecast the future of the target variable. This method assumes a link of cause and effect between the predictors and the target variable. Linear regression is a basic algorithm for machine learning that is utilized to forecast continuous outcomes based on input features. It is predicated on the variables having a linear relationship, as shown by the equation

$$Y = 0 + 1{*}X \tag{5.1}$$

During training, the goal is to find the values of coefficients 0 and 1 that minimize the variations between predicted and actual values. The linear regression line is positioned to minimize the vertical distances between observed and predicted values, providing the best-fit line to the data. Linear regression is frequently used for tasks like trend prediction and variable relationship understanding in a variety of fields.

Linear regression aids predictive maintenance by analyzing historical data, identifying patterns in equipment health, and predicting remaining useful life. This enables organizations to optimize maintenance schedules, reduce downtime, and allocate resources efficiently. By quantifying relationships between operational parameters and failure likelihood, linear regression facilitates data-driven decision-making, supporting a proactive maintenance approach for cost reduction and enhanced operational efficiency.

5.2.3.2 Lasso and Ridge Regression

Lasso (least absolute shrinkage and selection operator) regression is a regularization method used in linear regression to prevent overfitting and perform feature selection. It adds a penalty term to the ordinary least squares(OLS) objective function, introducing a constraint on the absolute values of the coefficients. λ is the regularization parameter controlling the strength of the penalty term. By effectively bringing some coefficients to zero, lasso performs feature selection by excluding less relevant variables from the model. This sparsity-inducing property makes lasso particularly useful when dealing with multidimensional datasets with many arbitrary, unimportant features.

Application in predictive maintenance:

- **Identifying critical features**: lasso regression can assist in determining the most important features in predicting equipment failures or maintenance needs. Less significant features' coefficients are probably going to be shrunk to zero.
- **Feature selection**: by eliminating irrelevant features, lasso regression simplifies the model and reduces the risk of overfitting.
- **Handling sparse datasets**: lasso can handle situations where many features are irrelevant or have negligible impact on the predictive maintenance task.

Ridge regression is another regularization method designed to address multicollinearity in linear regression models. Like lasso, it includes a penalty value to the OLS objective function, but instead of using the absolute values of coefficients, it uses the squared values. The regularization parameter λ controls the shrinkage of coefficients, and

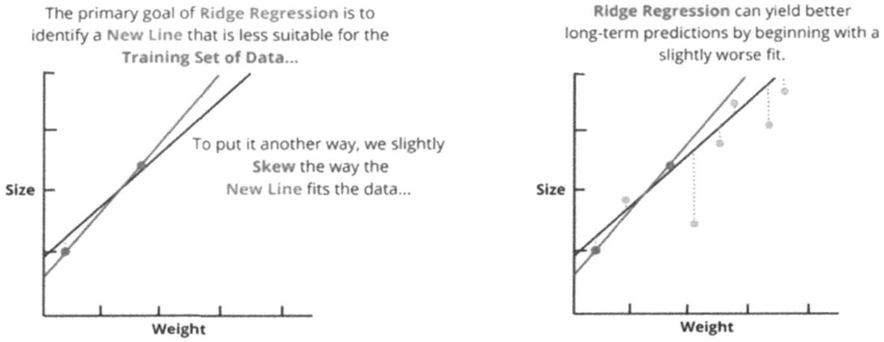

FIGURE 5.2 Identification of New Line using Ridge Regression.

ridge regression tends to shrink all coefficients toward zero, but without eliminating any entirely. Ridge regression is effective in stabilizing coefficient estimates when multicollinearity is present, reducing the impact of highly correlated predictors.

Figure 5.2 shows a concept diagram about ridge regression with text boxes explaining that the goal is to find a line that is less sensitive to the training data. Application in predictive maintenance:

- **Multicolinearity**: when dealing with sensors or features that may be highly correlated, ridge regression helps in stabilizing and improving the numerical stability of the model.
- **Generalization**: ridge regression can enhance the model's capacity for generalization by preventing overfitting when associated features are present.
- **Robustness**: ridge regression is less prone to outliers compared to ordinary least squares regression.

Regressions such as lasso and ridge help with feature selection and multicollinearity mitigation, which supports predictive maintenance. Lasso's sparsity-inducing property automatically chooses essential features, while ridge stabilizes coefficients, boosting model resilience. By enhancing generality, these regularization strategies guarantee precise predictions on fresh data. They support more effective and understandable models in predictive maintenance, directing proactive tactics for economical maintenance scheduling and resource allocation.

5.2.4 Time Series Models

Time series models are machine learning and statistical methods for examining and forecasting patterns seen in sequential data that are time-based. Time series data is made up of measurements or observations that are gathered over a period of time, with a specific timestamp assigned to each data point. Numerous disciplines, including finance, signal processing, economics, and predictive maintenance, use time series models. A commonly used time series model is the autoregressive integrated moving average predictive model.

5.2.4.1 Autoregressive Integrated Moving Average Predictive Model

Autoregressive integrated moving average is referred to as ARIMA. This method of statistical analysis makes use of time series data to comprehend a dataset or forecast future trends. ARIMA models use historical data to forecast future values, making them autoregressive. These types of models are commonly utilized in technical analysis to predict future outcomes of securities. The underlying assumption is that the future will resemble the past. ARIMA can be broken down into its constituent parts: moving average (MA), autoregression (AR), and differencing (I). In ARIMA, the order of the differencing, the moving average, and the autoregressive are marked by the notation ARIMA(p, d, q).

p is the order of the autoregressive component,
d is the degree of differencing,
q is the order of the moving average component.

Time series data with distinct trends, patterns, and seasonality are good candidates for the ARIMA model. In a variety of fields, including finance, economics, and environmental science for tasks like stock price prediction, economic forecasting, and climate modeling.

ARIMA models are estimated using historical time series data, and their parameters are optimized to reduce the variations between observed and predicted values. While ARIMA is a powerful tool, it has certain drawbacks, like the linearity and stationarity assumptions. In actual use, the model might require adaptation or combination with other methods to address data that contain more intricate patterns.

ARIMA is a useful tool in predictive maintenance. It uses historical time series data to forecast equipment health and potential failures. The stepwise application of ARIMA involves data collection, preprocessing, stationarity checks, model identification, training, validation, and continuous monitoring. Historical data includes sensor readings, maintenance logs, and failure events. Preprocessing ensures data cleanliness. ARIMA requires stationarity, achieved through differencing. Model identification involves analyzing autocorrelation and partial autocorrelation functions to determine components (p, q, and d). The model is then trained, validated, and adjusted based on performance metrics. ARIMA predictions provide insights into equipment health, triggering proactive maintenance actions. Continuous monitoring ensures the model remains adaptive.

The flowchart in Figure 5.3 shows the process in assessing an ARIMA model.

5.2.5 Random Forest

The ensemble learning technique random forest integrates the forecasts from several decision trees. Using a random subset of the features and data, each tree is built separately. The last step is to combine all of the individual trees' predictions into

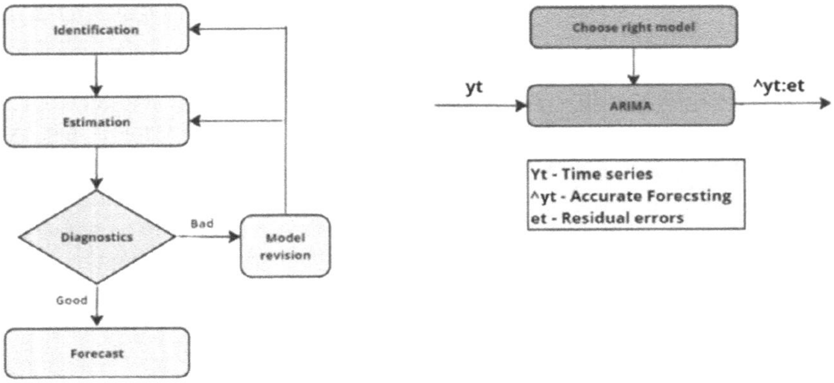

FIGURE 5.3 (a) ARIMA Algorithm Schematic. (b) ARIMA Model Assumptions.

one final prediction, usually by averaging (regression) or classifying the results by majority vote.

Application in predictive maintenance:

- **Equipment failure prediction**: random forests are widely used for predicting equipment failures. They can analyze sensor data, operational parameters, and historical maintenance records to identify patterns indicative of potential issues.
- **Remaining useful life (RUL) estimation**: random forests can estimate the remaining useful life of equipment, helping organizations plan maintenance activities more effectively and prevent unexpected failures.
- **Versatility across equipment types**: random forests are versatile and is applicable to various kinds of equipment across various industries. They adapt well to diverse maintenance scenarios and data structures.

Challenges and considerations:

- **Interpretability**: while random forests offer feature importance analysis, the model as a whole may lack interpretability compared to simpler models. Understanding the decision-making process across multiple trees can be challenging.
- **Computational complexity**: random forests can be computationally demanding, particularly when dealing with big datasets or lots of trees. However, advancements in hardware and optimization techniques have addressed some of these challenges.
- **Overfitting**: random forests are susceptible to overfitting, capturing noise in the training data. Techniques such as limiting tree depth, adjusting the minimum samples per leaf, or using feature selection can help mitigate overfitting.

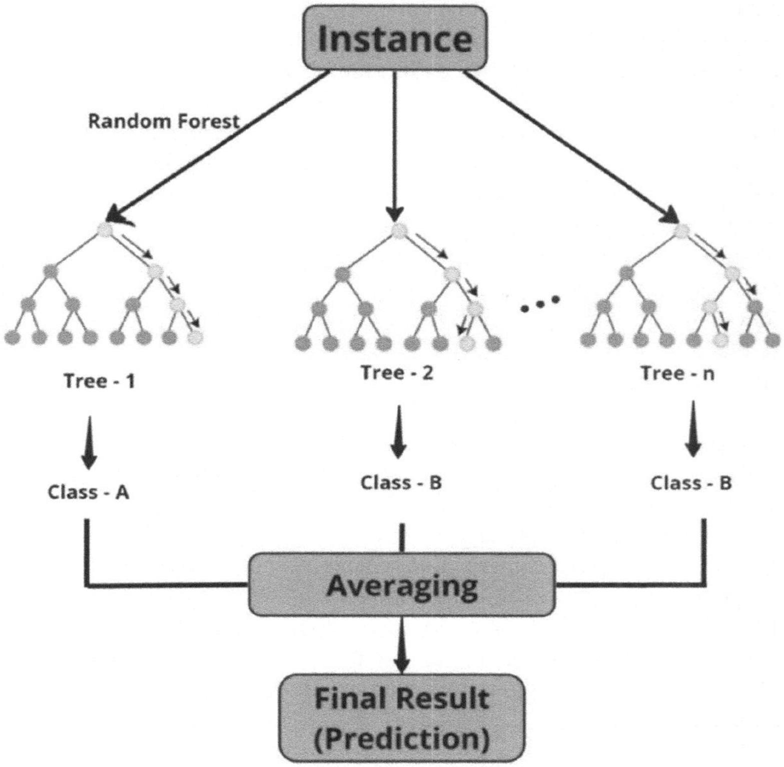

FIGURE 5.4 Simplified Random Forest.

- **Handling missing data**: random forests handle missing data by imputing missing values or using surrogate splits. However, the imputation strategy can impact model performance, and careful consideration is needed when dealing with missing data.

A diagram depicting the random forest algorithm is shown in Figure 5.4.

5.2.6 Gradient Boosting

Gradient boosting is an ensemble learning method that produces a sequence of weak learners. Every tree fixes the mistakes of its predecessor, resulting in a strong predictive model. Gradient boosting is effective in predictive maintenance due to its ability to capture complex relationships and improve predictive performance.

Application in predictive maintenance:

- **Failure prediction**: gradient boosting is widely used for predicting equipment failures. It can analyze diverse data sources such as sensor readings, operational parameters, and historical maintenance records to identify patterns indicative of potential issues.

- **Remaining useful life (RUL) estimation**: gradient boosting can estimate the remaining useful life of equipment. This is crucial for planning maintenance activities, optimizing resources, and avoiding unexpected downtime.
- **Anomaly detection**: gradient boosting excels at anomaly detection by capturing deviations from normal behavior. It can identify irregularities or unexpected patterns in equipment data, signaling potential issues.
- **Handling non-linearity**: gradient boosting is effective at handling non-linear relationships in data. In predictive maintenance, where failure patterns may not follow linear trends, this capability is crucial for accurate predictions.

Challenges and considerations:

- **Computational intensity**: training a large number of decision trees sequentially can be computationally intensive. However, optimizations and parallel processing can mitigate this challenge.
- **Overfitting**: gradient boosting is prone to overfitting, especially when the model complexity is high. Regularization techniques and hyperparameter tuning are essential to prevent overfitting and improve model generalization.
- **Handling imbalanced data**: in cases where maintenance events are rare, leading to imbalanced datasets, gradient boosting may require techniques like adjusting class weights or using specific algorithms designed to handle imbalanced data.

5.3 DEEP LEARNING MODELS

A game-changing tactic for businesses looking to improve the dependability and effectiveness of their machinery and equipment is predictive maintenance. Conventional maintenance techniques frequently depend on preset plans or reactive fixes for malfunctions, which leads to downtime and higher operating expenses. Conversely, predictive maintenance makes use of cutting-edge technologies, especially deep learning models, to evaluate enormous volumes of historical data and identify any equipment problems before they arise. Different deep learning architectures are essential to predictive maintenance in this situation. Autoencoders, generative adversarial networks (GANs), deep reinforcement learning (DRL), recurrent neural networks (RNNs), and convolutional neural networks (CNNs) are some of the deep learning models used in predictive maintenance. RNNs are good at processing sequential time series data, especially LSTM and GRU variations, which makes them useful for forecasting equipment breakdowns over time. CNNs are skilled at spotting patterns in spatial data, particularly when it comes to identifying wear or damage in image-based data that has been taken by sensors or cameras. [5] In order to detect anomalies, autoencoders learn effective representations of typical data and look for departures from the patterns they have learned. In situations where there is a lack of labeled failure data, GANs help generate synthetic data so that the model can train on fictitious defective scenarios. DRL is used to help the model learn the best practices for extending the lifespan of equipment when making maintenance choices in dynamic conditions.

5.3.1 Convolutional Neural Network (CNN)

Convolutional neural networks have become pivotal in modern deep learning, transforming computer vision and pattern recognition. Originally designed for image classification, CNNs now surpass visual data analysis by autonomously learning hierarchical features, mirroring the intricacies of the human visual system. Beyond their roots in computer vision, CNNs exhibit remarkable adaptability, extending their applications to predictive analysis in diverse domains. In predictive modeling, CNNs are indispensable, proving effective in tasks such as time series forecasting, healthcare analytics, and predictive maintenance in industrial settings. The automatic learning of intricate patterns and representations from varied datasets positions CNNs as versatile tools, enhancing accuracy and efficiency in predictions from complex data sources.

CNNs are powerful for predictive maintenance, particularly with time series sensor data or images. For time series data, a 1D CNN architecture with convolutional layers captures local patterns, pooling layers reduce dimensions, and fully connected layers facilitate global feature learning. In the case of images, a 2D CNN extracts hierarchical features through convolution and pooling layers, followed by dense layers for classification or regression. [6] Transfer learning with pre trained models like VGG or ResNet enhances performance, especially when data is limited. Data augmentation techniques, such as flipping and rotation, improve model robustness. Evaluation metrics like accuracy and lead time to failure are essential. Preprocessing, normalization, and partitioning data for training and testing are crucial steps. Hyperparameter tuning and cross-validation optimize CNNs for predictive maintenance tasks, making them effective tools for analyzing equipment health and predicting potential failures.

Figure 5.5 shows a diagram of the process of data collection and processing for predictive maintenance.

5.3.1.1 ResNet

Residual networks, or ResNets, have revolutionized predictive analysis in deep learning. They introduce the concept of residual learning, allowing networks to

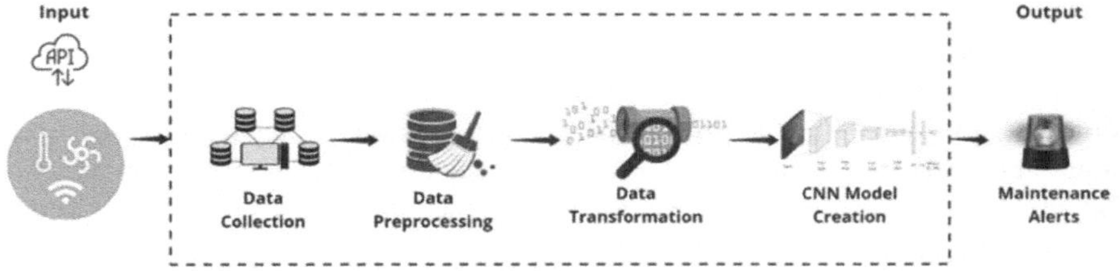

FIGURE 5.5 CNN Framework for Predictive Maintenance.

efficiently learn and extract hierarchical features. Traditional deep neural networks struggle with the vanishing gradient problem as depth increases, but ResNet overcomes this by using residual blocks. These blocks include a shortcut connection that permits the network to understand the distinction between input and output directly. This approach allows for the training of deep networks with smooth information flow during backpropagation. By passing input through a series of residual blocks and adding the learned residual to the original input, ResNet can capture intricate features even with increasing depth.
Architecture:

The core concept behind ResNet is the reformulation of the learning process from predicting the desired output to predicting the residual. Mathematically, the output of a residual block is represented as:

$$\text{Output} = F(\text{Input}) + \text{Input} \tag{5.2}$$

Here, F(Input) denotes the residual mapping learned by the layers within the block. This skip connection enables the network to easily learn identity mappings and facilitates the flow of gradients during backpropagation. ResNets are valuable in predictive maintenance for their ability to handle deep neural networks effectively. In the context of machinery health prediction, by solving the vanishing gradient issue, ResNets make it possible to train extremely deep networks. Their skip connections facilitate the information flowing via the network, enabling the modeling of intricate dependencies in time series sensor data or complex patterns in machinery images. In predictive maintenance tasks, a ResNet architecture can be used to examine sensor periodical data by employing 1D convolutional layers with residual connections. For image data, 2D convolutional layers can capture hierarchical features, and residual blocks enhance feature learning. With massive datasets, transfer learning with trained ResNet models can be advantageous when the available maintenance data is limited. By leveraging ResNets, predictive maintenance models gain the ability to learn intricate representations, enhancing the accuracy of fault predictions. Standardization, hyperparameter tuning, and data preprocessing are essential for best results. [7] Overall, ResNets contribute to robust and efficient predictive maintenance solutions, ensuring the reliability and longevity of machinery.

5.3.1.2 VGGNet

VGGNet, a groundbreaking CNN architecture, has revolutionized deep learning in image classification. Its meticulous layering, with options of 16 or 19 layers exclusively using 3×3 filters, showcases the power of depth in learning hierarchical features. Despite the computational demands, the benefits in feature learning and representation outweigh the costs. VGGNet's sequential arrangement of convolutional and max-pooling layers has become a benchmark and reference architecture for subsequent neural network designs. It processes input data through 3×3 convolutional layers and max-pooling layers for spatial down-sampling, capturing a wide range of high and low-level features for intricate image analysis. The mathematical formulations underpinning VGGNet involve convolutions, activations, and pooling operations. The convolution operation, symbolized by *, can be expressed as

$$H_i = \sigma\left(\sum_{m=1}^{3} \sum_{n=1}^{3} W_{mn} * X_{(i+m-1)(j+n-1)}\right),$$

where H_i represents the output feature map, W_{mn} denotes the filter weights, X_{ij} is the input matrix, and σ is the rectified linear unit (ReLU) activation function.

Beyond image classification, VGGNet's prowess extends far beyond image classification. Its depth and feature learning capabilities make it adaptable to a wide range of computer vision tasks.

- **Object recognition**: VGGNet can identify and localize objects in images with remarkable accuracy, finding applications in autonomous vehicles, robotics, and medical imaging.
- **Image segmentation**: by segmenting images into distinct regions, VGGNet facilitates scene understanding and object manipulation, playing a crucial role in medical diagnosis and autonomous navigation.
- **Transfer learning**: VGGNet's pre-trained weights can be transferred to other tasks, significantly reducing training time and improving effectiveness in tasks involving anomaly detection and facial recognition.
- **Ethical considerations**: VGGNet's power comes with ethical responsibility. Its use in image recognition raises concerns about bias, fairness, and transparency. Ensuring diverse datasets and responsible use is crucial to avoid perpetuating societal inequalities.

In predictive maintenance, VGG (visual geometry group) networks, celebrated for their straightforward yet potent architecture, prove beneficial for analyzing visual data pertinent to machinery health. Engineered for image classification tasks, multiple convolutional layers with 3×3 filters are a feature of VGG architectures, followed by max-pooling layers, creating a deep and hierarchical feature learning structure. Adapting VGG for predictive maintenance involves representing machinery health images in the required format (height, width, channels). The network design follows a sequence of convolutional and pooling layers, maintaining a consistent pattern seen in VGG models. This repetitive structure enables the network to discern intricate patterns and features in machinery images, crucial for identifying potential faults. By utilizing VGG networks, predictive maintenance models gain the capability to extract and comprehend complex visual information, enhancing the accuracy of fault predictions. Robust preprocessing of the data, normalization, and careful hyperparameter tuning are vital to harness the full potential of VGG in predictive maintenance applications. The adaptability of VGG networks makes them valuable tools for efficiently assessing and predicting the health of machinery based on visual data.

5.3.1.3 LeNet

LeNet is a groundbreaking architecture in convolutional neural networks. Originally designed for recognizing handwritten digits, it has influenced the wider field of predictive analysis. With seven layers, including convolutional and subsampling layers, LeNet stands out for its use of local receptive fields and shared weights. This allows the network to effectively learn features invariant to translation. LeNet's workflow involves passing input through convolutional layers to extract local features, followed by subsampling layers to enhance computational efficiency. Non-linear activation functions like sigmoid and tanh aid in capturing complex relationships within the data. The convolution operation generates a weighted sum using the kernel weights and input values, while average pooling calculates the average value within a region. LeNet's advantages lie in its hierarchical feature learning and weight sharing, contributing to parameter efficiency and reducing overfitting risk.

Mathematical formulation:

- **Convolution operation**: given an input feature map X and a learnable kernel W, the convolution operation Y is computed as:

$$Y[i, j] = \Sigma_m \Sigma_n X[i+m, j+n] \cdot W[m, n] \tag{5.3}$$

- **Average pooling**: the average pooling operation for a region R in the input feature map is computed as:

$$Y[i, j] = 1 /(| R| \Sigma_m \Sigma_n X[i+m, j+n]) \tag{5.4}$$

- **Sigmoid activation**: the sigmoid activation function for an input x is defined as:

$$\sigma(x)= 1 / 1+e^{-x} \tag{5.5}$$

- **Tanh activation**: the hyperbolic tangent (tanh) activation function for an input x is defined as:

$$\tanh(x) = e^x - e^{-x} / e^x + e^{-x} \tag{5.6}$$

LeNet is highly skilled in predictive analysis, particularly in identifying complex patterns in multidimensional data. It excels in NLP tasks like sentiment analysis and language translation, as well as time series forecasting, healthcare analytics, and industrial predictive maintenance. Its ability to learn intricate patterns greatly improves prediction accuracy from complex data sources. LeNet is a valuable framework for understanding complex patterns in data, contributing to the advancement of machine learning in various domains. Its historical contributions and ongoing relevance make it a fundamental reference point in the evolution of deep learning and its applications in predictive analytics.

5.3.2 Recurrent Neural Network (RNN)

Industry predictive maintenance procedures are being revolutionized by recurrent neural networks(RNNs). RNNs are used to analyze historical performance data of machinery

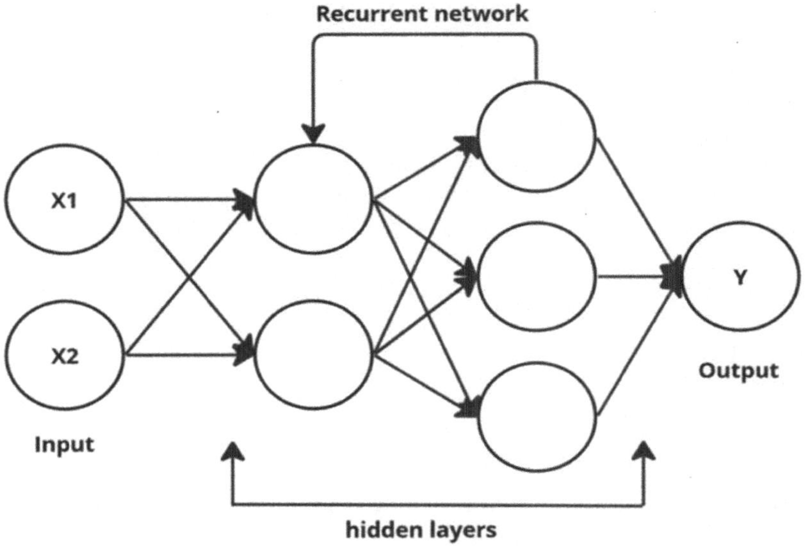

FIGURE 5.6 RNN Architecture.

and equipment because of their capacity to process sequential data. This makes it possible to anticipate possible problems before they happen, giving organizations the opportunity to plan maintenance tasks in advance. RNNs are especially good at processing time series data in the context of predictive maintenance, where the timing and order of events are crucial. RNNs may preserve information from prior time steps thanks to their recurrent connections, which gives them a way to record temporal dependencies. Forecasting equipment lifespans, estimating when maintenance is likely to be required, and seeing patterns suggestive of possible problems are just a few of the predictive maintenance applications for RNNs. Through the utilization of RNNs, industries can shift from reactive to proactive maintenance approaches, resulting in decreased downtime, lower operating expenses, and ultimately improved dependability of vital machinery and infrastructure. Recurrent neural networks(RNNs) come in several types, and each variant addresses certain challenges or limitations inherent in the basic RNN architecture. Figure 5.6 shows the RNN architecture.

Some key types of RNNs are vanilla RNN, long short-term memory, and gated recurrent unit.

5.3.2.1 Vanilla RNN

Vanilla recurrent neural networks(RNNs) also known as simple or Elman RNN, constitute a pivotal tool in sequential data analysis, offering a versatile solution to capture temporal dependencies. Unlike traditional RNNs, the vanilla variant incorporates a straightforward loop mechanism, facilitating the transfer of data transferring between time steps. This inherent architecture enables the model to maintain a hidden state that encapsulates insights from prior inputs. In its workflow, vanilla RNNs iteratively process sequential data, updating their hidden state for every time interval. This recurrent nature allows the network to learn and remember patterns over extended sequences. Mathematical formulation:

Applying a non-linear activation function (such as tanh or ReLU) to the sum of the product of the input at time t (x_t) yields the hidden state h at time t (h_t) and its associated weight (W_x), and the product of the hidden state at the previous time step ($h_{(t-1)}$) and its associated weight (W_h). A bias term b is also added. It is expressed as:

$$h_t = f(W_x * x_t + W_h * h_{(t-1)} + b) \tag{5.7}$$

The output y at time t is then calculated by applying a second linear transformation to h_t:

$$y_t = W_y * h_t + b_y \tag{5.8}$$

However, a notable challenge surfaces in the form of the problem of vanishing gradient during training. Gradients diminish exponentially as they propagate backward through time, impeding the network's capacity to effectively learn and retain data from distant past steps. This limitation poses constraints on the model's capacity to comprehend and utilize long-term dependencies within the data.

Vanilla recurrent neural networks play a crucial role in predictive analysis across various domains. In time series forecasting, they predict future values in financial markets, energy consumption, and more. They contribute to speech recognition systems, healthcare predictive modeling, fault detection in manufacturing, and weather prediction. Additionally, these networks are valuable in customer behavior analysis for marketing and e-commerce and play a key role in optimizing supply chain management by forecasting demand and optimizing inventory levels. The wide-ranging applications underscore the significance of vanilla RNNs in predictive analysis, where understanding sequential patterns is essential for making accurate predictions across various domains.

5.3.2.2 Long Short-term Memory

Long short-term memory(LSTM) networks have emerged as a crucial development in the realm of RNNs, demonstrating exceptional proficiency when simulating long-term dependent sequences. LSTM systems are integral to predictive analysis due to their prowess in modeling sequences with long-term dependencies. In time series forecasting, LSTMs excel at predicting future values in financial markets, energy consumption, and healthcare. [8] Their application in natural language processing extends to sentiment analysis and language translation. LSTMs also prove invaluable in speech recognition, customer behavior analysis, and optimizing supply chain management. Challenges include computational complexity, potential overfitting, and reduced interpretability due to their intricate architecture. However, their benefits lie in effectively capturing intricate temporal patterns, mitigating the vanishing gradient problem, and delivering superior performance in tasks demanding nuanced understanding of sequential data. The balance between challenges and benefits underscores the significance of LSTMs in predictive analysis across diverse domains.

5.3.2.3 Gated Recurrent Unit

The gated recurrent unit(GRU) stands as a powerful recurrent neural network variant, specifically designed to address the challenges posed by long-term dependencies in sequential data. GRUs share similarities with long short-term memory(LSTM) networks but introduce a simplified architecture with two crucial gates: the reset and the update gate.

Architecture:
At its core, a GRU comprises a periodically updated hidden state over sequential inputs. The update gate determines how much of the previous hidden state should be forgotten, while the update gate controls how much of the current input should be absorbed into the new hidden state. The simplicity of the GRU's architecture makes it computationally more efficient than LSTMs, rendering it particularly suitable for various applications.

The workflow of a GRU involves the following steps:
Reset gate operation: the reset gate is accountable for deciding portions of the previous hidden state should be reset based on the current input. Mathematically, it is expressed as:

$$r_t = \sigma\,(W_r\,[h_{t-1}, x_t] + b_r) \tag{5.9}$$

where r_t represents reset gate output, W_r represents the reset gate weights, and b_r is the reset gate bias.

Update gate operation: the update gate establishes the amount of new information from the current input that should be included in the new hidden state. It is calculated as:

$$z_t = \sigma\,(W_z\,[h_{t-1}, x_t] + b_z) \tag{5.10}$$

where z_t is the update gate output, W_z represents the update gate weights, and b_z is the update gate bias.

Candidate hidden state: a candidate hidden state, $\sim h_t$, is calculated to capture the new information from the current input:

$$h_t = \tanh\,(W\,[r_t \odot h_{t-1}, x_t] + b) \tag{5.11}$$

where $\odot$ denotes element-wise multiplication, W represents the candidate weights, and b is the candidate bias.

New hidden state: the new hidden state, h_t is determined by combining the previous hidden state with the candidate hidden state based on the update gate:

$$H_t = (1-z_t) \odot h_{t-1} + z_t \odot \sim h_t \tag{5.12}$$

The gated recurrent unit(GRU) has emerged as a key player in predictive analysis, demonstrating versatile applications across diverse domains. In time series forecasting, GRUs shine in predicting future values for dynamic variables like stock prices and weather patterns. Their proficiency lies in capturing temporal dependencies and discerning intricate patterns within sequential data. This makes them particularly valuable in applications requiring a nuanced understanding of historical context, such as natural language processing. GRUs can be employed to model equipment health trends over time. They excel in handling sequential patterns, making them suitable for predicting machinery failures based on sensor readings. GRUs learn to selectively update and reset information, allowing them to adapt to varying machinery conditions. To implement GRUs in predictive maintenance, the time series sensor data is fed into the network, and the GRU layers capture temporal dependencies. The model can then predict potential failures based on patterns learned from historical sensor readings. GRUs are advantageous when dealing with irregularly sampled data or when the importance of certain time steps varies.

The benefits of GRUs in predictive analysis are noteworthy. Their efficient learning of temporal dependencies, reduced computational complexity, and parameter reduction compared to more complex models like LSTM networks make them computationally efficient and resource-friendly. The simplified architecture of GRUs facilitates faster training, requiring less memory and computation power. However, challenges accompany the use of GRUs in predictive analysis. Their limited capacity to capture highly complex patterns, especially in comparison to more intricate models like LSTMs, poses a challenge in tasks demanding a high degree of intricacy. Additionally, while excelling in short-term dependencies, GRUs may underperform in tasks dominated by intricate long-term dependencies. The simplified architecture of GRUs also introduces interpretability challenges, as understanding their decision-making process may be less straightforward compared to models with more intricate structures.

5.3.3 Autoenocders

An artificial neural network type called an autoencoder is utilized for unsupervised learning, particularly in the field of feature learning and dimensionality reduction. The autoencoder's architecture is designed to encode the original input as precisely as possible after producing a lower-dimensional representation from the incoming data. The network consists of an encoder and a decoder, and it is trained to reduce the discrepancy between the reconstructed output and the input. The key elements of an autoencoder are broken down here:

- **Encoder**: the input data is mapped by the encoder to a lower-dimensional representation known as the "encoding" or "latent space". In this step, the input data is compressed into a representation that captures its key characteristics.
- **Decoder**: by using the encoded representation, the decoder recreates the original input data. The objective is to produce an output that closely resembles the input, ensuring that the network captures the most relevant information during the encoding process.

- **Latent space**: the lower-dimensional representation created by the encoder is called the latent space. This space contains a condensed form of the input data, where each point represents a potential configuration of the input features.
- **Loss function**: during training, an autoencoder minimizes a loss function that determines the difference between the input and the rebuilt output. Depending on the type of data, common loss functions include binary cross-entropy and mean squared error (MSE).

The training process typically entails providing the same data through the network during both the encoding and decoding steps. Autoencoders are versatile and can be adapted for various tasks, including dimensionality reduction, denoising, anomaly detection, and even generative modeling. A variety of autoencoder types exist, each with its specific variations, such as sparse autoencoders, variational autoencoders(VAEs), and denoising autoencoders. These variations add additional constraints or modifications to the basic autoencoder architecture, expanding its capabilities for different applications. Figure 5.7 shows prediction using autoencoders.

Types of Autoencoders:

- **Vanilla autoencoder**: the basic form of an autoencoder, consisting of an encoder to compress the input data and a reconstruction decoder. The goal is to reduce the variation between the original output and the reconstruction.
- **Sparse autoencoder**: introduces sparsity constraints during training, encouraging the autoencoder to acquire a minimal depiction of the input data. This can be useful for feature selection and learning more meaningful representations.
- **Denoising autoencoder**: capable of reassembling a clean, original input from a noisy or distorted one. This type of autoencoder is effective for denoising data and extracting robust features by learning to ignore noise during the encoding process.
- **Variational autoencoder (VAE)**: combines the principles of autoencoders with probabilistic modeling. VAEs learn a probabilistic distribution in the latent

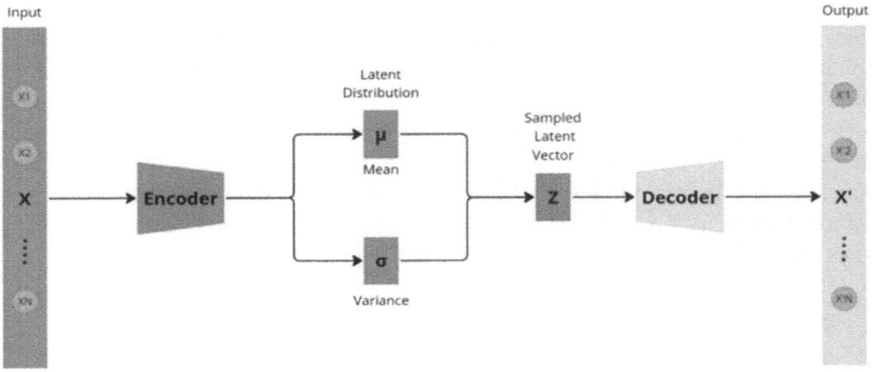

FIGURE 5.7 Prediction using Autoencoders.

space, enabling the generation of new data samples. This makes them useful for generative tasks.

- **Contractive autoencoder**: augments the basic autoencoder with a loss function penalty term that mandates a stable and reliable encoding for the model. This enhances the generalization of the learned representation and helps avoid overfitting.
- **Sequence-to-sequence autoencoder**: suitable for tasks where the input and output are sequences of varying lengths. It employs recurrent neural networks to handle sequences and can be used for tasks like language translation.

Autoencoders are a powerful tool for predictive maintenance, as they can learn meaningful representations from high-dimensional sensor data and help in identifying anomalies indicative of potential equipment failures. Here's how autoencoders can be applied in the context of predictive maintenance:

- **Normal operation modeling**: train an autoencoder using historical data collected during normal operating conditions. The autoencoder learns to encode the typical patterns and features present in the sensor data.
- **Anomaly detection**: during real-time operation, feed current sensor data through the trained autoencoder. If the reconstructed output significantly deviates from the input, it suggests an anomaly or deviation from normal behavior. This can be a signal for potential equipment failure or degradation.
- **Feature learning**: without the need for human feature engineering, autoencoders can automatically extract pertinent features from unprocessed sensor data. Especially when working with large datasets and complex sensor data, this can be very helpful.
- **Dimensionality reduction**: with the aid of auto encoders, one can reduce the dimensionality of sensor data while maintaining crucial information. This can be especially useful for efficient storage, processing, and visualization of data.
- **Time series analysis**: if the data involves time series information, recurrent autoencoders can be employed to recognize patterns and temporal connections in the sensor data. This is crucial for understanding how equipment conditions evolve over time.

Implementing autoencoders for predictive maintenance demands close examination of the unique qualities of the equipment, the nature of the data, and the available labeled data for training and validation. Regular model monitoring and updating are essential to ensure continued effectiveness in a dynamic operating environment. In summary, autoencoders play an important role in enhancing effectiveness of predictive maintenance strategies by providing tools for anomaly detection, fault diagnosis, and feature learning from sensor data. Their versatility and adaptability make them valuable assets in optimizing industrial operations and minimizing unplanned downtime.

5.3.4 Generative Adversarial Network

Generative adversarial networks (GANs) can be employed to enhance the capabilities of predictive models and address challenges related to data scarcity and imbalance.

GANs can create artificial sensor data that closely resembles real-world equipment operation. When failure events are uncommon, this synthetic data can be added to the training dataset to assist balance the class distribution and strengthen the predictive maintenance models' resilience. By generating additional realistic instances of sensor data, GANs contribute to the training of more accurate and generalizable predictive models, ultimately leading to improved anomaly detection and failure prediction in industrial systems. [9] With reference to predictive maintenance, a GAN can be used to generate synthetic sensor data to augment training datasets and improve the robustness of predictive models. Here's a simplified explanation of the GAN architecture with formulas and notations, specifically tailored for predictive maintenance:

- **Generator**: the generator, denoted as *G* accepts as input random noise z and generates synthetic sensor data *G*(*z*).
  - Generator function:

$$L_G = \text{Error}(D(G(z)),1) \tag{5.13}$$

- **Discriminator**: the discriminator, denoted as *D* evaluates both real sensor data *x* and generated sensor data *G(z)*, outputting a probability indicating whether the input is real or fake. *D(x)* indicates the probability that x is real, and *D(G(z))* indicates the probability that *G(z)* is real.
  - Discriminator function:

$$L_D = \text{Error}(D(x),1) + \text{Error}(D(G(z)),0) \tag{5.14}$$

- **Adversarial training**: the training consists of a minimax game in which the discriminator maximises this probability and the generator minimises the log probability of the discriminator correctly classifying the produced data.
  - Generator loss (minimizing):

$$L_{gen} = -\log(D(G(z))) \tag{5.15}$$

  - Discriminator loss (maximizing):

$$L_{disc} = -\log(D(x)) - \log(1 - D(G(z))) \tag{5.16}$$

Here, *D(x)* represents the probability assigned by the discriminator to real sensor data being real, and *1 – D(G(z))* denotes the probability assigned to generated sensor data being real. The generator aims to reduce the discriminator's log probability correctly classifying generated sensor data, while the discriminator aims to maximize this probability.

- **Overall loss function**: the total loss is a combination of the generator and discriminator losses.

$$L_{total} = L_{gen} + L_{disc} \tag{5.17}$$

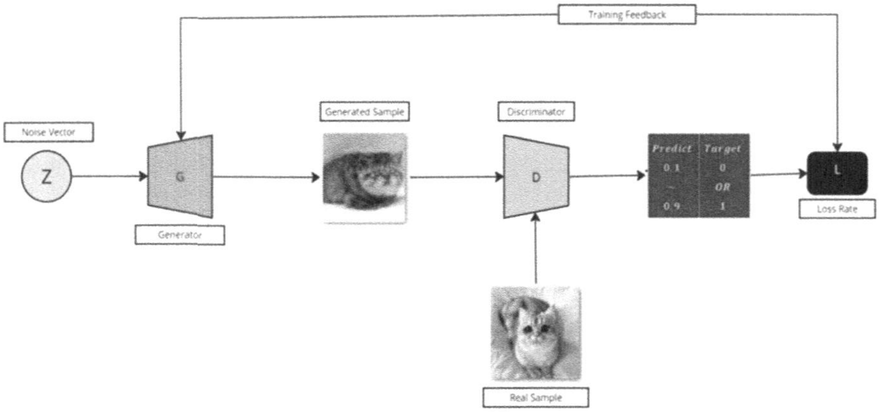

FIGURE 5.8 Training Algorithm of a GAN.

- **Training iterations**: the training involves alternating between updating the generator and discriminator parameters. The generator is updated to minimize its loss, and the discriminator is updated to maximize its loss.

In the predictive maintenance context, the generator is trained to create artificial sensor data that mimics real sensor data patterns during normal equipment operation. This synthetic data can then be combined with the real data for training predictive maintenance models, providing a more diverse dataset for improved model generalization. [10] The discriminator, in turn, learns to discern between actual and artificial sensor data, enhancing the quality of the generated samples.

Figure 5.8 shows the training algorithm of a GAN.

5.3.4.1 Types of GANs

GANs come in various types, each tailored to specific needs or challenges in different applications, including predictive maintenance. Here are some types of GANs that are applicable in the context of maintenance forecasting:

- **Standard GAN**: the basic GAN architecture, consisting of a generator and a discriminator. It is utilized for producing synthetic sensor data to augment training datasets for predictive maintenance models.
- **Conditional GAN (cGAN)**: extends the standard GAN by introducing additional information (conditions) during the generation process. In predictive maintenance, this could involve conditioning on specific equipment states, operating conditions, or other relevant parameters.
- **Wasserstein GAN (WGAN)**: an adaptation of the GAN architecture utilizing the Wasserstein distance for more stable training. WGANs can help mitigate training issues such as vanishing gradients and mode collapse, leading to an improved performance.

- **Deep convolutional GAN (DCGAN)**: the discriminator and generator both use convolutional layers to handle image data effectively. DCGANs are beneficial when dealing with sensor data that includes images or spatial information.
- **Time series GAN**: designed for generating synthetic time series data. In predictive maintenance, where temporal patterns are crucial, time series GANs can be used to generate realistic sequences of sensor data representing equipment behavior over time.
- **Recurrent GAN (RNN-GAN)**: integrates RNNs into the GAN architecture to record temporal relationships in data that is sequential. This is especially helpful for predicting equipment failures over time.
- **InfoGAN**: extends GANs to learn disentangled representations of data. It can be useful in predictive maintenance for learning interpretable and meaningful features from sensor data.
- **Adversarial autoencoder (AAE)**: combines the concepts of autoencoders and GANs, incorporating an encoder in addition to the generator and discriminator. AAEs can be useful for learning representations of sensor data and generating synthetic samples.

These variations of GANs offer flexibility and customization for different aspects of predictive maintenance, whether it involves different data types, temporal dependencies, or interpretability.

The particulars of the current predictive maintenance issue will determine which kind of GAN to use.

5.3.5 Deep Reinforcement Learning

Deep reinforcement learning (DRL) blends deep neural networks with reinforcement learning to revolutionize decision-making in complex environments. Intelligent agents use trial-and-error processes to navigate and learn from their surroundings, optimizing actions for maximum rewards over time. DRL involves agents interacting with environments, perceiving states, selecting actions, receiving rewards, and adjusting strategies. Deep neural networks, like Deep Q Networks (DQN), process sensory data to make informed decisions in high-dimensional input spaces. DRL relies on concepts such as the Bellman equation and Q-learning, which update Q-values to indicate expected future rewards. DQN approximates these values, handling complex decision spaces in real-world scenarios. DRL's advantages include learning from raw data and excelling in sequential decision- making tasks, as seen in gaming environments like Go and chess. [11] However, DRL faces challenges such as computational intensity, training instability, and the exploration-exploitation trade-off. The "reality gap" is a significant hurdle in transferring knowledge from simulated environments to the real world. [12] DRL includes various algorithms designed for specific scenarios.

5.3.5.1 Policy Gradient Methods

- **Objective**: directly optimize the policy function, which defines the probability distribution over actions given a certain state. [13]

- **Behavior determination**: policies are iteratively adjusted to maximize expected cumulative rewards, determining the agent's behavior.
- **Suitability**: suited for scenarios where an explicit policy needs optimization, often in environments with complex, non-linear decision spaces.

5.3.5.2 Value-Based Methods (e.g. DQN – Deep Q Networks)

- **Objective**: estimate the value function, representing the anticipated total benefits of performing a particular action in a particular state.
- **Decision guidance**: focus on evaluating the value of actions, aiding decision-making by prioritizing actions with higher expected rewards.
- **Effectiveness**: particularly effective in tasks where the value of actions is crucial, helping agents prioritize actions that lead to higher rewards.

5.3.5.3 Actor-Critic Methods

- **Integration**: integrate aspects of value-based and policy gradient approaches.
- **Actor role**: the actor component represents the policy function, determining the agent's behavior.
- **Critic role**: the critic component evaluates the value function, providing feedback on the expected rewards. Performance
- **Enhancement**: by integrating both approaches, actor-critic methods aim for enhanced performance and stability in learning.

5.3.5.4 Model-Based Methods

- **Approach**: involve constructing a model of the environment, capturing the dynamics and potential outcomes of actions.
- **Decision support**: the constructed model aids decision-making by simulating potential scenarios and their consequences.
- **Flexibility and adaptability**: tailored solutions. Each algorithm is tailored to specific scenarios and problem types, offering specialized solutions.
- **Domain adaptability**: adaptable across diverse applications in various domains, showcasing the versatility of DRL in addressing a wide range of challenges.

Figure 5.9 shows the DQN framework.

Deep Reinforcement Learning (DRL) is a transformative tool in predictive analysis, revolutionizing intelligent decision-making in complex scenarios. [14] DRL can learn directly from raw, unstructured data, unlike older approaches, which eliminates the need for human feature engineering. Its proficiency in handling time series data is particularly valuable in domains like finance, healthcare, and manufacturing. DRL excels at identifying temporal patterns, capturing

dependencies, and predicting future trends with exceptional granularity. It is especially effective in scenarios where understanding evolving data is crucial, such as predicting stock prices, patient outcomes, or equipment failures. [15] DRL's adaptability to diverse datasets and complex decision spaces allows it to unravel intricate relationships and provide accurate predictions. Its ability to generalize knowledge to

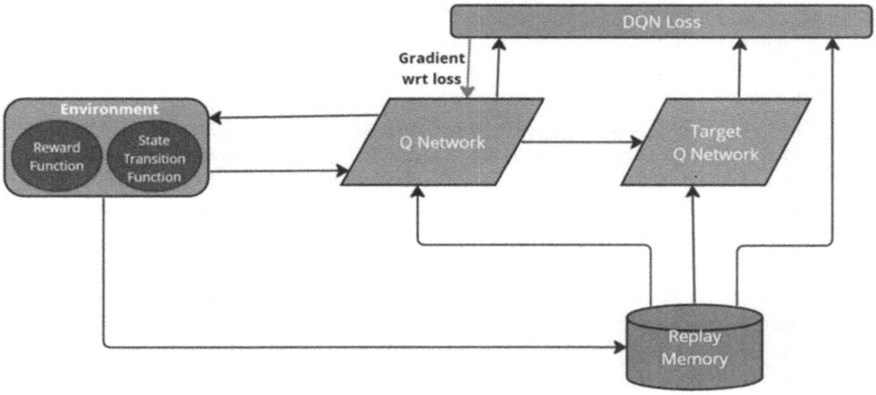

FIGURE 5.9 DQN Framework.

different scenarios enhances its utility across various predictive tasks. Whether in e-commerce, energy consumption, or supply chain logistics, DRL's capacity to extract nuanced insights makes it a game-changer. However, integrating DRL into predictive analysis requires careful consideration. [16] Training deep neural networks is challenging due to their high computational demands, especially with large datasets or real-time scenarios. Balancing prediction accuracy and computational resources is difficult. Additionally, understanding the decision- making process of DRL models, particularly in complex tasks, is crucial for trust and practical application.

REFERENCES

1. M. Guerroum, M. Zegrari, A.A. Elmahjoub, M. Berquedich and M. Masmoudi, "Machine learning for the predictive maintenance of a jaw crusher in the mining industry," 2021 IEEE International Conference on Technology Management, Operations and Decisions (ICTMOD), Marrakech, Morocco, 2021, pp.1–6, doi: 10.1109/ICTMOD52902.2021.9739338.
2. G. Makridis, D. Kyriazis and S. Plitsos, "Predictive maintenance leveraging machine learning for time-series forecasting in the maritime industry," 2020 IEEE 23rd International Conference on Intelligent Transportation Systems (ITSC), Rhodes, Greece, 2020, pp. 1–8, doi: 10.1109/ITSC45102.2020.9294450.
3. T.P. Carvalho, F.A.A.M.N. Soares, R. Vita, R. da P. Francisco, J.P. Basto and S.G.S. Alcalá, "A systematic literature review of machine learning methods applied to predictive maintenance," *Computers & Industrial Engineering*, vol. 137, no. August, p. 106024, 2019, doi: 10.1016/j.cie.2019.106024.
4. M. Paolanti, L. Romeo, A. Felicetti, A. Mancini, E. Frontoni and J. Loncarski, "Machine learning approach for predictive maintenance in Industry 4.0," 2018 14th IEEE/ASME International Conference on Mechatronic and Embedded Systems and Application, MESA, 2018, pp. 1–6, 2018, doi: 10.1109/MESA.2018.8449150.
5. C. Liu, D. Tang, H. Zhu and Q. Nie, "A novel predictive maintenance method based on deep adversarial learning in the intelligent manufacturing system," *IEEE Access*, vol. 9, pp. 49557–49575, 2021, doi: 10.1109/ACCESS.2021.3069256.

6. H. Wang, W. Zhang, D. Yang and Y. Xiang, "Deep-learning-enabled predictive maintenance in industrial internet of things: Methods, applications, and challenges," *IEEE Systems Journal*, vol. 17, no. 2, pp. 2602–2615, June 2023, doi: 10.1109/JSYST.2022.3193200.
7. J. Wang et al., "Deep heterogeneous GRU model for predictive analytics in smart manufacturing: Application to tool wear prediction," *Computers in Industry*, vol. 111, pp. 1–14, Oct. 2019.
8. W. Zhang, D. Yang and H. Wang, "Data-driven methods for predictive maintenance of industrial equipment: A survey," *IEEE System Journal*, vol. 13, no. 3, pp. 2213–2227, Sep. 2019.
9. K.S.H. Ong, W. Wang, D. Niyato and T. Friedrichs, "Deep-reinforcement-learning-based predictive maintenance model for effective resource management in industrial IoT," *IEEE Internet of Things Journal*, vol. 9, no. 7, pp. 5173–5188, 1 Apr. 2022, doi: 10.1109/JIOT.2021.3109955.
10. Y. Ran, X. Zhou, P. Lin, Y. Wen and R. Deng, "A survey of predictive maintenance: Systems, purposes and approaches," 2019. [Online]. Available: arXiv:1912.07383.
11. M. Compare, P. Baraldi and E. Zio, "Challenges to IoT-enabled predictive maintenance for industry 4.0," *IEEE Internet Things Journal*, vol. 7, no. 5, pp. 4585–4597, May 2020.
12. K.S.H. Ong, D. Niyato and C.Yuen, "Predictive maintenance for edge-based sensor networks: A deep reinforcement learning approach," in Proceedings of IEEE 6th WF-IoT, New Orleans, LA, USA, 2020, pp. 1–6.
13. Z. Chen, M. Wu, R. Zhao, F. Guretno, R. Yan and X. Li, "Machine remaining useful life prediction via an attention-based deep learning approach," IEEE *Transactions on Industrial Electronics*, vol. 68, no. 3, pp. 2521–2531, Mar. 2021.
14. G.R. Mode, P. Calyam and K.A. Hoque, "Impact of false data injection attacks on deep learning enabled predictive analytics," NOMS 2020 - 2020 IEEE/IFIP Network Operations and Management Symposium, Budapest, Hungary, 2020, pp. 1–7, doi: 10.1109/NOMS47738.2020.9110395.
15. T. Huuhtanen and A. Jung, "Predictive maintenance of photovoltaic panels via deep learning," in 2018 IEEE Data Science Workshop (DSW), IEEE, 2018, pp. 66–70.
16. R. Caponetto, F. Rizzo, L. Russotti and M. Xibilia, "Deep learning algorithm for predictive maintenance of rotating machines through the analysis of the orbits shape of the rotor shaft," in International Conference on Smart Innovation, Ergonomics and Applied Human Factors, Springer, 2019, pp. 245–250.

6 Role of Machine Learning and Deep Learning Models for Predictive Maintenance

Gaurav Kumar Gautam, Nishant Pushpak Koshti, Kesha Miralbhai Desai, Nirav Bhatt, Nikita Bhatt, and Hiren Mewada

6.1 INTRODUCTION

Machine learning (ML), a component of artificial intelligence, operates on the premise that machines, when provided with relevant data, can autonomously learn to solve specific problems. ML employs advanced mathematical and statistical tools to empower machines to independently execute intellectual tasks traditionally handled by humans. The concept of automating intricate tasks through ML has sparked considerable interest in the networking domain, with the expectation that activities related to communication network design and operation can be delegated to machines. Notable applications of ML in networking, such as intrusion detection, traffic classification, and cognitive radios, exemplify instances where ML has proven effective in meeting or surpassing expectations [1].

6.2 HOW DOES MACHINE LEARNING IMPACT OUR DAILY LIVES?

Machine learning is used when you ask Siri what the weather forecast is for today. You may praise machine learning when you use it to perform tasks more effectively or when you Google something at work. Our spam folders are another common example; a machine learning algorithm is used to distinguish between emails that belong in our inbox and those that are spam and don't require attention. Similar to this, Netflix uses an algorithm to choose which show to recommend depending on your tastes [2], [3].

Nearly everything we do involves machine learning, from self-driving vehicles to TV recommendations. The goal of these algorithms and machine learning, in general, is to enhance and simplify human lives significantly. The director of research and technology at Interactions, Srinivas Bangalore, states that effective machine learning

DOI: 10.1201/9781003480860-6

"should not be in your face". It needs to operate in the background, tracking and assisting in the faster and more effective accomplishment of objectives.

6.3 THE ESSENCE OF MACHINE LEARNING

Machine learning is fundamentally independent of conventional programming paradigms. It gives machines the capacity to draw conclusions from past evidence, adjust to changing circumstances, and make defensible choices. Understanding machine learning becomes essential in the field of predictive maintenance, where the operating efficiency of machinery is determined by its health.

Figure 6.1 shows different types of machine learning we will be discussed in the following section.

6.4 TYPES OF MACHINE LEARNING

The three main classes of machine learning techniques are:

1 Supervised learning
2 Unsupervised learning
3 Reinforcement learning.

FIGURE 6.1 Types of Machine Learning.

6.4.1 Supervised Learning

In supervised learning, input and predicted output pairings are used to train our model. Teaching the model to comprehend the connection between inputs and outputs is the goal. Consider the following scenario: we wish to educate a computer to identify which label on a graph of points corresponds to which point: the points are labeled with circles or crosses. This method is comparable to training a youngster to identify animals or shapes from photos. Decoding messages via communication channels and categorizing emails into spam and non-spam folders using historical data are two real-world applications of supervised learning [1], [4].

Figure 6.2 illustrates the many supervised learning approaches that fall into two groups that will be covered in more detail below.

There are two types of supervised learning techniques:

1 Classification
2 Regression.

Classification separates the data and regression fits the data.

6.4.1.1 Classification

Classification is the process of attempting to predict an output based on an input, and it is a subset of supervised learning. Classification attempts to categorize an unknown collection of elements into more established groups [1]. A series of labeled examples, such as a picture, text, or speech, is necessary for learning. Thousands or even millions of samples of data are needed to train a classifier to achieve high accuracy as the number of classes increases. Classification can be applied in scenarios where the aim is a structure or a sequence, such as in natural language processing, even though it usually focuses on simple categories.

6.4.1.2 Regression

In machine learning, regression is a potent method for creating models that can forecast continuous values. It learns from labeled data, where each data point has an input (independent variables) and a matching output (dependent variable). This is a form of supervised learning. The objective is to teach the model to recognize the correlation

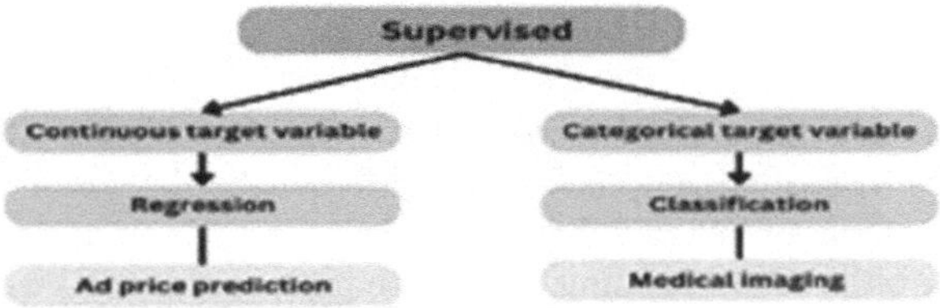

FIGURE 6.2 Types of Supervised Machine Learning Techniques.

between these variables and then apply that understanding to forecast the result for fresh, unobserved data [1].

Here are some key points about regression in machine learning:

- **Predicts continuous values:** regression works with continuous outputs like numerical values, prices, or temperatures, in contrast to classification, which predicts discrete categories.
- **Variety of algorithms:** though there are numerous alternative algorithms, each with pros and cons of its own, linear regression is the most popular and straightforward kind of regression. Polynomial regression, for instance, is capable of handling more intricate correlations between variables. Afterward, decision trees are an excellent tool for identifying non-linear patterns in data. Another choice is support vector machines, which are renowned for their capacity to manage big datasets and high-dimensional areas. These algorithms are appropriate for a variety of issues since each has particular benefits and drawbacks [4].

6.4.1.3 Various Applications

- **Logistic regression:** ideal for binary classification (Yes/No) but can handle multiple classes with adjustments.
- **Support vector machines (SVM):** establishes robust decision boundaries across classes; useful for data with several dimensions.
- **Decision trees:** simple to understand and effective with a variety of data formats, but if left unchecked, it can become overfit.
- **K-nearest neighbors (KNN):** sorts data points according to the training set's closest neighbors; useful for non-linear correlations.
- **Random forest:** combines several decision trees to minimize overfitting and increase accuracy.
- **Linear regression:** based on a linear relationship with input variables, predicts continuous values. Easy to understand and confined to linear interactions.
- **Polynomial regression:** captures nonlinear interactions by raising the values of the input variables.
- **Decision tree regression:** predicts continuous values by dividing the input according to predetermined criteria. Interpretable, yet unstable at times.
- **Support vector regression (SVR):** although it predicts continuous values based on a margin, it is classified similarly to SVM.
- **Gradient boosting:** combines several weak learners to increase accuracy (e.g. decision trees).

6.4.1.4 Advantages

- **High accuracy:** supervised models, particularly for classification and regression issues, can attain high accuracy on certain tasks when trained with an adequate amount of labelled data.
- **Interpretability:** the reasoning behind the predictions may be easy to comprehend depending on the method used, especially for simpler models like decision trees or linear regression.

6.4.1.5 Disadvantages

- **Reliance on labeled data:** the cost and time required to gather and annotate labeled data can limit the use of supervised learning in situations where there is a lack of data.
- **Overfitting:** excessively intricate models may overfit (adjust too much to the training set) and perform poorly on unobserved data. To lessen this, careful model selection and regularization strategies are essential.
- **Bias:** if supervised models do not carefully address their inherited biases from the training data, they may produce unfair or biased results.
- **Black box models:** unfair or biased outcomes may be produced by supervised models if their inherent biases from the training data are not adequately addressed.

6.4.2 Unsupervised Learning

Unlabeled data is used in unsupervised learning. In this case, the machine learns new patterns without being aware of any previous information or data. This kind of learning is particularly suited to clustering, the process of classifying data into sets of related data.

Unsupervised learning problems often fall into three categories: dimensionality reduction, association rules, and clustering. Unsupervised learning problems often fall into three categories: dimensionality reduction, association rules, and clustering [1].

Figure 6.3 illustrates unsupervised learning – we will delve further into its categorization in the following discussion.

6.4.2.1 Clustering

- Clustering algorithms include K-means, hierarchical clustering, and density-based spatial clustering of applications with noise (DBSCAN).
- They are useful for applications such as customer segmentation, anomaly detection, and image analysis.
- The objective of clustering is to group similar data points together based on their inherent characteristics.

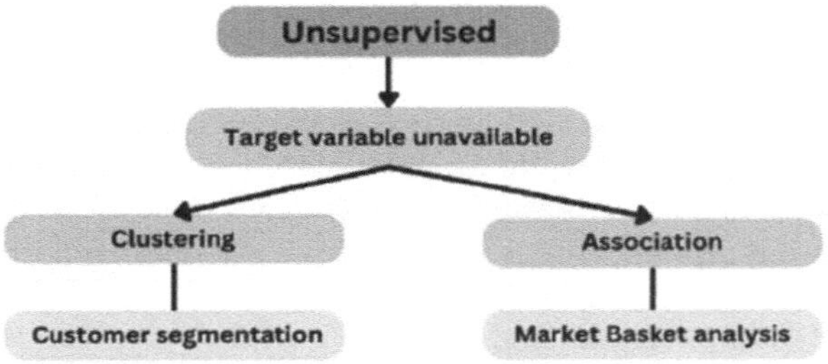

FIGURE 6.3 Types of Unsupervised Machine Learning.

6.4.2.2 Association Rule Mining

- In huge datasets, finds recurring patterns and correlations between variables
- This offers perceptions into consumer behavior, industry developments, and other relevant fields.
- Algorithms for association rule mining include FP-growth and the Apriori algorithm.

6.4.2.3 Dimensionality Reduction

- Minimizes the number of features in a dataset while maintaining the crucial data.
- When working with high-dimensional data, which can be costly to compute and prone to overfitting, this is useful.
- Principal component analysis (PCA) and independent component analysis (ICA) are two instances of dimensionality reduction techniques.

6.4.2.4 Techniques of Unsupervised Machine Learning

Clustering:

- K-means: easy to use and effective, it divides data into "k" predefined clusters according to closeness. well-liked for big databases.
- Hierarchical clustering: creates a hierarchy of clusters, facilitating flexible cluster discovery, although it may incur significant processing costs.
- Density-based spatial clustering of applications with noise (DBSCAN): identifies groups according to density; useful for datasets with noise and different densities.
- Self-organizing maps (SOMs): maintains linkages while visualizing high-dimensional data by projecting it onto a lower-dimensional grid.

Association rule mining:

- Apriori algorithm: finds association rules and recurring item groupings in transactional data, which is frequently utilized for market basket analysis.
- FP-growth: effectively finds recurring patterns by creating a little tree structure; for large datasets, this method is faster than Apriori.

Dimensionality reduction:

- Principal component analysis (PCA)**:** finds the data's principal components, or the directions with the most variance, and uses fewer dimensions to capture the majority of the information.
- Independent component analysis (ICA)**:** divides combined signals into independent parts, which is helpful for noise reduction and feature extraction.
- Non-negative matrix factorization (NMF): shows information as the product of non-negative matrices; helpful for identifying text data's hidden themes.

6.4.2.5 Benefits of Unsupervised Learning

- **Cost-effective:** not dependent on labeled data, which might be costly to gather and annotate.

- **Exploratory analysis:** reveals buried structures and patterns in data, producing surprising revelations.
- **Scalability:** effectively manages substantial volumes of unlabelled data.

6.4.2.6 Applications of Unsupervised Machine Learning

Customer segmentation:

- To enable customized marketing campaigns and product suggestions, clustering algorithms such as K-means group clients based on attributes like purchase history, demographics, or other characteristics.
- For instance: using listening patterns to categorize music streaming users and tailor playlists and promotions.

Anomaly detection:

- By identifying anomalous data points, methods such as the local outlier factor (LOF) can be used to detect fraudulent transactions, equipment malfunctions, or network breaches.
- As an illustration, consider analyzing expenditure trends to spot fraudulent credit card transactions.

Image and text analysis:

- Images with comparable visual characteristics are grouped together by clustering algorithms, which help with anomaly detection, content-based picture retrieval, and image organizing.
- Text summarization, document grouping, and sentiment analysis are made easier by latent topics found in text documents through topic modeling approaches such as latent dirichlet allocation (LDA).
- Clustering news items according to topics or organizing related medical photos for quick diagnosis are two examples.

Recommendation systems:

- Content-based strategies suggest products similar to those a user has previously enjoyed, while collaborative filtering algorithms forecast consumer preferences based on the behavior of similar users.
- As an illustration, suggesting movies to users based on their viewing preferences and reviews of related movies.

Scientific research:

- Unsupervised learning is useful for genetic variant discovery, protein structure identification, and astronomical data analysis.

Social network analysis:

- Online networks can be used to discover communities and significant users through clustering.

Market research:

- Consumer preferences and market trends can be found by analyzing product reviews and customer surveys.

6.4.3 Reinforcement Learning

In between supervised and unsupervised learning is where reinforcement learning falls. In contrast to unsupervised learning, some advice is provided, but not in the form of precise solutions for all scenarios. Reinforcement learning, on the other hand, functions by waiting to get input from the environment before acting on what it has observed. This feedback lets the learner know how well their actions match their objectives. When a learner must navigate a maze and make judgments step-by-step depending on what it observes and receives feedback after each turn, reinforcement learning is particularly helpful.

Figure 6.4 shows how an agent engages with the environment throughout the reinforcement learning process and gets rewarded for its efforts. The various forms of reinforcement learning are depicted in Figure 6.5, and we will delve into these in more detail in our discussion that follows.

6.4.3.1 Types of Reinforcement Learning

Positive: positive reinforcement is the process through which an event that results from a certain behavior intensifies and repeats the behavior. Put otherwise, it produces a favorable impact on behavior.

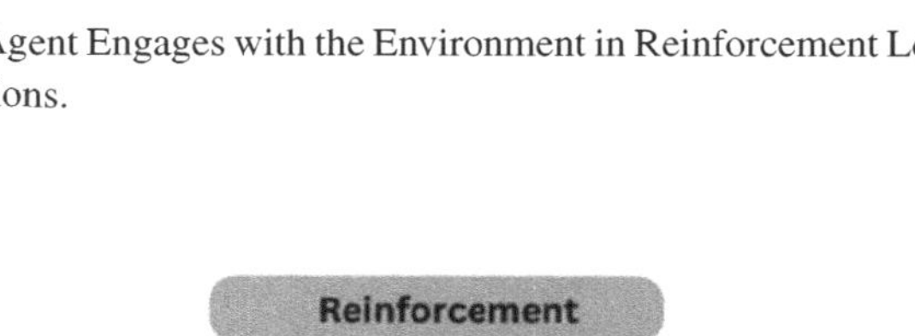

FIGURE 6.4 An Agent Engages with the Environment in Reinforcement Learning, Receiving Rewards for its Actions.

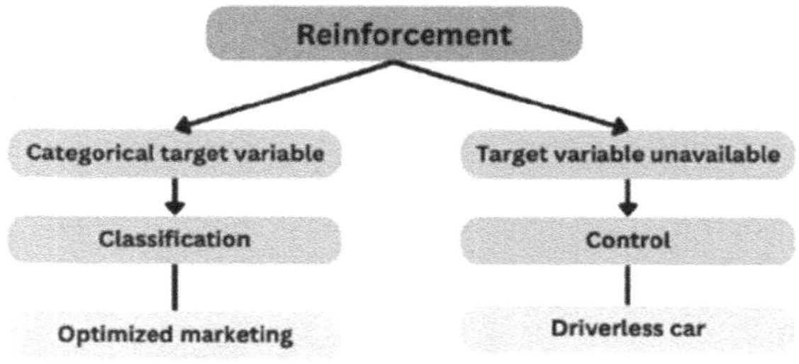

FIGURE 6.5 · Types of Reinforcement Machine Learning.

Positive reinforcement learning has the following benefits:

- increases performance
- holds change for a long time
- too much reinforcement might result in an overload of states, which can reduce the outcomes.

Negative: the definition of negative reinforcement is when a behavior gets stronger as a result of stopping or avoiding an unpleasant situation.

Advantages of negative reinforcement learning:

- enhanced action
- show disregard for a minimal level of performance
- it only offers enough to satisfy the bare minimum of behavior.

6.4.3.2 Elements of Reinforcement Learning

Element of reinforcement learning elements are as follows:

- **Guidelines**: the policy specifies the behavior of the learning agent for a specific duration. It is a mapping of perceived environmental states to appropriate behaviors to take in those states.
- **Reward system**: in a reinforcement learning problem, a goal is defined by the reward function. A reward function is a function that assigns a score based on the environment's condition.
- **Value function**: value functions define what is beneficial over the long term. The entire reward that an agent might anticipate earning over time, beginning from a certain state, is what determines the value of that state.
- **Environment model**: models are employed in planning.

6.4.4 Neural Networks

Artificial neural networks (ANNs), or deep neural networks (DNNs), are a collection of methods for creating robust learning systems. They incorporate several "hidden" layers that are utilized to derive intermediate representations, in contrast to methods like SVMs and Adaboost [4]. Despite being created in the 1980s, DNNs gained popularity around 2010 because of robust parallel hardware and user-friendly open-source software. DNNs encompass a wide variety of neural architectures, the most well-known of which are:

- **Recurrent neural networks (RNN):** a network in which neurons communicate with one another through feedback.
- **Convolutional neural networks (CNN):** typically used for visual and image identification, a feed-forward neural network.

6.4.5 Application of Machine Learning

Finance

Banking and finance are a perfect application for machine learning because of their quantitative character. Although there are many applications for the technology in the business, some of the more popular ones are credit and risk management, trading floors, and fraud [5].

Healthcare

- **Diagnoses:** by analyzing data, machine learning can spot patterns or warning signs in patients, which could result in earlier diagnoses and more effective therapies. Patient data: real-time health assessment is possible by gathering data from a patient's gadget [5], [6].
- **Drug discovery:** scientists may more accurately anticipate pharmacological side effects and the outcomes of drug tests without actually conducting them thanks to the data's capacity to identify trends.

Marketing and sales personalization

Online firms can use machine learning to advertise and offer products based on your search history and browser. Companies utilize the information they've gathered to provide clients with a special, tailored experience.

Oil and gas

- **Energy sources:** machine learning offers the possibility of finding new energy sources through the analysis of various minerals found in the earth.
- **Streamlining oil distribution:** oil distribution may be made more economical and efficient with the use of algorithms.
- **Reservoir modeling:** reservoir simulation, hydraulic fracturing optimization, and other areas can be the subject of specific machine-learning approaches.

Transportation

Data analysis can reveal patterns and trends that can be used to improve the efficiency of routes for delivery services, public transportation, and other purposes.

6.4.6 Challenges of Machine Learning

Even though machine learning has had a significant impact on all industries, there are still unknowns and difficulties with the technology.

Intelligent Assistant

The primary concern is that technology will eventually surpass humans. As we've discussed, technology isn't flawless and frequently needs human aid to guarantee accuracy. But there's still a lot of apprehension and anxiety about technology's potential to surpass human intelligence. Artificial Intelligence consists primarily of mathematical formulas and algorithms that need to be trained by humans. Thus, machine learning and artificial intelligence are only as intelligent as we educate them to be. Artificial intelligence (AI) can be a fantastic helper to increase human productivity when used correctly. Technology exists to help us and enhance our quality of life, not to subjugate and dominate us.

Issues with Unlabeled Data

The capacity of artificial intelligence and machine learning to handle unlabeled data is a more technical problem. Machine learning naturally requires a vast amount of labeled data to function at its best because it is a data-driven learning process [7].

Nonetheless, there are several instances where data is either unlabeled or not easily accessible. This increases the difficulty of designing algorithms. We're teaching these systems to become smarter and achieve human-level accuracy through ongoing research and new developments so that eventually unlabelled data will be just as sufficient as labeled data.

6.4.7 When can we use Machine Learning?

Machine learning finds its niche in scenarios where conventional engineering approaches encounter hurdles or constraints [1]. Let's dissect the circumstances and contexts where machine learning proves to be effective.

6.4.7.1 Model Deficit

- When the absence of physics-based mathematical models, owing to limited domain knowledge renders traditional model-based design approaches impractical.
- Machine learning steps in by directly learning patterns from available data, circumventing the need for explicit mathematical models.

6.4.7.2 Algorithm Deficit

- In cases where implementing existing algorithms based on robust mathematical models becomes overly intricate.
- Machine learning offers viable alternatives through hypothesis classes like scaled-down neural networks or tailored hardware implementations, presenting solutions with reduced complexity.

6.4.7.3 Availability of Training Data

- Machine learning shines when ample training datasets are accessible or can be generated, facilitating model learning through example instances.

6.4.7.4 Task Characteristics

- Tasks devoid of explicit reasoning or detailed explanation requirements for decision-making align well with machine learning paradigms.
- Machine learning models, often treated as opaque entities, efficiently map inputs to outputs without providing granular insights into the decision-making process.

6.4.7.5 Stationary Phenomenon

- Machine learning thrives in scenarios where the underlying phenomenon or function remains stable for significant durations, allowing for data accumulation and learning.

6.4.7.6 Requirement Constraints

- In situations with lenient requirement constraints or when performance assurances can be derived from numerical simulations, especially in the presence of algorithmic limitations.
- Machine learning is a feasible choice when weaker guarantees suffice, provided that the chosen hypothesis class is suitably comprehensive and the training data accurately represents the real-world distribution.

6.4.8 The Role of Data in Predictive Maintenance

An important tactic in contemporary manufacturing is predictive maintenance, which depends on efficient data use. Data is the foundation for proactive decisions made in predictive maintenance to maximize equipment performance, reduce downtime, and avert expensive breakdowns [4], [8].

Predictive maintenance fundamentally depends on gathering, evaluating, and interpreting enormous volumes of data produced by machinery and equipment. Numerous sources are included in this data, such as sensor readings, operational logs, maintenance records from the past, and environmental variables. Manufacturers can learn a great deal about the condition and behavior of their assets by utilizing this data. One of the primary difficulties in this field is the ability to precisely anticipate when a system will require repair [9].

The first step of a predictive maintenance process is project initiation, during which the issue is identified and the project's necessity is justified by projecting the total cost and possible return.

Data preparation, cleaning, normalization, transformation, and data reduction tasks including feature selection, instance selection, discretization, etc. are the primary activities of the second phase, which also involves data collecting and pre-processing. Next, create a predictive model that can identify malfunctions or anticipate when a malfunction may happen.

Furthermore, the industry's integration of the predictive maintenance concept with the production environment constitutes the fundamental step. Enabling predictive model monitoring for equipment dependability and business enhancement via manufacturing key performance indicator (KPI) monitoring are the two objectives. KPIs are therefore the primary source of information for the third phase, which entails tracking project performance [10].

Every phase has a significant challenge attached to it. It is very difficult to obtain the right data set for the project study phase, and once it is available, figuring out how to reduce it to a much more manageable size and improve its quality is a big challenge, especially in the case of companies that generate a large quantity and variety of data. In addition, decision-making based on data-driven could be inefficient.

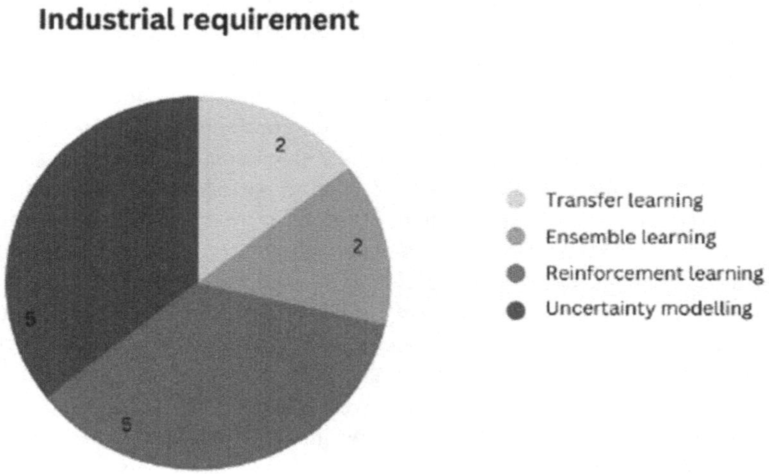

FIGURE 6.6 Industrial Requirements for ML Models in Maintenance.

The development of a machine learning model built on robust data mining is required to guarantee an efficient predictive maintenance system in terms of both cost and method. Additionally, in the event of unavailable or faulty data, the performance of the global predictive maintenance system could not be assessed.

Data management is, in fact, a need at every stage of the project.

Enabling condition monitoring is one of the fundamental functions of data in predictive maintenance. Data analytics algorithms can identify minute variations from standard operating conditions by continuously monitoring equipment characteristics including temperature, vibration, pressure, and fluid levels. These variances frequently function as early warning signs of approaching equipment malfunctions or performance deterioration, enabling maintenance crews to take preventative action before problems get worse [9].

Data assumes a paramount role in the predictive maintenance landscape. We explore how the meticulous collection and analysis of data fuel machine learning models, enabling them to forecast potential issues before they manifest. Figure 6.6 shows industrial requirements for predicted data in maintenance.

6.4.8.1 Machine Learning Methods Applied to Predictive Maintenance

Preventive maintenance(PvM), also known as time-based maintenance, scheduled maintenance, or PvM, is a type of maintenance that is carried out regularly according to a prearranged schedule to predict equipment or process problems. In general, it's a good strategy to stay out of trouble. On the other hand, needless corrective measures are implemented, raising operating expenses [8, 11].

PvM foresees maintenance interventions, which leads to a half-life spare part swap.

As a result, an effective maintenance plan should prolong the equipment's life, increase its condition, lower its failure rate, and save maintenance expenses.

Predictive technologies include machine learning techniques that analyze historical data, as well as integrity factors like visual aspects, wear and tear, and color changes from the original. Statistical inference methods and engineering approaches also aid in the early detection of potential issues. Predictive maintenance (PdM) is a maintenance management strategy that relies on continuously monitoring the condition of machinery or processes to determine when maintenance is necessary [11].

The goal of predictive maintenance, or PdM, is to foresee and stop mistakes or malfunctions before they happen. There are three primary approaches to maintenance: model-based, statistical, and artificial intelligence. These techniques support forecasting and diagnosis by keeping an eye on equipment issues. Artificial intelligence techniques have grown in popularity in PdM, even though statistical approaches need mathematical proficiency and model-based approaches depend on mechanical knowledge and equipment theory. For instance, to forecast future equipment failures, various artificial intelligence algorithms were tested with a statistical technique known as the life utilization model. Their results demonstrated that in this context, artificial intelligence techniques perform better than statistical ones.

6.4.8.2 Predictive Maintenance in the Industry Uses Machine Learning to Foresee Maintenance Needs and Increase Operational Efficiency

Predictive maintenance and condition monitoring is essential for averting unplanned electric motor and other industrial equipment failures, which can result in large financial losses and reduced system dependability [4], [8]. This research proposes a random forest approach-based machine learning system for predictive maintenance. Real-world industrial data was used to test the system through data collection and processing, machine learning application, and comparison with simulation tool outcomes.

The Azure cloud platform's data analysis tool was utilized to analyze the data. Data came from a variety of sources, such as communication protocols, sensors, and machine-programmable logic controllers (PLCs). Preliminary results show that the method accurately and efficiently forecasts various machine conditions.

Predictive maintenance and preventative maintenance are similar in that they both seek to schedule maintenance tasks in advance of machine problems. The method used to create these schedules differs, though. Predictive maintenance bases the scheduling of maintenance work on sensor data that is processed by algorithms [11].

In sectors where motors – such as induction motors, which account for around 70% of all electrical loads – power operations, assessing the health of these motors is a major priority. One of the most frequent problems is bearing failure, which can result in problematic maintenance and motor failures. The two primary issues that predictive maintenance seeks to address are decreasing unplanned downtime and increasing energy efficiency.

Two primary kinds of algorithms are used to accomplish these objectives:

- Those that concentrate on energy and efficiency, encompassing a range of assessment tools and techniques.
- Those who concentrated on keeping an eye on the system's condition, especially in identifying motor malfunctions using various equipment and techniques designed for fault detection.

Early discussions have looked into the application of intelligent decision support systems in the field of predictive maintenance. An essential component of predictive maintenance's effectiveness is algorithms. Various methods such as data processing, problem diagnosis, and prognostic prediction are used at different phases of predictive maintenance implementation.
There are three primary methods for predictive maintenance:

- **Data-driven methodology:** this method primarily uses data analysis from sensors and other sources to forecast maintenance requirements.
- **Model-driven methodology:** to forecast maintenance needs, models are developed here based on the fundamental physics or statistics of the machinery being observed.
- **Hybrid Approach:** using a hybrid approach to take advantage of each method's advantages and provide more accurate forecasts, this strategy blends aspects of model-based and data-driven approaches.

Using mathematical models, predictive maintenance is a useful technique for projecting future values of specific system characteristics, such as those of a machine or production process. This helps us identify any possible faults or anomalies before they become problematic. This is how it operates:

- We continuously monitor physical parameters in real-time, such as vibration or temperature.
- We estimate the parameters of the system for the near future based on these measurements.
- Next, we search for any indications that the system might be malfunctioning or behaving abnormally.
- Lastly, we design and execute remedial and preventive measures before the system reaches a critical state, which helps to avert unforeseen malfunctions and maintain system functionality.
- By using a variety of indicators, predictive maintenance can identify possible problems with machinery or equipment before they become serious ones. Here are a few instances:
  - Unusual vibrations in a machine may indicate that some of its mechanical components are bending or that the machine's bearings are wearing out.
  - A motor's temperature and current draw can be monitored to help identify any mechanical problems, such as friction, that may be harming the motor's performance.
  - The state of the rubbing contact components can be determined by looking at the particles in lubricants. To evaluate the general condition of the machine, sophisticated sensors can even examine the lubricating oils composition.
- The focus of these procedures' earliest stages is parameter estimation. This stage is important because it serves as the foundation for predictive maintenance (PdM), which uses technology to make precise predictions. It becomes difficult to identify anomalies and choose maintenance courses of action if the prediction algorithms provide erroneous predictions or have large reliability intervals.

To put it simply, there are two primary categories of forecasts:

- **Cross-sectional forecasting**: making an educated estimate about something we can't directly measure using the facts we do have is what forecasting is like. For example, we may use the amount of electric current flowing through an electronic component to determine how long it will endure under particular circumstances [2], [8].
- **Time series forecasting**: sequence of time predicting parameters that fluctuate over time is the task of forecasting. We wish to forecast the values of these parameters at a later time (t + dt) after we have measured them up to a specific point in time. Consider, for instance, estimating how many minutes your phone will last on a charge depending on how you have been using it.

In time series, we frequently search for the following:

- **Trends**: a sustained rise or fall in values over some time.
- **Seasonal phenomena**: these are periodic variations in values that exhibit a recurring pattern.
- **Secular occurrences**: these are values that fluctuate, going through ups and downs, but the length of these oscillations varies; they're not constant.

It's critical to comprehend the relationships between various measurements while analyzing data. Scatter plots are a useful tool for visualizing data and identifying patterns or dependencies among the data.

The underlying principle of linear regression is the straight-line behavior of the thing we're attempting to evaluate. We examine the residuals to see if the signal we are attempting to forecast follows this linear pattern or if there are additional variables at work that our model is missing.

To put it another way, residuals are the variations between the values that our model predicted and the actual values. We can determine if our model adequately accounts for the data or whether there are additional factors influencing the data that need to be taken into account by examining these residuals.

6.4.8.3 Analysing of the Residuals

Within our analysis, we focus on what is known as residuals. These are the discrepancies between the values our prediction line gives us and the actual measurements. To obtain the residual for each measurement, we deduct the expected value from the actual value [8].

$$ei = yi - y0$$

This above equation denotes the actual output (yi) and predicted output (y0), the difference between that is known as residuals.

To ensure that our regression model is performing as intended, we now employ the useful method of autocorrelation. This aids in determining whether there is a correlation between the residuals. By determining the proportion of residuals that lie inside a given range, we may compute the autocorrelation.

$$\mathrm{Int} = \pm 2 / \sqrt{N}$$

If over 95% of the residuals are contained within this range, we regard the residuals as uncorrelated and presume that the remaining noise is really "white noise". In the subject of predictive maintenance, in particular, this technique is very helpful for developing algorithms that employ linear regression to anticipate signals. It also aids in our real-time monitoring of any modifications to the system we're studying.

We can also ascertain whether linear regression is the best option for our prediction model or whether we need to look at other possibilities by testing the residuals.

6.4.9 Deep Learning

6.4.9.1 Introduction

Determining which characteristics to employ while addressing a problem is one of the major issues in the field of statistical machine learning. Consider that you are attempting to train a computer to distinguish between two categories of objects, such as cats and dogs. It's easiest if you can connect them with a straight line in "feature space". However, occasionally the items you wish to distinguish aren't so cleanly divided [12], [13].

This is the role of deep learning. Rather than attempting to manually choose the most relevant aspects, deep architectures – elegant networks consisting of numerous layers – allow the data to automatically identify the most relevant features. Even in cases when the interactions between features are extremely complicated, these deep networks do quite well at collecting complex patterns in the data.

However, training these deep networks isn't a walk in the park. Traditional methods that work well for simpler networks often struggle with deep ones. Adding more layers doesn't always make things better; in fact, it can sometimes make training even harder. The algorithms we use to teach these networks, like gradient descent, can get stuck in ruts or take forever to improve the model.

6.4.9.2 Deep Learning Model

Because neural networks are being used so widely, deep learning has taken the lead in machine learning over the last ten years. Even though model performance has significantly improved, especially in areas like text and image categorization, these gains have not come without a price. Network complexity has grown, resulting in more parameters and higher training resource requirements. For example, a model such as GPT-3, with its enormous 175 billion parameters, needs a significant outlay of funds just for training. Additionally, experimenting with various settings – known as hyperparameters – increases the computing cost [13, 14].

These models' inefficiency may make them impractical for immediate implementation in real-world circumstances, even with their remarkable performance on certain tasks [15], [16]. When training or deploying models, deep learning practitioners frequently run across issues like efficiently managing computing resources and maximizing performance for practical applications. When training or deploying a model, a deep learning practitioner may encounter the following difficulties.

Long-term server-side scaling
Large-scale deep learning model deployment and training can be very costly. It can be rather expensive upfront to train a model, particularly if we're beginning from scratch. However, there are situations when using pre-trained models can save money. The true continuous expense arises during model deployment and long-term operation, a process known as "inference". This entails applying the model to real-world applications like as data analysis or prediction. It still consumes a lot of resources, such as CPU and RAM (computer memory), which can build up over time.

Enabling deployment on-device
On tiny devices like IoT (internet of things) gadgets and smart devices, some deep learning applications must launch immediately. This is crucial for many reasons, including protecting sensitive data, avoiding the need for internet connections, and expediting processes.

Thus, the models we employ for these devices must have a high level of efficiency. This implies that they must be compact enough to fit on the gadget and quick enough to operate without depleting the battery too quickly. The secret to ensuring that everything functions well and swiftly in the real world is to optimize the models for these devices.

Data sensitivity and privacy
Reducing the amount of data used for training is essential when working with sensitive user data. This is due to privacy concerns associated with gathering and retaining user data. We can lessen the requirement for substantial data collection by training models effectively with only a tiny portion of the available data. This guarantees that we are managing data ethically and responsibly in addition to helping to preserve user privacy.

Fresh programs
Some new applications have specific requirements or limitations, whether it's about the quality of the model or how much space it takes up. Off-the-shelf models that are already available might not be able to meet these unique needs or restrictions. This means we might have to develop custom models tailored to these specific constraints to make sure the application works as intended.

Model explosion
Even if a single model operates flawlessly, colocation – the practice of training or deploying many models concurrently on the same infrastructure –may put a burden on the available resources. This suggests that there's a chance the infrastructure won't be able to support the demands of running several models concurrently. This could therefore result in ineffectiveness or even failures when performing the different apps.

6.4.9.3 Deep Learning Model for Predictive Maintenance

Prognostic health management focuses on predicting how much longer a machine or system will operate before needing maintenance, using data from sensors or monitoring tools. These predictions play a crucial role in optimizing maintenance planning through predictive maintenance [17], [18], [11].

Predicting when a system or component may malfunction is the main goal of prognostics. This prediction is based on a concept called the remaining useful life (RUL), which calculates how long a system will last by analyzing its behavior over time. With these RUL estimates in hand, we can better plan maintenance – a process known as health management.

We refer to this entire process as prognostic health management (PHM). We frequently utilize a stochastic (random) approach to produce these predictions because there are a lot of unknowns. This means that when we estimate RUL, we take into account every possible outcome and its probability. Then, considering this uncertainty about the potential failure date of the system, we are essentially making decisions regarding the next course of action when it comes to maintenance planning.

Model-based and data-driven methods are the two primary categories of approaches used in the field of prognostics. These techniques were categorized into four groups by a recent review paper: hybrid approaches, artificial intelligence (AI), statistics model-based, and physics model-based. Data-driven AI techniques have grown more and more popular recently due to their exceptional ability to handle complicated degradation patterns that are difficult to depict with just statistics or physics.

One problem with applying machine learning to prognostics, though, is that it's not always obvious how to take uncertainty into account when making forecasts. To successfully plan maintenance based on the estimates of remaining useful life (RUL), this uncertainty is essential.

The prognostic health management (PHM) community has developed multiple datasets including data from continuous monitoring of systems/components as they deteriorate to aid academics in developing and testing these prognostic approaches. These datasets cover a wide range of topics, including industrial equipment, batteries, turbofan engines, and rolling bearings. Several of these datasets are offered by the NASA Prognostics Center of Excellence for public use.

These databases have made it possible for academics to create and hone a wide range of data-driven prognostic algorithms. But up until now, the majority of research has concentrated on estimating the remaining useful life and hasn't given as much thought to what comes after, namely how to manage the system's health in light of those forecasts.

6.4.9.4 Predictive Maintenance Decision Policies Based on RUL Predictions

We need to assess how well prognostic algorithms support maintenance decisions to determine which ones are most effective in predicting remaining useful Life (RUL). We test them using various policies to do this. A policy is similar to a set of guidelines that instruct us on what to do given the information at hand, such as historical monitoring data and previous actions conducted [19], [11].

A policy could assist us in determining, for instance, whether or not to replace a component before it fails. It might state something like, "Should the component be replaced beforehand? Either yes or no."

6.4.9.5 Decision-oriented Metric for Prognostics Performance Evaluation

This section presents a metric that assesses how well the maintenance decisions prompted by the algorithm's predictions match the intended outcomes in a given decision scenario, as part of our attempts to develop an organized framework for assessing and optimizing prognostic algorithms.

6.4.9.6 Utilizing Data Analysis to Evaluate a Typical Predictive Maintenance (PdM) Approach

What matters most to us when evaluating a predictive maintenance (PdM) policy is the average maintenance cost per unit hour over an indefinite period. We can compute this average cost by taking the predicted maintenance cost within one cycle of the component and dividing it by the expected length of that cycle, according to a theory known as renewal theory, specifically the renewal-reward theorem. Essentially, we are calculating the average lifetime maintenance expenses of a component and dividing that figure by the anticipated length of the component's lifecycle. This provides us with a ratio that aids in our comprehension of the maintenance policy's cost-effectiveness. Furthermore, we frequently employ models to simulate a wide range of lifecycle scenarios to provide precise estimations of these expectations [11].

The several phases of predictive maintenance (PdM), including preprocessing, feature engineering, anomaly detection, diagnosis, prognosis, and mitigation, are shown in Figure 6.7. All of these phases work together to improve the system's maintenance processes' efficacy.

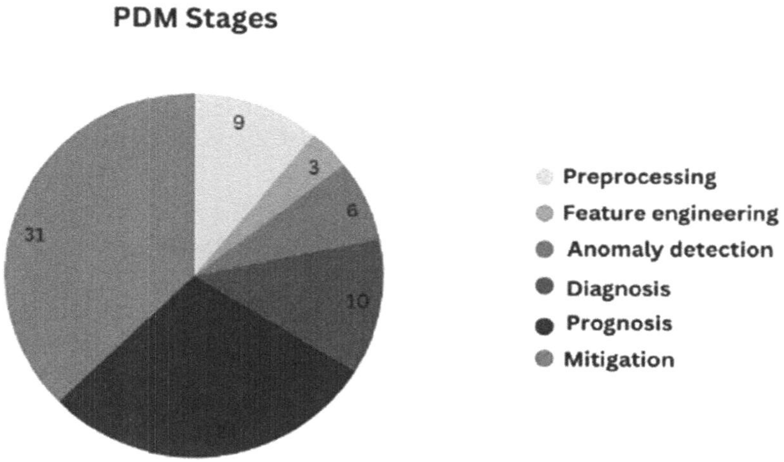

FIGURE 6.7 The Different PdM Stages.

6.5 CONCLUSION

This chapter "Machine Learning and Deep Learning Models for Predictive Maintenance" concludes by exploring the critical role that artificial intelligence and advanced data analytics play in transforming predictive maintenance procedures in the industrial sector. We have analyzed the efficacy of machine learning and deep learning models in anticipating equipment failures, enhancing maintenance schedules, and reducing expensive downtime through our investigation of these models.

Manufacturers can save maintenance costs and increase operational efficiency by proactively identifying possible faults before they worsen by utilizing these advanced models [20]. Every step of the predictive maintenance process – from feature engineering and preprocessing to anomaly detection, diagnosis, prognosis, and mitigation – is essential to guaranteeing the dependability and durability of crucial equipment.

To fully utilize predictive maintenance solutions, the chapter stresses the importance of integrating cutting-edge technology like cloud computing, big data analytics, and IoT sensors. Manufacturers now have unparalleled possibilities to achieve predictive maintenance excellence and maintain their competitive edge in the highly competitive market scenario of today, thanks to ongoing breakthroughs in data analytics and artificial intelligence [21].

This chapter serves as a comprehensive guide for manufacturers looking to harness the power of machine learning and deep learning models for predictive maintenance. By embracing these innovative techniques, organizations can transform their maintenance strategies, optimize resource allocation, and ultimately drive greater productivity and profitability in the manufacturing sector [22].

REFERENCES

1 Simeone, O. (2018). A very brief introduction to machine learning with applications to communication systems. *IEEE Transactions on Cognitive Communications and Networking*, *4*(4), 648–664.

2 Alemdar, H., Tunca, C., & Ersoy, C. (2015). Daily life behaviour monitoring for health assessment using machine learning: bridging the gap between domains. *Personal and Ubiquitous Computing*, *19*, 303–315.

3 Ferrario, A., & Demiray, B. (2024). Understanding reminiscence and its negative functions in the everyday conversations of young adults: A machine learning approach. *Heliyon*, *10*(1), e23825.

4 Ouadah, A., Zemmouchi-Ghomari, L., & Salhi, N. (2022). Selecting an appropriate supervised machine learning algorithm for predictive maintenance. *The International Journal of Advanced Manufacturing Technology*, *119*(7-8), 4277–4301.

5 Ghatge, S. K., & Parasar, A. (2023). Application of Artificial Intelligence (AI), Internet of Things (IoT), and Big Data in Healthcare, Finance, and Transportation. *PriMera Scientific Medicine and Public Health*, *2*(2023), 27–36.

6 Beck, N. (2001). Time-series–cross-section data: What have we learned in the past few years? *Annual review of political science*, *4*(1), 271–293.

7 Ando, R. K., Zhang, T., & Bartlett, P. (2005). A framework for learning predictive structures from multiple tasks and unlabeled data. *Journal of Machine Learning Research*, *6*(11), 1817–1853.

8 Paolanti, M., Romeo, L., Felicetti, A., Mancini, A., Frontoni, E., & Loncarski, J. (2018, July). Machine learning approach for predictive maintenance in industry 4.0. In *2018 14th IEEE/ASME International Conference on Mechatronic and Embedded Systems and Applications (MESA)* (pp. 1–6). IEEE.

9 Zhang, W., Yang, D., & Wang, H. (2019). Data-driven methods for predictive maintenance of industrial equipment: A survey. *IEEE systems journal*, *13*(3), 2213–2227.

10 Sajid, S., Haleem, A., Bahl, S., Javaid, M., Goyal, T., & Mittal, M. (2021). Data science applications for predictive maintenance and materials science in context to Industry 4.0. *Materials today: proceedings*, *45*, 4898–4905.

11 Carvalho, T. P., Soares, F. A., Vita, R., Francisco, R. D. P., Basto, J. P., & Alcalá, S. G. (2019). A systematic literature review of machine learning methods applied to predictive maintenance. *Computers & Industrial Engineering*, *137*, 106024.

12 Lauzon, F. Q. (2012, July). An introduction to deep learning. In *2012 11th International Conference on Information Science, Signal Processing and their Applications (ISSPA)* (pp. 1438–1439). IEEE.

13 Choi, R. Y., Coyner, A. S., Kalpathy-Cramer, J., Chiang, M. F., & Campbell, J. P. (2020). Introduction to machine learning, neural networks, and deep learning. *Translational vision science & technology*, *9*(2), 14–14.

14 Singh, A., Sengupta, S., & Lakshminarayanan, V. (2020). Explainable deep learning models in medical image analysis. *Journal of imaging*, *6*(6), 52.

15 Menghani, G. (2021). Efficient Deep Learning: A Survey on Making Deep Learning Models Smaller. *Faster, and Better. arXiv*, *2106*.

16 Bhatt, C., Kumar, I., Vijayakumar, V., Singh, K. U., & Kumar, A. (2021). The state of the art of deep learning models in medical science and their challenges. *Multimedia Systems*, *27*(4), 599–613.

17 Serradilla, O., Zugasti, E., Rodriguez, J., & Zurutuza, U. (2022). Deep learning models for predictive maintenance: a survey, comparison, challenges and prospects. *Applied Intelligence*, *52*(10), 10934–10964.

18 Zonta, T., Da Costa, C. A., da Rosa Righi, R., de Lima, M. J., da Trindade, E. S., & Li, G. P. (2020). Predictive maintenance in the Industry 4.0: A systematic literature review. *Computers & Industrial Engineering*, *150*, 106889.

19 Kamariotis, A., Tatsis, K., Chatzi, E., Goebel, K., & Straub, D. (2024). A metric for assessing and optimizing data-driven prognostic algorithms for predictive maintenance. *Reliability Engineering & System Safety*, *242*, 109723.

20 Jhaveri, R. H., Revathi, A., Ramana, K., Raut, R., & Dhanaraj, R. K. (2022). A review on machine learning strategies for real-world engineering applications. *Mobile Information Systems*, *2022*, 2–4.

21 Nguyen, K. T., & Medjaher, K. (2019). A new dynamic predictive maintenance framework using deep learning for failure prognostics. *Reliability Engineering & System Safety*, *188*, 251–262.

22 Mathur, P. (2018). *Machine learning applications using Python: Case studies from healthcare, retail, and finance*. Apress.

7 Data Analytics and AI for Predictive Maintenance in Pharmaceutical Manufacturing

Rati Kailash Prasad Tripathi

7.1 INTRODUCTION

The pharmaceutical manufacturing industry operates in an environment where precision, reliability, and adherence to regulatory standards are paramount. In this dynamic landscape, the integration of advanced technologies such as data analytics and artificial intelligence (AI) has emerged as a transformative force, particularly in the domain of predictive maintenance [1]. As pharmaceutical facilities are characterized by complex machinery and critical processes, the proactive identification and mitigation of equipment failures play a pivotal role in ensuring continuous production, product quality, and compliance with stringent regulatory requirements [2].

Data analytics and AI in the context of predictive maintenance empower pharmaceutical manufacturers to move beyond traditional, reactive maintenance approaches. By harnessing real-time sensor data, historical performance records, and sophisticated machine learning (ML) algorithms, these technologies enable the anticipation of equipment failures before they occur, allowing for targeted and timely interventions. This paradigm shift from preventive to predictive maintenance not only minimizes unplanned downtime but also optimizes resource utilization, reduces operational costs, and enhances overall manufacturing efficiency [3].

The present chapter explicates into the intricacies of how these technologies are reshaping the industry. From the collection and integration of diverse data sources to the deployment of advanced ML models, the key components of a predictive maintenance framework tailored to the unique needs of pharmaceutical production facilities are navigated. Furthermore, the seamless integration of these technologies with existing maintenance systems have been examined, ensuring a cohesive and automated approach to addressing equipment health.

The tangible benefits experienced by pharmaceutical manufacturers who have embraced predictive maintenance is also unveiled, as we journey through case studies

DOI: 10.1201/9781003480860-7

and success stories. Additionally, the regulatory landscape to underscore the importance of validation processes and compliance adherence has been addressed, ensuring that these innovative technologies align with the rigorous standards of the pharmaceutical industry.

The exploration concludes by contemplating the challenges faced in the implementation of predictive maintenance and forecasting future trends that hold the potential to further revolutionize the landscape. Through this comprehensive overview, the objective is to illuminate the transformative power of data analytics and AI in predictive maintenance, offering insights that resonate with pharmaceutical professionals seeking to enhance the reliability, efficiency, and compliance of their manufacturing processes.

7.1.1 Importance of Predictive Maintenance in Pharmaceuticals

The pharmaceutical industry is characterized by complex machinery and critical processes, making unplanned downtime a costly and potentially hazardous occurrence. Predictive maintenance mitigates these risks by leveraging real-time sensor data, historical performance records, and sophisticated ML algorithms. By analyzing patterns and anomalies, this approach allows for the identification of potential equipment failures, enabling timely interventions and minimizing disruptions to manufacturing processes [4].

The benefits of predictive maintenance extend beyond mere cost savings. By addressing issues before they escalate, pharmaceutical manufacturers can ensure the consistent production of high-quality medicines, reduce the risk of product recalls, and enhance overall operational resilience. The proactive nature of predictive maintenance also contributes to the optimization of resource utilization, as maintenance activities can be scheduled based on actual equipment health rather than arbitrary timelines [5].

Moreover, in an industry where compliance with regulatory standards is non-negotiable, predictive maintenance aligns seamlessly with the need for robust validation processes. Ensuring equipment reliability through predictive maintenance not only safeguards product quality but also supports pharmaceutical manufacturers in meeting the stringent requirements set forth by regulatory bodies [5].

Thus, the importance of predictive maintenance in pharmaceuticals lies in its ability to transform maintenance practices from reactive to proactive, providing a reliable and efficient way to manage complex manufacturing processes. By embracing data analytics and AI-driven predictive maintenance, pharmaceutical companies can safeguard product quality, reduce operational costs, and reinforce their commitment to regulatory compliance in an ever-evolving industry [6].

7.1.2 Objectives of Data Analytics and AI Integration

The integration of data analytics and artificial intelligence (AI) in various industries, including pharmaceutical manufacturing, is driven by specific objectives aimed at enhancing efficiency, decision-making, and overall business performance. In context of predictive maintenance in pharmaceuticals, the key objectives of integrating data analytics and AI has been outlined in Table 7.1 [6].

TABLE 7.1
Objectives of Integrating Data Analytics and AI in Predictive Maintenance in Pharmaceuticals

S. No.	Parameters	Objective	Rationale
1	Proactive equipment maintenance	Anticipate and prevent equipment failures before they occur	Minimize unplanned downtime, optimize production schedules, and ensure the continuous operation of critical pharmaceutical manufacturing processes.
2	Optimized resource utilization	Efficiently allocate resources for maintenance activities	Utilize personnel, materials, and time more effectively by focusing maintenance efforts on equipment that requires attention based on data-driven insights, reducing unnecessary interventions.
3	Data-driven decision-making	Utilize data analytics to inform strategic and operational decisions	Leverage insights from historical and real-time data to make informed decisions about equipment maintenance, process improvements, and resource allocation.
4	Cost reduction and operational efficiency	Reduce operational costs associated with unplanned downtime and reactive maintenance	By proactively addressing potential equipment failures, minimize the financial impact of unexpected disruptions, and optimize overall operational efficiency.
5	Enhanced product quality and compliance	Ensure the consistent production of high-quality pharmaceuticals while meeting regulatory standards	Mitigate the risk of product recalls and uphold compliance with stringent regulatory requirements by maintaining equipment reliability and consistency in manufacturing processes.
6	Continuous improvement and innovation	Foster a culture of continuous improvement through feedback loops and innovation	Embrace emerging technologies, refine predictive models based on performance feedback, and stay at the forefront of advancements in data analytics and AI to continually enhance predictive maintenance strategies.
7	Alignment with industry 4.0 principles	Embrace the principles of Industry 4.0 for smart manufacturing	Integrate data analytics and AI as part of a larger digital transformation strategy, aligning with Industry 4.0 principles to create a connected and intelligent manufacturing ecosystem.

7.2 DATA COLLECTION IN PHARMACEUTICAL MANUFACTURING

Data collection in pharmaceutical manufacturing is a critical process that involves gathering a diverse range of information to monitor, control, and optimize various aspects of the production environment to ensure product quality, safety, and adherence to regulatory standards. Key components of data collection include the utilization of sensors to capture real-time data from equipment and processes. These sensors monitor critical parameters such as temperature, pressure, humidity, and other relevant variables, providing a comprehensive view of the manufacturing environment [7]. Additionally, data collection extends beyond sensor-based inputs to encompass batch records, equipment logs, and historical performance data. This holistic approach allows pharmaceutical manufacturers to establish a robust dataset that serves as the foundation for advanced analytics and AI applications, particularly in predictive maintenance and process optimization [8].

The integration of sophisticated data collection methods not only facilitates compliance with regulatory requirements but also empowers manufacturers to make informed decisions. By leveraging comprehensive and accurate data, pharmaceutical companies can enhance operational efficiency, identify potential areas for improvement, and ensure the consistent production of high-quality pharmaceutical products.

7.2.1 Sensor Data

Sensor data refers to information collected by sensors, which are devices designed to detect and measure physical changes or conditions in the surrounding environment. In various industries, including manufacturing and healthcare, sensors play a crucial role in capturing real-time data to monitor and control processes. In context of pharmaceutical manufacturing, sensors are employed to measure critical parameters such as temperature, pressure, humidity, pH levels, and other relevant variables [8]. These sensors generate continuous streams of data, providing insights into the dynamic conditions of manufacturing equipment and processes. The data collected from sensors is invaluable for ensuring product quality, optimizing operational efficiency, and meeting regulatory requirements [9]. In pharmaceutical manufacturing, where precision is essential, sensor data is used to maintain optimal conditions for the production of pharmaceuticals, prevent deviations, and support compliance with stringent industry standards [10].

The integration of sensor data into advanced analytics and AI applications allows for predictive maintenance, process optimization, and real-time monitoring. By harnessing the power of sensor data, industries can enhance decision-making, minimize downtime, and ensure the reliability and consistency of their manufacturing processes, contributing to improved overall operational performance and product quality [11].

7.2.2 Process Data

Process data refers to information generated during the various stages of a manufacturing or industrial process. In pharmaceutical manufacturing, process data encompasses a wide array of details related to the production of pharmaceutical products. This data includes parameters such as reaction times, ingredient quantities, flow rates, and other factors relevant to the manufacturing process [12].

Capturing and analyzing process data is crucial for ensuring the reproducibility, quality, and compliance of pharmaceutical products. Manufacturers use sophisticated monitoring and control systems to collect real-time data from equipment and instruments involved in the production process. This comprehensive dataset serves as a valuable resource for optimizing manufacturing conditions, identifying potential issues, and maintaining consistency in product quality [13]. By leveraging process data, pharmaceutical manufacturers can make informed decisions, enhance efficiency, and adhere to regulatory requirements. The utilization of process data in advanced analytics and AI applications further empowers manufacturers to streamline operations and respond proactively to deviations, contributing to the overall excellence of pharmaceutical manufacturing processes [14].

7.2.3 Data Integration and Storage

Data integration and storage are essential components of managing and leveraging information effectively within an organization. In context of pharmaceutical manufacturing, where vast amounts of diverse data are generated, proper integration and storage mechanisms are critical for ensuring accessibility, accuracy, and the ability to derive meaningful insights [15].

7.2.3.1 Data Integration

Data integration involves combining and unifying information from diverse sources within an organization to provide a unified view of the data. In pharmaceutical manufacturing, data integration ensures that information from various systems, such as sensors, laboratory equipment, and enterprise resource planning (ERP) systems, is consolidated into a cohesive and interconnected dataset. Integrated data allows for a comprehensive understanding of the manufacturing process, facilitates real-time monitoring, and supports advanced analytics and decision-making [16].

7.2.3.2 Data Storage

Data storage refers to the systematic organization and retention of data for future use. In pharmaceutical manufacturing, data storage is crucial for preserving historical records, batch information, and sensor data. It ensures data integrity, traceability, and compliance with regulatory requirements. Pharmaceutical companies often employ secure and scalable data storage solutions, such as cloud-based storage or on-premises databases, to accommodate the volume and complexity of manufacturing data. Efficient data storage enables easy retrieval and analysis of historical information, supporting activities such as audits, regulatory submissions, and continuous improvement initiatives [17].

7.2.3.3 Advantages of Data Integration and Storage in Synergy

a **Interconnected systems**: integrated data is seamlessly stored in a structured manner, creating a foundation for reliable and efficient storage systems.

b **Real-time updates**: data integration ensures that storage systems receive real-time updates, maintaining the relevance and accuracy of stored information.
c **Adaptability**: as pharmaceutical manufacturing processes evolve, integrated and well-organized storage systems provide the flexibility to adapt to new data sources and technologies.

7.2.4 Data Cleaning and Pre-processing

Data cleaning and pre-processing are integral stages in the data analysis process, crucial for ensuring the accuracy and reliability of the information used for further analysis or modeling. In pharmaceutical manufacturing, where precision and data integrity are paramount, these steps play a vital role in optimizing datasets for meaningful insights and decision-making [18].

Data cleaning involves the identification and rectification of errors, inconsistencies, and missing values within the dataset. This process addresses issues such as typos, duplicate entries, and outliers that could distort analytical results. In the pharmaceutical context, data cleaning is essential to maintain the quality of information, reducing the risk of inaccuracies in critical decision-making processes [18].

Data pre-processing transforms raw data into a format suitable for analysis or modeling. This includes tasks such as normalization to ensure consistent scales for numerical variables, encoding categorical variables for ML compatibility, handling imbalanced datasets, and selecting relevant features. These steps are vital for enhancing the performance of ML models and preventing biases or inaccuracies in the outcomes [18].

In pharmaceutical manufacturing, these processes contribute to maintaining data integrity, complying with regulatory standards, and supporting accurate decision-making. By systematically handling missing data, outliers, and optimizing the dataset's structure, data cleaning and pre-processing lay the foundation for robust analyses, ensuring that insights derived from the data are both meaningful and trustworthy in the complex and highly regulated pharmaceutical industry [18].

7.3 ML MODELS FOR PREDICTIVE MAINTENANCE

Predictive maintenance in pharmaceutical manufacturing relies on advanced ML models to anticipate and prevent equipment failures, ultimately ensuring continuous operational efficiency. The integration of ML models has emerged as a transformative approach to predictive maintenance. These models leverage historical data, real-time sensor information, and advanced algorithms to predict potential equipment failures before they occur. Several ML models are commonly applied in the context of predictive maintenance. These includes failure prediction models, anomaly detection models, and feature engineering models [19].

7.3.1 Failure Prediction Models

In the dynamic landscape of predictive maintenance, failure prediction models stand as the cornerstone, leveraging advanced technologies to foresee and prevent equipment breakdowns before they occur. In pharmaceutical manufacturing, where precision and reliability are paramount, these models play a pivotal role in ensuring uninterrupted production, product quality, and regulatory compliance.

The primary goal of failure prediction models is to forecast when equipment or machinery is likely to fail. By proactively identifying potential failures, these models empower organizations to schedule maintenance activities strategically, minimizing downtime, and optimizing operational efficiency [20].

7.3.1.1 Types of Failure Prediction Models

7.3.1.1.1 Regression Models

Regression models, such as linear regression, polynomial regression, and time-to-failure models, are employed to predict the remaining useful life (RUL) of equipment components. They estimate the time until a particular machine or part is likely to fail, allowing for proactive maintenance scheduling [21].

7.3.1.1.2 Classification Models

Classification models, including support vector machines (SVM), random forests or decision trees, are utilized for binary outcomes, that is it categorize equipment health into failure or non-failure classes within defined time frames. This assists in prioritizing maintenance tasks based on the urgency of potential failures [21].

7.3.1.1.3 Time Series Models

Time series models, like autoregressive integrated moving average (ARIMA), and long short-term memory (LSTM) networks, are applied to analyze patterns and trends in historical data to predict future equipment behavior. These models capture the temporal dependencies of equipment performance, aiding in predicting future behavior [21].

7.3.1.1.4 Deep Learning Models

Deep learning architectures, such as convolutional neural networks (CNN) or recurrent neural networks (RNN), utilize neural networks to capture intricate relationships within data for improved accuracy. They excel in tasks like predicting equipment failures by learning from the nuances present in large and diverse datasets [21].

7.3.1.1.5 Ensemble Learning Models

Ensemble learning techniques, such as random forests or gradient boosting, combine the strengths of multiple models to improve predictive accuracy. These models are effective in handling diverse and complex data sources common in pharmaceutical manufacturing [21].

7.3.1.1.6 Prognostic Models

Prognostic models use both historical and real-time data to forecast the future health and performance of equipment components. They go beyond predicting failures by providing insights into the degradation and wear-and-tear of machinery [21].

7.3.1.1.7 Applications of Failure Prediction Models

Failure prediction models play an important role in data integration, real-time data monitoring, and maintenance scheduling [21].

- **Data integration**: failure prediction models integrate data from various sources, including sensor data, process records, and historical performance data.
- **Real-time monitoring**: by continuously analyzing real-time data, these models provide timely warnings, allowing for immediate corrective actions.
- **Optimized maintenance scheduling**: pharmaceutical manufacturers benefit from optimized maintenance schedules, minimizing disruptions and ensuring equipment reliability.

7.3.1.2 Benefits of Failure Prediction Models

- **Operational efficiency**: failure prediction models contribute to enhanced operational efficiency by reducing unplanned downtime and optimizing resource utilization.
- **Cost reduction**: proactive maintenance strategies, guided by these models, result in cost savings associated with reactive repairs and equipment replacement.
- **Continuous improvement**: ongoing refinement based on feedback and evolving data ensures these models remain at the forefront of predictive maintenance, aligning with the industry's evolving needs.

7.3.2 Anomaly Detection

Anomaly detection, a powerful technique in data analysis, is designed to identify patterns or observations that deviate significantly from the expected or normal behavior within a dataset. This method plays a crucial role in various industries, including pharmaceutical manufacturing. In pharmaceutical manufacturing, anomaly detection is instrumental in maintaining product quality, ensuring equipment reliability, and preventing unexpected deviations from operational standards [8].

7.3.2.1 Techniques in Anomaly Detection

7.3.2.1.1 Statistical Methods

Statistical models determine anomalies by analyzing data distributions and identifying observations significantly distant from the mean or median. For example, Z-score analysis, Grubbs' test, and percentile-based methods [22–23].

7.3.2.1.2 Machine Learning Algorithms

Supervised and unsupervised machine learning algorithms, including isolation forests, one-class SVM, and autoencoders, learn normal patterns and detect anomalies based on deviations. For instance, isolation forests excel in isolating anomalies, one-class

SVM detects outliers, and autoencoders reconstruct normal data, highlighting anomalies in the process [22–23].

7.3.2.1.3 Time Series Analysis

Analyzing temporal data sequences helps identify anomalies based on unusual trends or patterns over time. Time series anomaly detection is crucial for monitoring equipment behavior and predicting potential failures in pharmaceutical manufacturing [22–23].

7.3.2.2 Applications of Anomaly Detection Models

Anomaly detection models play an important role in sensor data monitoring, quality control, and equipment health monitoring [22–23].

- **Sensor data monitoring**: anomaly detection is applied to real-time sensor data, ensuring that unusual variations in temperature, pressure, or other critical parameters are promptly identified.
- **Quality control**: in pharmaceutical production, anomaly detection is utilized to identify anomalies in product quality, preventing defective batches from reaching the market.
- **Equipment health monitoring**: anomalies in equipment behavior or performance can signal potential failures, allowing for proactive maintenance and minimizing unplanned downtime.

7.3.2.3 Benefits of Anomaly Detection Models

- **Early issue identification**: anomaly detection enables the early identification of deviations, preventing potential disruptions or quality issues.
- **Predictive maintenance**: by recognizing anomalies in equipment performance, pharmaceutical manufacturers can implement predictive maintenance strategies, reducing maintenance costs and downtime.

7.3.2.4 Challenges of Anomaly Detection Models

- **False positives**: striking the right balance to avoid unnecessary alarms for normal variations.
- **Data quality**: ensuring the accuracy and reliability of data inputs to avoid misleading anomaly detection results.

7.3.3 Feature Engineering

Feature engineering is a transformative process in ML where raw data is refined and augmented to create meaningful features that enhance the predictive capabilities of models. This critical step in the data preparation pipeline is particularly relevant in predictive maintenance models in pharmaceutical manufacturing, contributing to the optimization of performance and the accuracy of predictions.

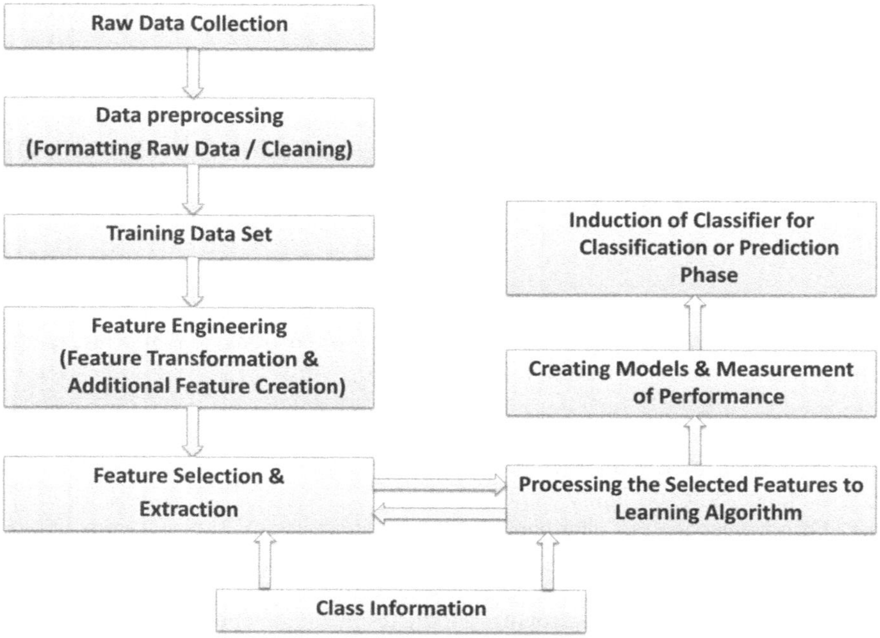

FIGURE 7.1 Comprehensive Framework for Feature Engineering.

Feature engineering involves the creation and transformation of variables (features) to extract relevant information, improve model performance, and enhance the interpretability of the data. The primary goal is to provide machine learning models with well-crafted, informative features that capture the essential patterns and relationships within the data [24–29]. Figure 7.1 represents a comprehensive framework for feature engineering in classification.

7.3.3.1 Techniques in Feature Engineering

7.3.3.1.1 Variable Transformation

It includes techniques like logarithmic transformation, square root transformation, or Box-Cox transformation to normalize data distributions and improve model robustness.

7.3.3.1.2 Binning or Discretization

It consists of grouping continuous variables into discrete bins or categories to capture non-linear relationships or patterns.

7.3.3.1.3 Interaction Features

It consists of creating new features by combining or interacting existing features to capture synergistic effects that may impact the target variable.

7.3.3.1.4 Time-based Features

It consists of extracting temporal information such as day of the week, month, or time of day to capture trends or patterns over time in predictive maintenance models.

7.3.3.1.5 Aggregation

It consists of combining multiple related features by aggregating or summarizing data to provide a holistic view of a system's behavior.

7.3.3.1.6 One-hot Encoding

It consists of converting categorical variables into binary vectors to ensure compatibility with machine learning algorithms.

7.3.3.2 Applications of Feature Engineering Techniques

Feature engineering in the pharmaceutical industry often involves tailoring features to the unique characteristics of manufacturing processes, ensuring that the created features capture relevant information about equipment health and performance [25–27]. The various applications of feature engineering techniques include:

- **Sensor data enhancement**: feature engineering can involve creating lag features or statistical summaries of sensor data, providing additional context and capturing trends in equipment behavior.
- **Time-to-failure features**: features related to the time remaining until equipment failure (RUL) can be engineered to enhance the predictive capabilities of models.
- **Event-based features**: incorporating features related to maintenance events, downtimes, or historical failures to capture the impact of past events on future equipment behavior.
- **Quality control metrics**: creating features that quantify product quality deviations or anomalies in order to enable models to link equipment performance with product quality, contributing to comprehensive predictive maintenance strategies.

Feature engineering in the pharmaceutical industry must align with regulatory standards, ensuring that models comply with good manufacturing practice (GMP) and other relevant guidelines. Additionally, rigorous validation processes are essential to confirm that the features engineered contribute to the reliability of predictive maintenance models without compromising product quality or regulatory compliance [27].

7.3.3.3 Benefits of Feature Engineering Techniques

- **Improved model performance**: well-engineered features contribute to more accurate predictions, enhancing the overall performance of predictive maintenance models.
- **Interpretability**: carefully crafted features can make models more interpretable, providing insights into the factors influencing predictions [28–29].

7.3.3.4 Challenges of Feature Engineering Techniques

- **Data complexity**: in complex datasets, identifying the most relevant features and creating meaningful transformations can be challenging.
- **Overfitting**: feature engineering, if not carefully executed, can lead to overfitting, where models perform well on training data but poorly on new, unseen data [28–29].

7.4 PREDICTIVE ANALYTICS FOR EQUIPMENT HEALTH

Predictive analytics has emerged as a transformative force in the pharmaceutical industry, particularly in the realm of equipment health monitoring. The primary goal of predictive analytics for equipment health is to proactively identify potential failures or deviations from normal operating conditions. By leveraging historical data, real-time sensor information, and sophisticated algorithms, predictive analytics empowers pharmaceutical manufacturers to transition from reactive to proactive maintenance strategies [14].

7.4.1 Remaining Useful Life (RUL) Prediction

Remaining useful life (RUL) prediction is a crucial aspect of predictive maintenance, aiming to estimate the remaining operational lifespan of equipment or components before they are likely to fail. This predictive modeling technique is widely employed across various industries, including manufacturing, aviation, and healthcare, to optimize maintenance strategies and minimize downtime. RUL prediction seeks to anticipate the remaining time until a piece of equipment or a component reaches the end of its useful life. By forecasting the RUL, organizations can proactively schedule maintenance activities, reduce unplanned downtime, and optimize resource utilization [30].

The key components include (a) data collection, where RUL prediction relies on historical and real-time data, including sensor information, maintenance records, and performance metrics; (b) using ML models, such as regression models, time series analysis, and other ML techniques to analyze the data and predict the remaining useful life of equipment; (c) using feature engineering techniques, which includes crafting informative features from raw data to enhance the accuracy of the predictive models by capturing relevant information about the equipment's health and performance [31].

Historical data is processed and cleaned to ensure accuracy and reliability in predicting the remaining useful life. ML models are trained on the processed data, learning patterns and relationships that correlate with the equipment's degradation over time. The trained model is then applied to real-time data, continuously monitoring the equipment's health and predicting its remaining useful life as new information becomes available [32].

7.4.1.1 Advantages of RUL

- **Proactive maintenance**: RUL prediction enables proactive maintenance scheduling, allowing organizations to address potential issues before equipment failure occurs.

- **Downtime reduction**: by anticipating failures and planning maintenance in advance, organizations can minimize unplanned downtime and optimize overall operational efficiency.
- **Resource optimization**: efficient use of resources by targeting maintenance efforts where they are most needed, based on the predicted remaining useful life.
- **Integration with Industry 4.0**: RUL prediction can be integrated seamlessly with Industry 4.0 initiatives, incorporating advanced technologies like the internet of things (IoT) for enhanced monitoring and analysis.
- **AI and ML advances**: ongoing advancements in AI and ML will likely lead to more accurate and adaptive RUL prediction models.

7.4.1.2 Challenges of RUL

- **Data quality**: RUL prediction relies heavily on the quality and reliability of the data used for training and monitoring.
- **Model complexity**: developing accurate RUL prediction models may require sophisticated algorithms, and choosing the appropriate model can be challenging.
- **Adaptability**: models need to adapt to evolving equipment conditions, necessitating continuous refinement and updates based on feedback and changing data patterns.

7.4.2 Real-time Monitoring with IoT

Real-time monitoring with the internet of things (IoT) involves the continuous tracking, analysis, and visualization of data from connected devices and sensors. This technology enables instantaneous insights into processes, facilitating proactive decision-making. In real-time IoT monitoring, data is collected, transmitted, and analyzed in real-time, offering immediate visibility into various applications such as predictive maintenance, healthcare monitoring, and supply chain optimization [33–34]. The various stages of real-time IoT monitoring has been depicted in Figure 7.2.

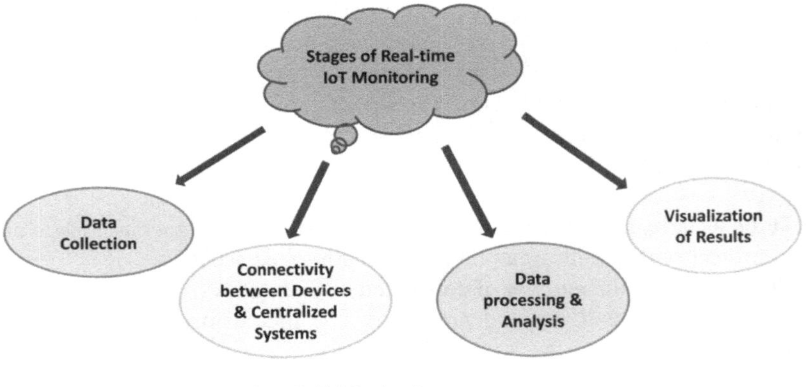

FIGURE 7.2 Stages of Real-time IoT Monitoring.

Stages of real-time IoT monitoring:

1. **Data collection**: sensors embedded in devices collect real-time data, capturing information such as temperature, humidity, pressure, or equipment status.
2. **Connectivity**: the collected data is transmitted in real-time through IoT networks, allowing seamless communication between devices and centralized systems.
3. **Analysis**: advanced analytics and ML algorithms process the real-time data, identifying patterns, anomalies, or trends that can be crucial for decision-making.
4. **Visualization**: results are presented through intuitive dashboards or interfaces, providing stakeholders with instant and actionable insights into the monitored processes.

7.4.2.1 Applications of Real-time Monitoring

- **Predictive maintenance**: real-time monitoring aids in predicting equipment failures or maintenance needs by analyzing continuous data streams.
- **Healthcare monitoring**: in healthcare, IoT enables real-time tracking of patient vital signs, ensuring immediate response to critical situations.
- **Supply chain visibility**: monitoring goods in transit in real-time helps optimize logistics, enhance security, and streamline supply chain operations.

7.4.2.2 Benefits and Challenges of Real-time Monitoring

The benefits include improved operational efficiency, cost savings through optimized workflows, and the ability to respond swiftly to emerging issues [34].

- **Proactive decision-making**: immediate access to real-time data allows for quick and informed decision-making, preventing potential issues or optimizing processes.
- **Efficiency**: real-time monitoring improves operational efficiency by reducing delays, minimizing downtime, and enhancing overall productivity.
- **Cost savings**: early identification of issues and efficient resource allocation result in cost savings through optimized maintenance and improved operational workflows.

Challenges include data security concerns, but ongoing advancements, including 5G integration and edge computing, continue to enhance the capabilities of real-time monitoring with IoT [34].

- **Data security**: ensuring the security of real-time data transmissions is crucial to protect sensitive information from unauthorized access.
- **Scalability**: as the number of connected devices grows, scalability challenges may arise in managing and processing the increasing volume of real-time data.

7.5 INTEGRATION WITH MAINTENANCE SYSTEMS

Integration with maintenance systems refers to the seamless incorporation of advanced technologies, data analytics, and insights into traditional maintenance processes. Integration with maintenance systems represents a strategic evolution in how organizations manage and optimize their maintenance processes. By embracing modern technologies and data-driven approaches, businesses can transition from reactive to proactive maintenance strategies, resulting in improved efficiency, reduced costs, and enhanced overall operational reliability.

7.5.1 CMMS Integration

The integration of computerized maintenance management system (CMMS) in the pharmaceutical industry is instrumental in streamlining maintenance processes, ensuring compliance, and optimizing overall operational efficiency. Major characteristics of CMMS integration in the pharmaceutical industry are illustrated in Figure 7.3 [35].

- **Equipment maintenance**
  Predictive maintenance: CMMS integration with predictive maintenance tools enables pharmaceutical companies to anticipate equipment failures, schedule maintenance proactively, and reduce unplanned downtime.

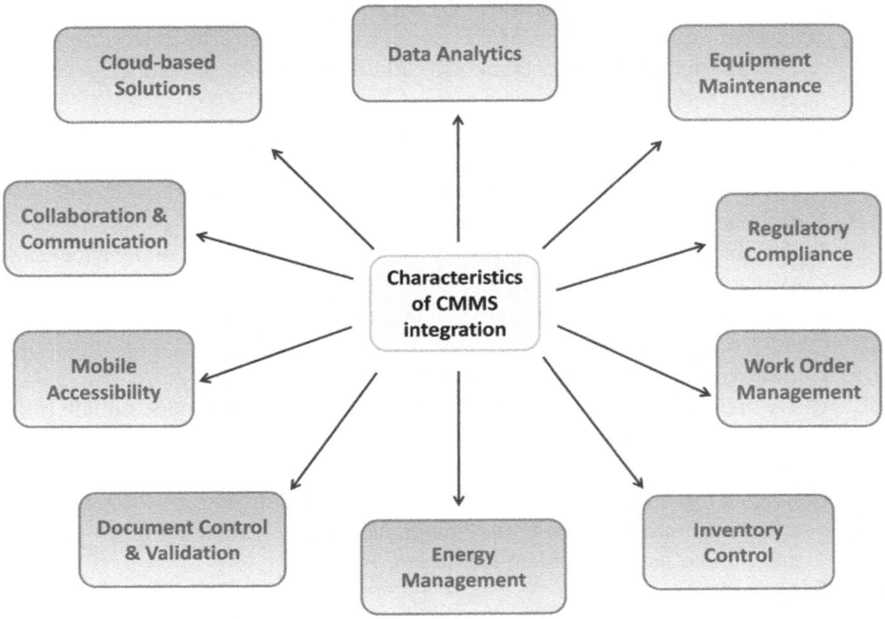

FIGURE 7.3 Characteristic Features of CMMS Integration.

Condition monitoring: real-time data from connected sensors and equipment can be fed into CMMS, allowing for continuous monitoring of equipment health and performance.

- **Regulatory compliance**

  Documentation and auditing: CMMS assists in maintaining comprehensive records of maintenance activities, equipment calibration, and compliance documentation. this ensures adherence to good manufacturing practice (GMP) regulations and simplifies the auditing process.

- **Work order management**

  Efficient workflows: CMMS integration streamlines work order processes, FACILITATING efficient planning, scheduling, and tracking of maintenance tasks. This results in improved workforce productivity and resource utilization.

  Asset tracking: integration allows for accurate tracking of pharmaceutical assets, including equipment and spare parts. This contributes to better inventory management and timely procurement of necessary components.

- **Inventory control**

  Spare parts management: CMMS integration aids in maintaining an up-to-date inventory of spare parts. This ensures that critical components are available when needed, minimizing downtime and optimizing maintenance efficiency.

- **Energy management**

  Resource optimization: CMMS integration with energy management systems enables pharmaceutical companies to monitor and control energy consumption in manufacturing processes. This contributes to sustainability goals and cost reduction.

- **Document control and validation**

  Documentation Integrity: CMMS ensures the integrity and accessibility of essential maintenance documents, including standard operating procedures (SOPs) and validation documentation.

  Change control: CMMS assists in managing change control processes, ensuring that any modifications to equipment or processes are documented, validated, and compliant with regulatory standards.

- **Mobile accessibility**

  Remote monitoring: mobile integration with CMMS allows maintenance teams to access real-time data, create work orders, and monitor equipment health remotely. This is particularly beneficial for on-the-go maintenance activities and troubleshooting.

- **Collaboration and communication**

  Cross-departmental integration: CMMS integration facilitates collaboration between maintenance, production, and quality assurance departments, ensuring that all stakeholders have access to relevant information.

  Communication channels: integration with communication tools allows for efficient communication within maintenance teams, ensuring quick response times and effective problem resolution.

- **Data analytics**

  Performance analysis: CMMS integration enables the collection and analysis of data related to equipment performance and maintenance activities. This data-driven approach supports informed decision-making and continuous improvement initiatives.

- **Cloud-based solutions**
  Scalability and accessibility: integration with cloud-based CMMS solutions enhances scalability, accessibility, and data security. It enables remote access to maintenance data, promoting flexibility in operations.

Thus, by leveraging the capabilities of CMMS, pharmaceutical companies can optimize maintenance strategies, reduce downtime, and enhance overall manufacturing reliability.

7.5.2 IoT Connectivity

In the pharmaceutical industry, IoT connectivity plays a pivotal role in enhancing efficiency, ensuring compliance, and improving overall operational processes. Some key applications and considerations for IoT connectivity in pharmaceutical sector are illustrated in Figure 7.4 [33].

- **Supply chain monitoring**
  Cold chain management: IoT sensors can be deployed to monitor temperature-sensitive pharmaceuticals during transportation and storage. Real-time

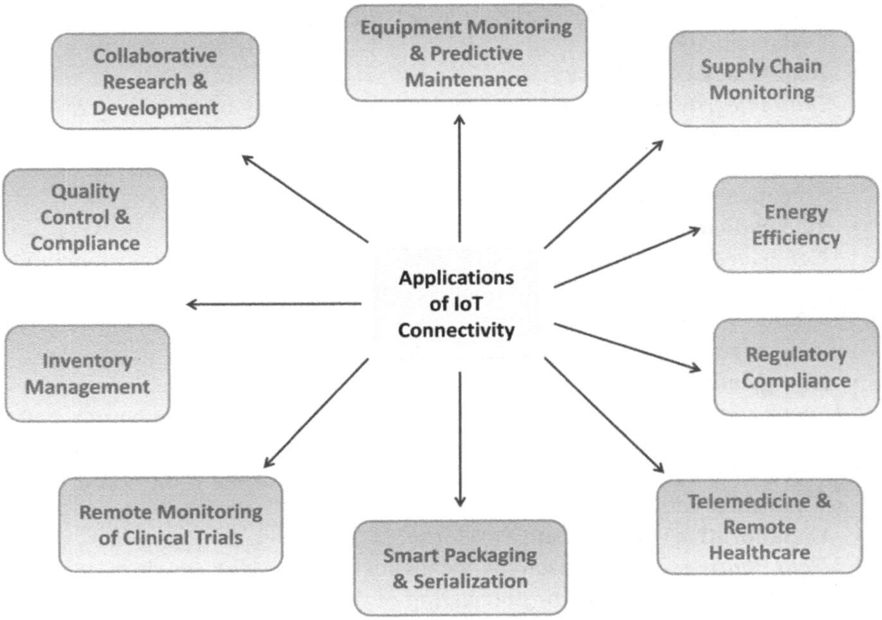

FIGURE 7.4 Key Applications of IoT Connectivity in Pharmaceutical Sector.

tracking ensures adherence to regulatory requirements and maintains the integrity of sensitive medications.

- **Equipment monitoring and predictive maintenance**
  Condition monitoring: IoT-enabled sensors on manufacturing equipment provide real-time data on performance metrics. This facilitates predictive maintenance, reducing downtime, and optimizing the efficiency of pharmaceutical production processes.
- **Quality control and compliance**
  Environmental monitoring: IoT sensors monitor environmental conditions in manufacturing facilities, ensuring compliance with good manufacturing practice (GMP) regulations. This includes monitoring factors such as temperature, humidity, and air quality.
- **Inventory management**
  Smart warehousing: RFID tags and IoT devices help in real-time tracking of pharmaceutical inventory within warehouses. This ensures accurate inventory levels, minimizes errors, and expedites order fulfilment processes.
- **Remote monitoring of clinical trials**
  Patient monitoring devices: IoT devices and wearables facilitate remote patient monitoring during clinical trials. This enables real-time data collection on patient health, improving the efficiency of trials and reducing the need for physical site visits.
- **Smart packaging and serialization**
  Track and trace: IoT-enabled smart packaging with serialization technology helps track pharmaceutical products throughout the supply chain. This enhances traceability, reduces counterfeiting risks, and ensures product authenticity.
- **Regulatory compliance**
  Data integrity: IoT connectivity ensures data integrity by providing real-time monitoring and recording of critical parameters. This is essential for meeting regulatory requirements and maintaining the highest standards in pharmaceutical manufacturing.
- **Telemedicine and remote healthcare**
  Connected medical devices: IoT devices in healthcare settings, including connected medical instruments and monitoring devices, facilitate remote patient consultations, data collection, and real-time healthcare management.
- **Energy efficiency**
  Smart energy management: IoT sensors help optimize energy consumption in pharmaceutical facilities. This includes monitoring and controlling lighting, HVAC systems, and other energy-intensive processes, contributing to sustainability efforts.
- **Collaborative research and development**
  Data sharing: IoT connectivity facilitates collaborative research efforts by enabling secure and efficient sharing of data between pharmaceutical companies, research institutions, and regulatory bodies.

While implementing IoT connectivity in the pharmaceutical industry offers numerous benefits, it is crucial to address security and data privacy concerns. Robust cybersecurity measures must be in place to safeguard sensitive information and ensure the integrity of pharmaceutical processes. Additionally, compliance with industry regulations, such as FDA guidelines, is essential to uphold the highest standards of quality and safety in pharmaceutical manufacturing and distribution.

7.6 CONTINUOUS MONITORING AND FEEDBACK LOOP

Continuous monitoring and feedback loops are essential components of a dynamic and responsive system that ensures ongoing improvement, optimization, and adaptability. In various industries, including pharmaceuticals, this approach is crucial for maintaining operational excellence, enhancing product quality, and meeting regulatory standards. Table 7.2 illustrates continuous monitoring and feedback loops contributing to the pharmaceutical industry. Creating continuous monitoring and feedback loops ensures that organizations can quickly respond to changes, minimize risks, and maintain high standards in an evolving industry landscape [36].

7.7 OPTIMIZATION OF PREDICTIVE MAINTENANCE STRATEGY

The optimization of predictive maintenance strategy is a critical initiative aimed at maximizing the effectiveness and efficiency of maintenance practices within an organization. Leveraging advanced technologies and data-driven insights, this approach involves refining the predictive maintenance processes to minimize downtime, reduce costs, and enhance overall operational performance. By continuously analyzing and adjusting strategies based on real-time data, organizations can ensure the timely identification of potential equipment issues, optimize maintenance schedules, and proactively address maintenance needs. This iterative optimization process not only extends the lifespan of assets but also contributes to improved resource utilization, increased operational reliability, and a more streamlined maintenance workflow. Table 7.3 enlists the various strategies to optimize predictive maintenance [4, 6].

Thus, optimizing a predictive maintenance strategy involves a holistic and strategic approach to maintenance planning, execution, and analysis. By harnessing the power of data and technology, organizations can achieve higher levels of operational efficiency, reduced costs, and increased reliability of critical assets.

7.8 REGULATORY COMPLIANCE IN PHARMACEUTICAL PREDICTIVE MAINTENANCE

7.8.1 Validation Processes

The validation processes in pharmaceutical predictive maintenance involve systematic procedures to ensure that the predictive maintenance system meets predefined requirements and operates effectively within the regulated environment of the pharmaceutical industry. Validation is a critical step to demonstrate that the predictive maintenance processes, software, and technologies are reliable, accurate, and

TABLE 7.2
Continuous Monitoring and Feedback Loops Contributing to Pharmaceutical Sector

S. No.	Parameters	Continuous Monitoring	Feedback Loop
1	Quality Control and Compliance	Utilizing real-time data and sensor technologies to monitor production processes, ensuring adherence to Good Manufacturing Practice (GMP) standards	Rapid detection of deviations triggers immediate corrective actions, preventing non-compliance issues and ensuring high product quality.
2	Process Optimization	Tracking key parameters in manufacturing processes, such as temperature, pressure, and reaction rates, to identify opportunities for optimization	Analyzing data to make informed adjustments to processes, improving efficiency, reducing waste, and enhancing overall operational performance.
3	Predictive Maintenance	Employing IoT sensors and condition monitoring to continuously assess the health of manufacturing equipment	Predictive analytics and machine learning models analyze equipment data, predicting potential failures and optimizing maintenance schedules for minimal downtime.
4	Supply Chain Management	Tracking the movement and storage conditions of pharmaceutical products throughout the supply chain	Immediate alerts and adjustments based on real-time data ensure the integrity of pharmaceuticals during transportation and storage.
5	Regulatory Adherence	Keeping a constant check on processes and documentation to comply with regulatory requirements	Iterative reviews and adjustments based on regulatory feedback, ensuring continuous compliance with evolving standards.
6	Data Analytics and Reporting	Collecting and analyzing data from various sources, including manufacturing processes and quality control measures	Data-driven insights inform decision-making, facilitate trend analysis, and support continuous improvement initiatives through informed adjustments.
7	Risk Management	Identifying and assessing risks associated with manufacturing processes and supply chain activities	Implementing risk mitigation strategies based on real-time data and continuous assessment, ensuring a proactive approach to risk management.

TABLE 7.2 (Continued)
Continuous Monitoring and Feedback Loops Contributing to Pharmaceutical Sector

S. No.	Parameters	Continuous Monitoring	Feedback Loop
8	Environmental Monitoring	Monitoring environmental conditions within manufacturing facilities, including temperature, humidity, and air quality	Prompt responses to deviations from optimal conditions, maintaining a controlled environment for pharmaceutical production.
9	Collaborative Teams and Communication	Fostering collaboration among different departments and teams involved in pharmaceutical manufacturing	Regular feedback sessions, meetings, and communication channels ensure that insights from various stakeholders contribute to continuous improvement efforts.
10	Product Lifecycle Management	Managing the entire lifecycle of pharmaceutical products, from development to manufacturing and distribution	Iterative improvements based on feedback from each phase, ensuring that the product lifecycle is optimized for efficiency and quality.

compliant with regulatory standards. The key elements of the validation processes are outlined in Table 7.4 [4].

The validation processes in pharmaceutical predictive maintenance are part of a broader quality assurance framework and are essential for ensuring the reliability and compliance of predictive maintenance practices in pharmaceutical manufacturing. Regular and well-documented validation activities contribute to maintaining the highest standards of quality and safety within the industry.

7.8.2 Compliance with Industry Regulations

Compliance with industry regulations is a critical aspect of ensuring that businesses operate ethically, safely, and in accordance with legal standards. In various industries, including pharmaceuticals, compliance is vital due to the high level of scrutiny, safety considerations, and potential impact on public health. Several key principles and considerations for ensuring compliance with industry regulations has been depicted in Table 7.5 [3].

Compliance with industry regulations is a multifaceted and ongoing process that requires dedication, vigilance, and a commitment to upholding the highest standards. By embedding a culture of compliance within the organization, implementing robust processes, and staying informed about regulatory changes, businesses can navigate the regulatory landscape successfully and contribute to the overall integrity of their industry.

TABLE 7.3
Optimization Strategies for Predictive Maintenance

S. No.	Parameters	Optimization Strategies
1	Data-driven Decision-Making	Utilizing real-time and historical data to make informed decisions about maintenance activities. Analyzing data patterns to identify trends, anomalies, and potential equipment failures.
2	Advanced Analytics and Machine Learning	Implementing sophisticated analytics and machine learning models to improve the accuracy of failure predictions. Continuously refining algorithms based on ongoing data feedback for better performance.
3	Proactive Planning and Scheduling	Anticipating equipment failures and scheduling maintenance activities proactively to minimize downtime. Optimizing maintenance schedules to balance resource utilization and operational continuity.
4	Condition Monitoring Technologies	Deploying advanced sensors and monitoring technologies for real-time condition assessment of critical assets. Integrating internet of things (IoT) devices to enhance the scope and accuracy of condition monitoring.
5	Risk-based Prioritization	Prioritizing maintenance tasks based on the risk and criticality of equipment. Allocating resources to high-priority assets to maximize the impact of maintenance efforts.
6	Continuous Improvement	Establishing a feedback loop for continuous improvement, incorporating lessons learned from past maintenance activities. Iteratively refining predictive maintenance processes based on performance analysis and evolving data trends.
7	Integration with CMMS and EAM Systems	Integrating predictive maintenance solutions with computerized maintenance management systems (CMMS) and enterprise asset management (EAM) systems. Streamlining workflows and ensuring seamless data exchange between predictive maintenance tools and broader asset management platforms.
8	Employee Training and Skill Development	Providing training for maintenance personnel to enhance their skills in interpreting predictive maintenance data Ensuring that the workforce is equipped to leverage advanced technologies and methodologies.

TABLE 7.3 (Continued)
Optimization Strategies for Predictive Maintenance

S. No.	Parameters	Optimization Strategies
9	Cost Optimization	Balancing the costs associated with predictive maintenance with the benefits of extended equipment life and reduced downtime. Analyzing the return on investment (ROI) to justify and optimize maintenance expenditures.
10	Continuous Monitoring and Reporting	Establishing robust systems for continuous monitoring of equipment health and performance. Generating regular reports on the effectiveness of predictive maintenance activities and using insights for further optimization.

TABLE 7.4
Key Aspects of the Validation Processes

S. No.	Key Aspect	Requirement	Mitigation
1	User Requirement Specification (URS)	Definition of User Requirements	Clearly defining the user requirements for the predictive maintenance system, including its functionalities, capabilities, and intended use within the pharmaceutical manufacturing environment.
2	Risk Assessment	Identification of Risks	Conducting a comprehensive risk assessment to identify potential risks associated with the predictive maintenance system. This includes risks to product quality, patient safety, and compliance with regulatory standards.
3	Design Specification	Detailed Design Documentation	Developing detailed design specifications that outline the architecture, functionalities, and features of the predictive maintenance system based on the user requirements.
4	Installation Qualification (IQ)	Verification of Installation	Verifying that the predictive maintenance system is installed correctly and according to specifications. This includes checking hardware, software, and network configurations.
5	Operational Qualification (OQ)	Functional Testing	Conducting functional testing to ensure that the predictive maintenance system performs as intended. This involves validating each function and feature of the system to confirm its operational capabilities.

(*continued*)

TABLE 7.4 (Continued)
Key Aspects of the Validation Processes

S. No.	Key Aspect	Requirement	Mitigation
6	Performance Qualification (PQ)	Real-world Performance Testing	Evaluating the real-world performance of the predictive maintenance system in the pharmaceutical manufacturing environment. This includes testing the system under normal operating conditions to validate its reliability and accuracy.
7	Validation Protocols	Documented Validation Protocols	Developing detailed validation protocols that outline the testing procedures, acceptance criteria, and documentation requirements for each stage of validation.
8	Data Integrity	Ensuring Data Integrity	Validating the integrity, security, and traceability of data generated by the predictive maintenance system. This includes confirming that data is accurately collected, stored, and retrievable.
9	Validation of Predictive Models	Model Validation	If predictive maintenance involves the use of machine learning models or algorithms, validating these models for accuracy, reliability, and compliance with regulatory requirements.
10	Change Control Procedures	Documentation of Changes	Implementing change control procedures to manage any modifications to the predictive maintenance system. Documenting and justifying changes is essential for maintaining compliance.
11	Validation Report	Comprehensive Validation Report	Compiling a comprehensive validation report summarizing the entire validation process, including test results, deviations, and any corrective actions taken.
12	Periodic Revalidation	Ongoing Revalidation	Establishing a plan for periodic revalidation to ensure that the predictive maintenance system continues to meet requirements and remains in compliance as the system evolves or changes are implemented.

7.9 CHALLENGES AND FUTURE TRENDS

7.9.1 Current Challenges in Implementing Predictive Maintenance

Implementing predictive maintenance can bring significant benefits to industries by optimizing equipment performance, reducing downtime, and minimizing maintenance costs. However, several challenges may be encountered during the implementation process. Table 7.6 highlights some current challenges in implementing predictive maintenance [37].

TABLE 7.5
Considerations for Ensuring Compliance with Industry Regulations

S. No.	Key Aspect	Requirement	Considerations
1	Understanding Regulatory Landscape	Stay Informed	Regularly monitor and stay informed about relevant industry regulations, guidelines, and standards applicable to the specific sector, such as good manufacturing practice (GMP) in pharmaceuticals.
2	Establishing a Compliance Framework	Create Policies and Procedures	Develop comprehensive policies and procedures that outline how the organization will comply with specific regulations. This includes detailing responsibilities, processes, and reporting mechanisms.
3	Regulatory Training	Employee Education	Ensure that employees are educated and trained on relevant regulations, compliance requirements, and the importance of adhering to established protocols.
4	Risk Assessment and Management	Identify and Mitigate Risks	Conduct regular risk assessments to identify potential areas of non-compliance. Develop and implement strategies to mitigate and manage these risks effectively.
5	Documentation and Record-Keeping	Maintain Accurate Records	Keep detailed and accurate records of all activities related to compliance. This includes documentation of processes, procedures, training sessions, audits, and any corrective actions taken.
6	Audit and Monitoring Programs	Regular Audits	Implement regular internal and external audits to assess compliance with regulations. Use these audits to identify areas for improvement and corrective actions.
7	Data Integrity	Ensure Data Accuracy	In industries where data is crucial, maintain data integrity by implementing systems and practices that ensure accurate and reliable data collection, storage, and retrieval.
8	Continuous Improvement	Learn from Incidents	Establish a culture of continuous improvement by learning from any incidents or non-compliance issues. Implement corrective and preventive actions to avoid recurring problems.

(*continued*)

TABLE 7.5 (Continued)
Considerations for Ensuring Compliance with Industry Regulations

S. No.	Key Aspect	Requirement	Considerations
9	Third-Party Vetting	Evaluate Suppliers and Partners	If applicable, ensure that suppliers, contractors, and business partners also adhere to industry regulations. Establish criteria for vetting and monitoring third-party entities.
10	Communication with Regulatory Bodies	Transparent Communication	Maintain open and transparent communication with regulatory bodies. Report incidents promptly and work collaboratively to address any concerns raised during inspections or audits.
11	Adaptability to Regulatory Changes	Stay Flexible	Regulations can change, so organizations must be adaptable. Stay vigilant to regulatory updates, assess their impact on operations, and make necessary adjustments to maintain compliance.
12	Ethical Conduct and Corporate Governance	Promote Ethical Behavior	Emphasize ethical conduct throughout the organization. Encourage a strong corporate governance framework that aligns with regulatory expectations.
13	Legal Counsel Involvement	Legal Guidance	Engage legal counsel to provide guidance on interpreting and complying with complex regulations. Legal experts can help ensure that organizational practices align with the legal landscape.
14	Crisis Management and Response Plans	Prepare for Contingencies	Develop crisis management and response plans to address potential non-compliance incidents swiftly and effectively. Be prepared to communicate with stakeholders, including the public, if necessary.

Addressing these challenges requires a holistic approach, involving careful planning, strategic investments, and ongoing collaboration between stakeholders. Organizations should prioritize understanding their specific challenges and tailoring solutions to ensure successful implementation of predictive maintenance strategies.

7.9.2 Emerging Trends in Data Analytics and AI for Pharmaceutical Manufacturing

Several emerging trends in data analytics and AI have been impacting pharmaceutical manufacturing. These trends contribute to enhanced efficiency, improved quality control, and better decision-making processes. Table 7.7 enlists some notable

TABLE 7.6
Challenges in Implementing Predictive Maintenance and Their Solutions

S. No.	Issue	Challenge	Solutions
1	Data Quality and Availability	Predictive maintenance relies heavily on data from sensors, equipment, and other sources. Inconsistent, inaccurate, or insufficient data can hinder the effectiveness of predictive maintenance models.	Establish robust data collection processes, invest in high-quality sensors, and ensure data accuracy and completeness. Implement data cleansing techniques to address issues with data quality.
2	Integration with Existing Systems	Integrating predictive maintenance solutions with existing systems, such as Computerized Maintenance Management Systems (CMMS) or Enterprise Resource Planning (ERP) systems, can be complex and may require significant adjustments.	Plan for integration challenges in advance, work closely with IT teams, and consider flexible solutions that can adapt to existing infrastructure.
3	Costs and Return on Investment (ROI)	Implementing predictive maintenance involves upfront costs for sensors, software, and infrastructure. Organizations may face challenges in demonstrating a clear and quick ROI.	Conduct a thorough cost-benefit analysis, prioritize critical assets for predictive maintenance, and communicate the long-term benefits of reduced downtime and maintenance costs.
4	Skilled Workforce Shortage	The implementation of predictive maintenance requires a skilled workforce, including data scientists, engineers, and technicians. There is a shortage of professionals with expertise in both maintenance and data analytics.	Invest in training programs, collaborate with educational institutions, and consider partnerships with external experts or service providers.
5	Change Management	Introducing predictive maintenance often requires a cultural shift within the organization. Resistance to change and lack of buy-in from employees can impede successful implementation.	Communicate the benefits of predictive maintenance, involve employees in the implementation process, and provide training and support to address concerns.

(*continued*)

TABLE 7.6 (Continued)
Challenges in Implementing Predictive Maintenance and Their Solutions

S. No.	Issue	Challenge	Solutions
6	Scalability Issues	Predictive maintenance solutions may face scalability challenges when expanding to cover a larger number of assets or across multiple facilities.	Choose scalable solutions, leverage cloud-based platforms, and design the implementation with future expansion in mind.
7	Security and Privacy Concerns	Predictive maintenance systems involve the collection and analysis of sensitive operational data. Ensuring the security and privacy of this data is a critical concern.	Implement robust cybersecurity measures, comply with data protection regulations, and regularly audit and update security protocols.
8	Complexity of Predictive Models	Developing accurate predictive models can be complex, especially for diverse and interconnected systems. Ensuring the models are reliable and adaptable to changing conditions is a challenge.	Collaborate with data scientists and subject matter experts, continuously refine models based on real-world performance, and leverage machine learning techniques for adaptability.
9	Regulatory Compliance	Predictive maintenance systems must comply with industry-specific regulations and standards, particularly in highly regulated sectors such as healthcare and pharmaceuticals.	Stay informed about regulatory requirements, design systems with compliance in mind, and document adherence to standards in the implementation process.
10	Lack of Historical Data	In cases where there is a lack of historical maintenance data, building accurate predictive models can be challenging.	Implement data collection strategies moving forward, consider hybrid models that leverage both historical and real-time data, and focus on building a comprehensive dataset over time.

emerging trends in the application of data analytics and AI in pharmaceutical manufacturing [38].

These trends underscore the transformative potential of data analytics and AI in pharmaceutical manufacturing, providing opportunities for innovation, efficiency gains, and advancements in patient care.

TABLE 7.7
Emerging Trends in the Application of Data Analytics and AI in Pharmaceutical Manufacturing

S. No.	Emerging Trends	Description	Impact
1	Advanced Predictive Analytics	Utilizing advanced predictive analytics models to foresee equipment failures, optimize production processes, and streamline supply chain management	Improved maintenance planning, reduced downtime, and enhanced overall operational efficiency.
2	Real-time Monitoring and Control	Implementation of real-time monitoring systems for manufacturing processes, enabling immediate adjustments and interventions based on live data	Enhanced quality control, increased responsiveness to deviations, and improved process optimization.
3	Digital Twins	Creating digital replicas (digital twins) of physical manufacturing processes or equipment, allowing for simulation, analysis, and optimization	Improved understanding of processes, predictive maintenance capabilities, and faster problem-solving.
4	IoT Integration	Integration of the Internet of Things (IoT) devices and sensors for comprehensive data collection and monitoring of equipment and processes	Increased connectivity, real-time data insights, and enhanced overall visibility into manufacturing operations.
5	AI-powered Quality Control	Implementation of AI algorithms for quality control processes, including the identification of defects or deviations in real-time	Higher accuracy in quality assurance, reduced waste, and compliance with stringent regulatory standards.
6	Supply Chain Optimization	Utilizing AI and data analytics to optimize the pharmaceutical supply chain, ensuring efficient inventory management and timely deliveries	Reduced lead times, minimized stockouts, and enhanced overall supply chain resilience.
7	Augmented Reality (AR) for Training and Maintenance	Adoption of AR technologies for training personnel and assisting in maintenance tasks, providing augmented information and guidance	Improved training efficiency, reduced human errors, and enhanced maintenance procedures.
8	Blockchain for Traceability	Implementation of blockchain technology for end-to-end traceability of pharmaceutical products throughout the supply chain	Increased transparency, reduced counterfeiting risks, and compliance with traceability regulations.

(continued)

TABLE 7.7 (Continued)
Emerging Trends in the Application of Data Analytics and AI in Pharmaceutical Manufacturing

S. No.	Emerging Trends	Description	Impact
9	Explainable AI (XAI)	Emphasis on developing AI models that provide clear and understandable explanations for their decisions, especially in critical areas like drug development and quality control	Increased trust in AI systems, better regulatory acceptance, and improved decision-making processes.
10	Personalized Medicine and Drug Discovery	Leveraging AI and data analytics for personalized medicine approaches, including patient-specific drug formulations and targeted therapies	Accelerated drug discovery, improved treatment outcomes, and a shift towards more patient-centric healthcare.
11	Regulatory Compliance Automation	Adoption of AI and data analytics to automate regulatory compliance processes, ensuring adherence to evolving standards	Enhanced efficiency in compliance reporting, reduced compliance-related risks, and improved audit readiness.

7.10 CONCLUSIONS

In conclusion, the integration of data analytics and artificial intelligence (AI) in pharmaceutical manufacturing represents a transformative paradigm shift, bringing forth a host of benefits ranging from enhanced operational efficiency to improved product quality and compliance. As the industry embraces emerging trends, it becomes clear that these technologies play a pivotal role in shaping the future of pharmaceutical manufacturing.

The utilization of advanced predictive analytics, real-time monitoring, and control mechanisms empowers manufacturers to proactively address challenges, optimize processes, and reduce downtime. Digital twins and the integration of the internet of things (IoT) offer unprecedented visibility into manufacturing operations, fostering a data-driven decision-making environment. The application of AI in quality control, supply chain optimization, and personalized medicine not only improves outcomes but also positions pharmaceutical companies at the forefront of innovation.

Moreover, the adoption of augmented reality for training and maintenance, as well as the incorporation of blockchain for traceability, exemplifies the industry's commitment to leveraging cutting-edge technologies for efficiency, safety, and compliance. Explainable AI (XAI) addresses concerns regarding transparency and trust, while regulatory compliance automation ensures that pharmaceutical manufacturers stay ahead of evolving standards.

As these trends continue to unfold, it is crucial for pharmaceutical companies to remain agile, adaptable, and committed to ongoing advancements. Collaboration with technology experts, investment in workforce training, and a proactive approach to regulatory adherence will be instrumental in realizing the full potential of data analytics and AI in pharmaceutical manufacturing. The journey towards a more connected, intelligent, and patient-centric industry is underway, promising a future where innovation and efficiency converge to drive sustainable success.

REFERENCES

1 Jagatheesaperumal, S.K., Rahouti, M., Ahmad, K., Al-Fuqaha, A., Guizani, M., The duo of artificial intelligence and big data for industry 4.0: applications, techniques, challenges, and future research directions, *IEEE Internet of Things Journal*, 9, 12861, 2022.

2 Ganesh, S., Su, Q., Nagy, Z., Reklaitis, G., Advancing smart manufacturing in the pharmaceutical industry, in *Smart Manufacturing*, Soroush, M., Baldea, M., Edgar, T.F., Eds., Elsevier, Amsterdam, 2020, chap. 2.

3 Jalundhwala, F., Londhe, V., A systematic review on implementing operational excellence as a strategy to ensure regulatory compliance: a roadmap for Indian pharmaceutical industry, *International Journal of Lean Six Sigma*, 14, 730, 2023.

4 Manchadi, O., Ben-Bouazza, F.-E., Jioudi, B., Predictive maintenance in healthcare system: a survey, *IEEE Access*, 11, 61313, 2023.

5 Marques, C.M., Moniz, S., de Sousa, J.P., Barbosa-Povoa, A.P., Reklaitis, G., Decision-support challenges in the chemical-pharmaceutical industry: findings and future research directions, *Computers & Chemical Engineering*, 134, 106672, 2020.

6 Ben-Bouazza, F.-E., Manchadi, O., Dehbi, Z.E.O., Rhalem, W., Ghazal, H. (2023). Machine learning based predictive maintenance of pharmaceutical industry equipment, in: *International Conference on Advanced Intelligent Systems for Sustainable Development. AI2SD 2022. Lecture Notes in Networks and Systems*, Kacprzyk, J., Ezziyyani, M., Balas, V.E., Eds., Springer, Cham, 714, 2023.

7 Silva, A.F., Vercruysse, J., Vervaet, C., Remon, J.P., Lopes, J.A., De Beer, T., Sarraguça, M.C., In-depth evaluation of data collected during a continuous pharmaceutical manufacturing process: a multivariate statistical process monitoring approach, *Journal of Pharmaceutical Sciences*, 108, 439, 2019.

8 Kammerer, K., Hoppenstedt, B., Pryss, R., Stokler, S., Allgaier, J., Reichert, M., Anomaly detections for manufacturing systems based on sensor data - insights into two challenging real-world production settings, *Sensors*, 19, 5370, 2019.

9 Eifert, T., Eisen, K., Maiwald, M. Herwig, C., Current and future requirements to industrial analytical infrastructure - part 2: smart sensors, *Analytical and Bioanalytical Chemistry*, 412, 2037, 2020.

10 Jelsch, M., Roggo, Y., Kleinebudde, P., Krumme, M., Model predictive control in pharmaceutical continuous manufacturing: a review from a user's perspective, *European Journal of Pharmaceutics and Biopharmaceutics*, 159, 137, 2021.

11 Coito, T., Firme, B., Martins, M.S.E., Vieira, S.M., Figueiredo, J., Sousa, J.M.C., Intelligent sensors for real-time decision-making, *Automation*, 2, 62, 2021.

12 Yu, L.X., Pharmaceutical quality by design: product and process development, understanding, and control, *Pharmaceutical Research*, 25, 781, 2008.

13 Tao, F., Qi, Q., Liu, A., Kusiak, A., Data-driven smart manufacturing, *Journal of Manufacturing Systems*, 48, Part C, 157, 2018.

14 Unal, P., Albayrak, O., Jomaa, M., Berre, A.J., Data-driven artificial intelligence and predictive analytics for the maintenance of industrial machinery with hybrid and cognitive digital twins, in *Technologies and Applications for Big Data Value*, Curry, E., Auer, S., Berre, A.J., Metzger, A., Perez, M.S., Zillner, S., Eds., Springer, Cham, 2022.
15 Dash, S., Shakyawar, S.K., Sharma, M., Kaushik, S., Big data in healthcare: management, analysis and future prospects. *Journal of Big Data*, 6, 54, 2019.
16 Wang, Y., Kung, L., Byrd, T.A., Big data analytics: understanding its capabilities and potential benefits for healthcare organizations, *Technological Forecasting and Social Change*, 126, 3, 2018.
17 Leal, F., Chis, A.E., Caton, S., et al., Smart pharmaceutical manufacturing: ensuring end-to-end traceability and data integrity in medicine production, *Big Data Research*, 24, 100172, 2021.
18 Kuhn, M., Johnson, K., Data pre-processing, in *Applied Predictive Modeling*, Kuhn, M., Johnson, K., Eds., Springer, New York, NY, 2013.
19 Javaid, M., Haleem, A., Singh, R.P., Suman, R., Rab, S., Significance of machine learning in healthcare: features, pillars and applications, *International Journal of Intelligent Networks*, 3, 58, 2022.
20 Ferencek, A., Borstnar, M.K., Data quality assessment in product failure prediction models, *Journal of Decision Systems*, 29, sup1, 79, 2020.
21 Achouch, M., Dimitrova, M., Ziane, K., Sattarpanah Karganroudi, S., Dhouib, R., Ibrahim, H., Adda, M., On predictive maintenance, in Industry 4.0: overview, models, and challenges, *Applied Science*, 12, 8081, 2022.
22 Patcha, A., Park, J.-M., An overview of anomaly detection techniques: existing solutions and latest technological trends, *Computer Networks*, 51, 3448, 2007.
23 Omar, S., Ngadi, A., Jebur, H.H., Machine learning techniques for anomaly detection: an overview, *International Journal of Computer Applications*, 79, 33, 2013.
24 Topolski, M., Application of feature extraction methods for chemical risk classification in the pharmaceutical industry, *Sensors*, 21, 5753, 2021.
25 Rawat, T., Khemchandani, V., Feature engineering (fe) tools and techniques for better classification performance, *International Journal of Innovations in Engineering and Technology*, 8, 169, 2017.
26 Heaton, J., An empirical analysis of feature engineering for predictive modeling, in *SoutheastCon 2016*, Norfolk, VA, USA, 1, 2016.
27 Huysentruyt, K., Kjoersvik, O., Dobracki, P., et al., Validating intelligent automation systems in pharmacovigilance: insights from good manufacturing practices, *Drug Safety*, 44, 261, 2021.
28 Meagher, D.P., Engineering automated systems for pharmaceutical manufacturing: quality, regulations and business performance, PhD thesis, Dublin City University, 2006.
29 Ahmed, S.F., Alam, M.S.B., Hassan, M., et al., Deep learning modelling techniques: current progress, applications, advantages, and challenges, *Artifical Intelligent Review*, 56, 13521, 2023.
30 Juodelyte, D., Cheplygina, V., Graversen, T., Bonnet, P., Predicting bearings degradation stages for predictive maintenance in the pharmaceutical industry, *KDD '22: Proceedings of the 28th ACM SIGKDD Conference on Knowledge Discovery and Data Mining*, 3107, 2022.
31 Ferreira, C., Goncalves, G., Remaining useful life prediction and challenges: a literature review on the use of machine learning methods, *Journal of Manufacturing Systems*, 63, 550, 2022.

32 Wang, Y., Zhao, Y., Addepalli, S., Remaining useful life prediction using deep learning approaches: a review, *Procedia Manufacturing*, 49, 81, 2020.
33 Singh, M., Sachan, S., Singh, A., Singh, K.K., Internet of things in pharma industry: possibilities and challenges, in *Emergence of Pharmaceutical Industry Growth with Industrial IoT Approach*, Balas, V.E., Solanki, V.K., Kumar, R., Eds., Academic Press, 195, 2020, chap. 7.
34 Sharma, A., Kaur, J., Singh, I., Internet of things (IoT) in pharmaceutical manufacturing, warehousing, and supply chain management, *SN Computer Sciences*, 1, 232, 2020.
35 Liggin, P., The research and implementation of maintenance excellence on clean utility systems in the pharmaceutical industry, Masters Dissertation, Dublin, DIT, 2008.
36 Brun, Y., Di Marzo Serugedo, G., Gacek, C., et al., Engineering self-adaptive systems through feedback loops, in *Software Engineering for Self-Adaptive Systems*. Lecture Notes in Computer Science, Cheng, B.H.C., de Lemos, R., Giese, H., Inverardi, P., Magee, J., Eds., Springer, Berlin, Heidelberg, 5525, 2009.
37 Dalzochio, J., Kunst, R., Pignaton, E., Binotto, A., Sanyal, S., Favilla, J., Barbosa, J., Machine learning and reasoning for predictive maintenance in Industry 4.0: current status and challenges, *Computers in Industry*, 123, 103298, 2020.
38 Maharjan, R., Lee, J.C., Lee, K., et al., Recent trends and perspectives of artificial intelligence-based machine learning from discovery to manufacturing in biopharmaceutical industry, *Journal of Pharmaceutical Investigation*, 53, 803, 2023.

8 Real-time Violence Detection in Video Streams

Exploiting ResNet-50 for Enhanced Accuracy

Surinder Kaur, Ajay Dureja, Mahesh Kumar, Mohit Dayal, and Shyla

8.1 INTRODUCTION

The rate of violent criminal activity has grown over the past few decades, motivating companies to utilize surveillance systems to spot risky areas and react swiftly to violent actions. CCTV cameras are the foundation of the conventional surveillance method, which is used to monitor and supervise people. A small group of security personnel were responsible for the prior assignment, which involved watching through vast amounts of CCTV material [1]. The typical approach is therefore unreliable due to a number of issues including worker weariness, boredom, and observational discontinuity. Violence detection, which must be able to function in a variety of situations to identify quarrels from security cameras and regulate violent occurrences, is one of the difficult problem that security systems identify in commercial and public infrastructure, such as shopping malls, have to deal with. Public violence has skyrocketed in the modern era, both on the streets and in high schools. As a result, surveillance cameras are used everywhere. This has made it easier for the government to recognize these incidents and take the appropriate action. However, the majority of systems in use today necessitate human inspection of these recordings in order to detect such occurrences, which is essentially ineffective [2]. Thus, having a useful system that can recognize and automatically monitor the security footage is essential. The field of computer vision has seen a historic shift as a result of the development of numerous deep learning algorithms made possible by the availability of massive data sets and processing resources. To solve issues like object detection, recognition, tracking, action recognition, and legend creation, a variety of solutions have been created. Because violent incidents can happen anywhere at any moment in a city, it is ineffective to rely on people to keep an eye out for and identify violent incidents. These activities typically result in highly unpleasant scenarios, which is why it is critical that automatic identification of such events occur through real-time

DOI: 10.1201/9781003480860-8

video footage, enabling the involved authority to make the necessary and important decision. Consequently, the notion of putting in place tools and systems to identify these kinds of occurrences through real-time monitoring and video retrieval has been presented. Eliminating the aforementioned practical restrictions and effectively bringing down crime rates are the main priorities.

The study's principal contributions are as follows: the study concurrently incorporates geographical and temporal information by utilising sophisticated deep learning architectures. As a result, a novel model is put out that performs better with somewhat lesser network depth. In order to further improve efficiency, the system uses a sophisticated motion representation technique inside the temporal stream, reducing the need for costly optical low and making use of motion attributes to accurately identify the behavior. To do this, a generic behavior recognition technique that already exists is expanded for the specific aim of detecting irregularities in crowds. In a similar vein, it is suggested to process only the useful portions of frames in order to avoid processing large amounts of unnecessary pixels [3].

In order to identify a common set of abnormalities that can encompass a broader range of abnormal behaviors in crowded environments, the study explores the most recent advancements in crowd abnormality identification. The two main kinds that are highlighted are aggressive confrontations and escape panics. Crowds moving in one direction, crowds dispersing from a single place, and crowd turmoil or evacuation actions are a few instances of the former kind of uncommon crowd behavior. Attacks, fights, and trampling are some frequent instances of violent encounters that occur in crowded settings.

This research integrates the identification of multiple often observed anomalous behaviors within a congested environment. In this manner, the most pertinent datasets are used to direct the abnormal event detection model's training process after determining the species of the chosen abnormal events. A few improvements are made throughout the study, primarily to increase the suggested detection system's efficiency.

This paper's primary goal is to suggest an automated method for identifying violent actions. Prior research has demonstrated that deep learning approaches out perform other approaches in terms of accuracy and error rates. Our objective is to improve, attain more accuracy, and out perform the existing techniques of violence detection and recognition.

8.2 LITERATURE REVIEW AND METHODOLOGY

In general, CNN is divided into two tiers. The input of every neuron in the first layer, known as the feature extraction layer, is linked to the local receptive fields of the layer before it, which extracts the local feature. Following the extraction of the local features, the positional relationship with other features will also be ascertained. The other is the feature map layer; each network computing layer is made up of several feature maps. Each feature map is a plane with equal weights for all of its neurons. The sigmoid function is used in the feature map structure as the activation function of the convolution network, giving the feature map shift invariance.Additionally, the network has fewer free parameters since neurons in the same mapping plane share weight. The computational layer that comes after each convolution layer in a convolution neural network computes the local average and the second extract; this special two-feature extraction structure lowers the resolution [4]. CNN is primarily used to detect zoom, displacement, and other types of two-dimensional graphic distortion invariance. When utilizing CNN, the feature detection layer learns implicitly from the training data instead of explicitly by extracting features from the data. Furthermore, the neurons in the same feature map plane have the same weight, allowing the network to study concurrently. This is a significant benefit of the convolution network over the mutually connected neural network. CNN has a distinct edge in speech recognition and image processing because to the unique structure of its local shared weights. Its structure is more akin to the real neural network found in biology. The network becomes less complex when shared weights are used. Specifically, multi-dimensional input vector images can be fed straight into the network, saving the laborious process of reconstructing data during the feature extraction and classification stages. This paper presents a prediction model for various purposes including satellite communications [5] in sub-tropical regions. The model is based on slant path rain exahaustion measurements in Singapore using beacon signals like WINDS and GE23 satellites. Cumulative distributions of rainfall rate and corresponding rain deplition are analyzed. Existing models, including Yamada, DAH, Karasawa, and Ramachandran, are compared with measurement data from nine countries. Results indicate room for improvement in existing models, leading to the proposal of a new slant path rain devitilization model tailored for the sub-tropical region.The assigned model outperforms existing ones, providing more accurate predictions for satellite links in tropical climates.The paper introduces a data-driven model [6] for the prediction of accurate rainfall, leveraging PWVfrom GPS signals along with various ground-based weather features. A systematic analysis of atmospheric parameters, including temperature with relative humidity, dew point, solar radiation, PWV, seasonal, and diurnal (S&D) variables, is conducted. The study identifies PWV, solar radiation, and S&D features as key contributors to rainfall prediction.

Using these optimal features, a machine learning algorithm is employed, achieving an experimental evaluation with a true detection rate (80.4%), a false alarm rate (20.3%), and an overall accuracy (79.6%) based on a four-year database from 2012 to 2015.The study focuses on rainfall prediction [7] in a watershed basin using a data-mining approach, specifically applying RR and TB data.

Various data-mining algorithmss, including impartial networks, RF, and support vector machines, are employed to construct prediction models. Three models are developed: Model I uses radar data from Oxford(OM), Model II predicts rainfall from both radar and local TB data in Oxford, and Model III incorporates radar and nearby TB data from cities such as South Amana and Iowa City. The results reveal that Model II outperforms the others in predicting both current and future rainfall, highlighting the significance of combining radar and local TB data for improved accuracy. The study emphasizes the potential application of data-mining algorithms in RF, providing valuable insights for water quality assessments in watershed management.

In this section, we outline the approach employed subsequent to data collection. Our methodology involved converting the videos from our training dataset into a series of images, which were then input into our two models. Following training, the models were applied to classify the videos in the test dataset for violence detection [8].

8.2.1 CNN Layers Used

- **Time distributed:** this wrapper class facilitates the application of a neural layer to each feature as an input. The input is expected to be at least 3D, and the dimension at index one of the first input is interpreted as the temporal dimension.
- **LSTM:** after extracting the frames output is sent to another network architecture having selective neurons called LSTM (long short-term memory). The videos can be used to analyse temporal information and retrieve data from these neurons. If the model notice any instances of violence, the video will be flagged as having violent content [9].
- **Dense:** the dense layer in a neural network is characterized by deep interconnections, where each and every neuron within the neural layer is inter and itra connected to every neuron in the preceding layers. It is widely employed in various models. The matrix incorporates trainable parameters, subject to updates through the process of backpropagation. The output generated by different layers is an 'm'-dimensional vector, enabling dimensionality transformation.
- **Dropout:** the dropout layer functions as a mask that selectively nullifies the contributions of certain neurons to the subsequent layer while leaving others unaltered.

8.2.2 Data Pre-processing

The data includes an average of 150 frames in a single video, the videos from the combined RWF-2000 and RLVS dataset were converted to frames. Owing to the internet-sourced nature of the dataset, the movies were limited to a 30 frames per second frame rate. After that, 10 frames were randomly chosen from every movie at regular intervals, and these frames were scaled to 112 × 112. In the CNN model BGR was changed to RGB. ResNet-50 requires a special type of preprocessing in which the RGB needs to be changed to BGR. Each of the three channels was normalized, which will zero-center each color channel with respect to the ImageNet dataset [10].

8.2.3 Data Augmentation

- A key method in deep learning is data augmentation, which improves the diversity and variability of training data. Data augmentation works to prevent overfitting of the model by applying various changes to the input data, which in turn helps to acquire more robust and generalized patterns. The following data augmentation methods were used in the scenario that was described [11]:
- **Rescaling pixel intensities:** each frame's pixel intensities underwent a factor of 1/255 rescaling. This rescaling operation ensures that the pixel values are within the range of 0 to 1, which deep learning models frequently employ. Rescaling guarantees constant input ranges between samples and aids in normalizing the data.

- **Horizontal and vertical flipping:** the dataset's rescaled photos were turned both vertically and horizontally. This transformation creates new samples by reflecting the original images along the horizontal and vertical axes. By introducing these flipped versions, the model learns to recognize patterns and objects from different orientations, enhancing its ability to generalize and perform well on unseen data.
- **Random rotation**: the rescaled images were randomly rotated during data augmentation. This rotation introduces variations in the orientation of objects, allowing the model to learn to detect and classify objects regardless of their rotation angle. By including randomly rotated images, the model becomes more robust to changes in object orientation and improves its performance under different viewing angles.
- **Random magnification:** the images in the dataset were randomly magnified as part of the data augmentation process. This augmentation technique involves scaling the images by different factors, either increasing or decreasing their size. By applying random magnification, the model learns to handle variations in object size and scale, making it more adaptable to objects of different sizes within the input data.

8.2.4 Training

One method for lessening model over-fitting is the neural network train-validate-test cycle. Another name for the method is early stopping. Though the train-validate-test process isn't conceptually challenging, it can be challenging to describe because it involves multiple interrelated concepts. A neural network's weights and biases are determined throughout the training phase. The train-test approach is typically employed to finish the training procedure. About 80% of the data that is accessible is divided into two sets: a test set and a training set [12].

The values of the accessible data's input and output are known. The training data set is used to train the neural network. Various settings for the weights and biases are examined in order to identify the set of values that will allow the computed output values to most closely approximate the known, correct output values.

Stated differently, training is the process of figuring out which weights and biases to use in order to minimise error.

There are many training methods available, such as back-propagation and particle swarm optimization. When training, the test data is not utilised in any way. After training, the final neural network model's weights and biases are only applied to the test data once. The model's performance on the test data gives you a very rough sense of how accurate it will be when new, unseen data is added to it.

8.2.5 Overfitting

Overfitting is a key issue when utilising neural networks. An excessively long training regimen may cause the model to become overfit. A set of weight and bias values that yield outputs that almost precisely match the training set of data is the final result.

But when similar weight and bias values are applied to new data, the model performs horribly. The train-validate-test methodology aims to help spot the point at which model overfitting occurs so that training can be stopped. From the available data, three sets are created: a test set (20%), a validation set (20%), and a training set (typically 60% of the data). This is carried out instead of dividing the data into train and test. Following pre-processing and augmentation, the frames are passed through each layer of the ResNet-50 architecture, which is renowned for its cutting-edge performance. This architecture makes use of residual connections to help solve the vanishing gradient issue and make deep network training easier. The generated feature maps are subjected to batch normalisation after the ResNet-50 layers. By lowering internal covariate shift, this normalisation strategy aids in both stabilising and speeding up the training process.

The feature maps derived from the ResNet-50 layers are subjected to average pooling in order to minimise the number of parameters and accelerate computation [13].

By preserving significant information, average pooling reduced the spatial dimensions of the feature maps while maintaining computing efficiency. A different CNN model with eight layers is used in parallel. The convolution layers in this CNN model are sequential, and the number of filters increases from 64 to 128 to 256 to 512, in that order. To add non-linearity, ReLU activation functions are applied after every convolution layer. In the CNN model, following each of the two convolution layers, by decreasing spatial dimensions and regularising the network, respectively, both strategies aid in the mitigation of overfitting. An LSTM layer is used at the end to process the feature map sequences from the CNN and ResNet-50 models in order. Long short-term memory (LSTM) type recurrent neural networks (RNNs) effectively extract relevant information from sequential input data, allowing the model to identify temporal correlations and produce predictions based on the background of previous feature maps. Figure 8.1 presents a flow chart that illustrates the entire process.

Figure 8.1 represents the methodology adapted for the predictions made for the detection of violent videos. The dataset has been gathered and collected from open

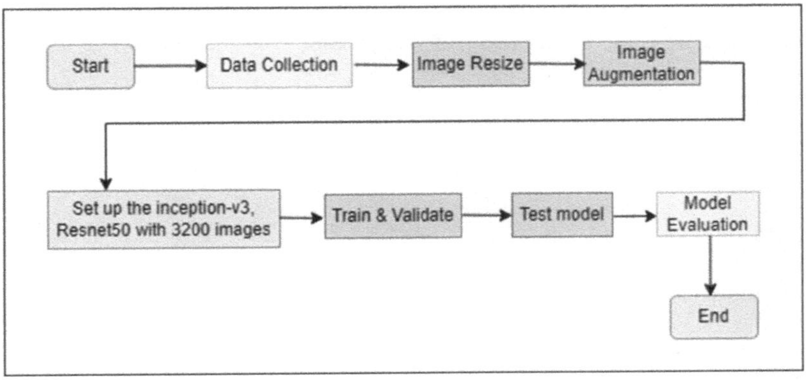

FIGURE 8.1 Methodology Flowchart [14].

source environment after that the image is resized and augmented for further processing to train, test, and validate the model.

8.3 RESULT AND ANALYSIS

Six performance metrics in all were used to confirm the suggested model's functionality. These metrics, which are defined as follows, are recall (Rc), accuracy (Ac), precision (Pr), F1 score (F-one score), false positive rate (Fpr), and false negative rate (Fnr) [15]:

- **Ac**

  The Ac is determined by dividing the number of subjects with accurate labels by the total number of subjects .

$$Ac = \frac{Tpr + Tnr}{Tpr + Tnr + Fpr + Fnr} \tag{8.1}$$

- **Pr**
  The proportion of accurately labeled "Violent Videos" to all videos that have been given that classification.

$$Pr = \frac{Tpr}{Tpr + Fnr} \tag{8.2}$$

- **Rc**
  The proportion of "Violent Videos" that are genuinely violent to those that are appropriately labeled as such.

$$Rc = \frac{Tpr}{Tpr + Fnr} \tag{8.3}$$

- **F-one score-**
  F-one score is the harmonic mean of precision (Pr) and recall (Rc).

$$F\text{-one score} = \frac{2^*(\text{Rc}^*\text{Pr})}{(\text{Rc} + \text{Pr})} \tag{8.4}$$

- **Fpr**
  The ratio of videos incorrectly labeled as violent to all "Non-Violent Videos" is known as the Fpr.

$$F\text{pr} = \frac{Fpr}{Tn + Fp} \tag{8.5}$$

- **Fnr**
  Fnr refers to the ratio of videos that are actually violent to those that are mistakenly classified as "non-violent".

$$Fnr = \frac{Fn}{Tp + Fn} \tag{8.6}$$

A comparison is made between the two models' performance in classifying videos for the purpose of detecting violence: CNN and ResNet-50. The results produced a realistic performance on the entire data set used by averaging across five folds of cross validation.

The CNN model's performance metrics across all folds are displayed in Table 8.1. 74% is the average accuracy achieved on a real dataset.

Figure 8.2 represents the training and validation accuracy of the trained model based on distinct epochs for five fold validation using CNN model.

Figure 8.3 represents the training and validation loss graph of the trained model based on distinct epochs for five fold validation using CNN model.

The performance metrics for the ResNet-50 model across all folds are displayed in Table 8.2. Average accuracy after applying ResNet-50 model on real dataset is 83%.

Figure 8.4 shows accuracy graph for ResNet-50 model by representing model train and validation accuracy per epoch considering all categorization levels.

Figure 8.5 shows loss graph for ResNet-50 model by representing model train and validation loss per epoch considering all categorization levels depending on predicted and actual value.

Figure 8.6 represents the AUC curve for five folds considering two parameters: Tpr and Fpr. The mean ROC curve for RESNET-50 is shown having mean value of 0.83 and standard deviation 1.

The confusion matrix is a powerful method for analyzing classification problems [5]. It describes how the data belonging to a single class can be assigned to multiple possible classes. Normalized confusion matrix is used for representing the traffic

TABLE 8.1
Metrics Measuring the CNN Model's Performance over Each Fold

	Ac	F-One Score	Rc	Pr	Fpr	Fnr
Fold1	0.7575	0.779043	0.855	0.715481	0.34	0.145
Fold2	0.7325	0.748826	0.7975	0.705752	0.3325	0.2025
Fold3	0.6575	0.718686	0.875	0.609756	0.56	0.125
Fold4	0.7575	0.765133	0.79	0.741784	0.275	0.21
Fold5	0.77875	0.778473	0.7775	0.779449	0.22	0.2225
Average	0.73675	0.758032	0.819	0.710444	0.3455	0.181
Standard Deviation	0.0422463	0.0225504	0.0386199	0.0564434	0.115698	0.0386199

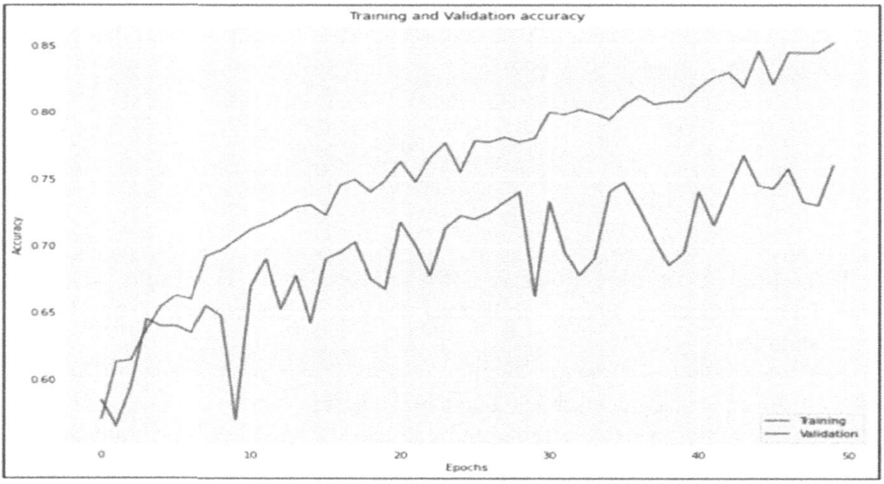

FIGURE 8.2 Training and Validation.

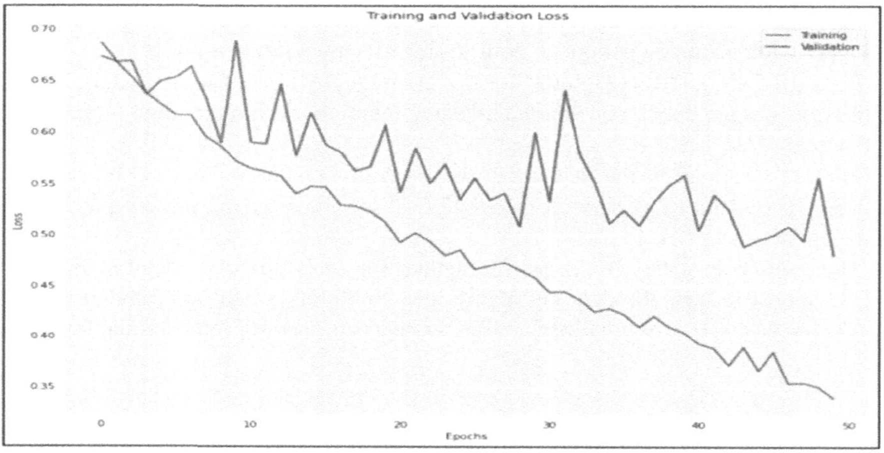

FIGURE 8.3 Training and Validation Loss Graph.

congestion's classes. The normalized confusion matrix that has been used for simulation is shown in Figure 8.7.

By leveraging the powerful capabilities of deep learning and the ResNet-50 model, it is possible to detect and classify violent content in images or video frames with a high degree of accuracy. ResNet-50, with its deep layers and residual connections, enables the model to learn intricate patterns associated with violence. The architecture's ability to capture both levels of features which contributes to its success in violence detection tasks.

TABLE 8.2
Measures of the ResNet-50 Model's Performance over Each Fold

	Ac	F-one Score	Rc	Pr	Fpr	Fnr
Fold1	0.83125	0.827145	0.8075	0.847769	0.145	0.1925
Fold2	0.81625	0.818294	0.8275	0.809291	0.195	0.1725
Fold3	0.83875	0.835249	0.8175	0.853786	0.14	0.1825
Fold4	0.8225	0.824691	0.835	0.814634	0.19	0.165
Fold5	0.85375	0.856089	0.87	0.842615	0.1625	0.13
Average	0.8325	0.832294	0.8315	0.833619	0.1665	0.1685
Standard Deviation	0.0130862	0.0130784	0.0213659	0.0181115	0.022561	0.0213659

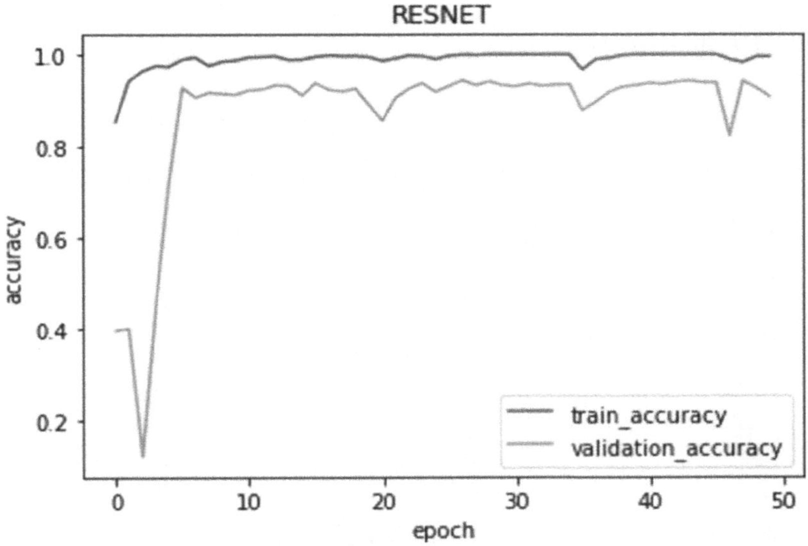

FIGURE 8.4 Accuracy Graph.

8.4 CONCLUSION AND FUTURE SCOPE

Violence detection using the ResNet-50 architecture has proven to be an effective approach. By leveraging the powerful capabilities of deep learning and the ResNet-50 model, it is possible to detect and classify violent content in images or video frames with a high degree of accuracy. ResNet-50, with its deep layers and residual connections, enables the model to learn intricate patterns associated with violence. Continued research and exploration of novel architectures and techniques will contribute to further advancements in violence detection using deep learning. With

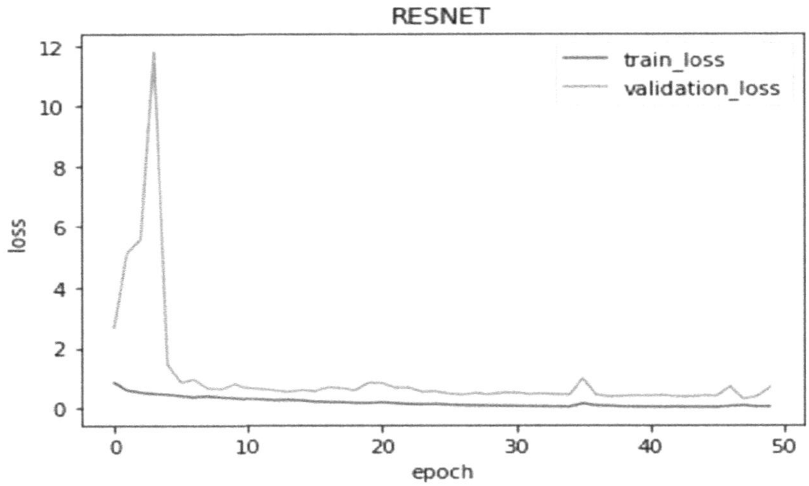

FIGURE 8.5 Loss Graph.

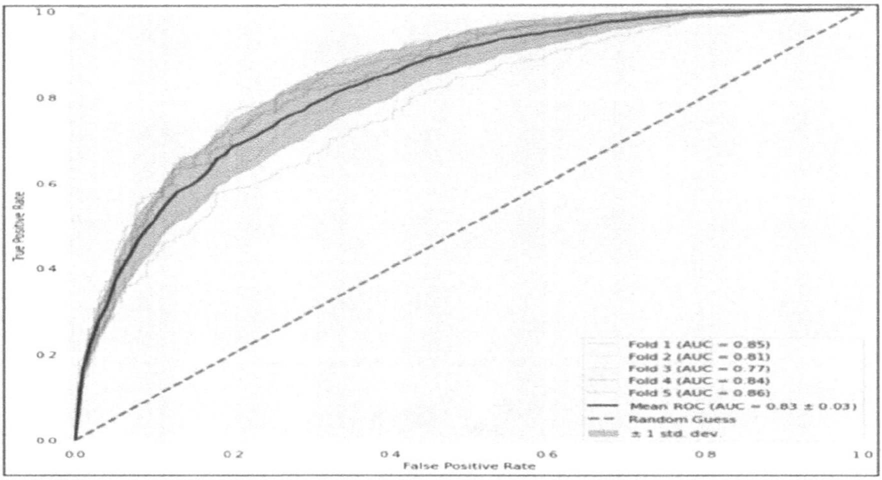

FIGURE 8.6 Average ROC for RESNET-50 Model.

ongoing developments in the field, we can expect even more accurate and efficient models for detecting violence in visual media, which will have significant implications for various domains, including content moderation, security, and public safety.

- The work can yet be improved, and this can be approached from the following angles: (1) by resolving the present issues with the suggested system, and (2) by adjusting the parameters to increase accuracy [16].

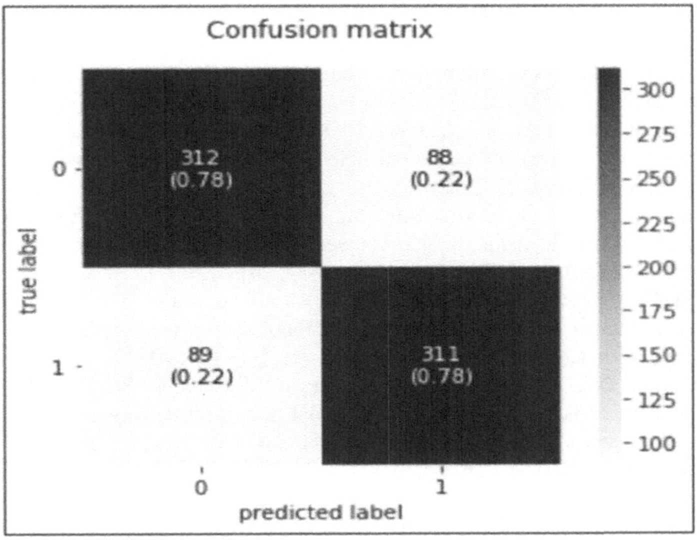

FIGURE 8.7 Confusion Matrix for the Final Fold.

- The authors will use more data to train the model and test different data augmentation strategies in the future to create a more resilient model. The authors will expand research to include a range of action recognition tasks for more flexible use.

REFERENCES

1 Abto Software, "Violence detection for smart surveillance systems," online at: www.abtosoftware.com/blog/violence-detection, reached on May 26, 2021.

2 Cheng, M., K. Cai and M. Li, "RWF-2000: An open large scale video database for violence detection," 2021, doi: 10.1109/icpr48806.2021.9412502.

3 Roman, D. G. C. and G. C. Chavez, "Violence detection and localization in surveillance video," 2020, doi: 10.1109/SIBGRAPI51738.2020.00041.

4 https://vitalflux.com/accuracy-precision-recall-f1-score-python-example/

5 Dinesh Jackson, S. R. et al., "Real-time violence detection for football stadium using big data analysis and deep learning via bidirectional LSTM," *Comput. Networks*, vol. 151, pp. 191–200, 2019.

6 Butt, U. M., S. Letchmunan, F. H. Hassan, S. Zia and A. Baqir, "Detecting video surveillance using VGG19 convolutional neural networks", *Int. J. Adv. Comput. Sci. Appl.*, vol. 11, no. 2, pp. 674–682, 2020.

7 Lin, W., Gao, Q. Wang and X. Li, "Learning to detect anomaly occurrences in crowd scenes from synthetic data," *Neurocomputing*, vol. 436, pp. 248–259, May 2021.

8 Sabokrou, M., M. Fayyaz, M. Fathy, Z. Moayed and R. Klette, "Deepanomaly: Fully convolutional neural network for fast anomaly detection in crowded scenes," *Comput. Vis. Image Understand.*, vol. 172, pp. 88–97, Jul. 2018

9 Hu, J., X. Liao, W. Wang and Z. Qin, Detecting compressed deepfake videos in social networks using frame-temporality two-stream convolutional network. *IEEE Trans. Circ. Syst. Video Technol.*, vol. 32, pp. 1089–1102, 2022.

10 Jalal, A., M. Mahmood and A. S. Hasan, Multi-features descriptors for human activity tracking and recognition in Indoor-outdoor environments. In: 2019 16th International Bhurban Conference on Applied Sciences and Technology (IBCAST), Islamabad, Pakistan, pp. 371–376, 2019.
11 Farooq, M. U., M. N. M. Saad and S. D. Khan, "Motion-shape-based deep learning approach for divergence behaviour detection in high-density crowd," *Vis. Comput.*, vol. 38, pp. 1553–1577, 2022.
12 Ajani, O. S. and H. El-Husseiny, An ANFIS-based Human Activity Recognition using IMU sensor Fusion. In: 2019 Novel Intelligent and Leading Emerging Sciences Conference (NILES), Giza, Egypt, pp. 34–37, 2019.
13 Afza, F., M. A. Khan, M. Sharif, S. Kadry, G. Manogaran, T. Saba, et al. A framework of human action recognition using length control features fusion and weighted entropy-variances based feature selection, vol. 106, 104090, 2021.
14 Hu, X., J. Dai, Y. Huang, H. Yang, L. Zhang, W. Chen, G. Yang and D. Zhang, "A weakly supervised framework for abnormal behaviour detection and localization in crowded scenes," *Neurocomputing*, vol. 383, pp. 270–281, Mar. 2020.
15 Vieira, J. C., A. Sartori, S. F. Stefenon, F. L. Perez, G. S. De Jesus and V. R. Q. Leithardt, "Low-cost CNN for automatic violence recognition on embedded systems," *IEEE Access*, vol. 10, pp. 25190–25202, 2022.
16 Kwan-Loo, K. B., J. C. Ortiz-Bayliss, S. E. Conant-Pablos, H. Terashima-Marin and P. Rad, "Detection of violent behavior using neural networks and pose estimation," *IEEE Access*, vol. 10, pp. 86339–86352, 2022.

9 The Analytics Advantage

Sculpting Tomorrow's Decisions Today

M Vaidya, S. Singh, and B. Jaisinghani

9.1 INTRODUCTION

In the modern era, data serves as a foundational element in our global and business landscapes, guiding us into an era where the significance of information is unparalleled and marked by unprecedented volume of data and rapid technological advancements has emerged as a pivotal force shaping the landscape of business, research, and governance. As organizations grapple with an abundance of information, the integration of data science methodologies has become instrumental in extracting actionable insights, fostering informed decision-making processes, and gaining a competitive edge in diverse industries. The allure for organizations to gather, store, and analyze abundant company data has been amplified by recent strides in information technology and an intensified commitment to digitalization, particularly post covid pandemic. As we navigate through the intricacies of this data-driven world, it becomes evident that the ability to harness and leverage data is not just a competitive advantage but a fundamental necessity for sustainable growth and innovation. According to projections published on Statista [1], the worldwide datasphere, encompassing data generated, represented, duplicated, and consumed globally, is projected to reach an astonishing 181 zettabytes by 2025.This underscores the vast scale and significance of data in our interconnected global landscape. Furthermore, a recent global survey by McKinsey & Company [2] has revealed several noteworthy trends:

- A considerable fraction of businesses, exceeding 20%, credit a noteworthy share of their earnings before interest and taxes (EBIT) to the effects of data science.
- Fifty percent of the surveyed organizations indicate that they have integrated data science into a minimum of one operational aspect.
- The survey brings attention to a noteworthy trend: individuals working in high-performing data science organizations are 2.3 times more likely than their counterparts to perceive their C-suite leaders as exceptionally effective.

In the realm of Industry 4.0 and smart manufacturing, data analytics takes on a transformative role, extending beyond mere numerical analysis. It encompasses the intricate processes of visualization, storytelling, and communication. The true essence of data analytics lies in its ability to decode complex datasets, providing organizations

DOI: 10.1201/9781003480860-9

with clear, meaningful, and actionable insights crucial for effective decision-making. This chapter aims to explore not only the tangible benefits of data analytics but also the cultural shifts required for organizations to fully embrace data-driven decision-making. Moreover, this chapter will shed light on the diverse personas that contribute to the creation of a holistic data analytics platform and different tools that can be used for analysis.

9.2 DATA AND TYPES OF DATA

The exponential growth of data, coupled with the evolution of sophisticated data analytics tools, has bestowed upon decision-makers an unprecedented capacity to harness the power of information. From predicting market trends and optimizing operational processes to enhancing healthcare outcomes and guiding policy formulations, data analytics in addition to data science has become a linchpin in the decision-making fabric of modern societies. We find ourselves in an era defined by "data science and advanced analytics", where virtually every aspect of our daily lives is digitally documented [3]. Consequently, the contemporary electronic landscape is a rich repository of diverse data types, which includes commercial information, financial records, health-related data, multimedia content, insights from the internet of things (IoT), and data pertaining to cybersecurity [4]. The data can be manifested in structured, semi-structured, or unstructured formats, as its volume continues to grow steadily. Data science is essentially a conceptual framework that integrates statistics, data analysis, and associated methodologies, aiming to comprehend and analyze real-world phenomena through the lens of data. To go ahead with this we need to understand what is data? Data refers to raw facts, figures, or information that can be collected, stored, and processed. It is the foundation of any meaningful analysis or decision making process. Data can take various forms, including text, numbers, visuals, sound, and more. Data is often classified into different types determined by its format and attributes. Types of data include:

1 **Structured data**

Structured data is indicative of highly organized information and formatted in a way that is easily searchable and queryable by traditional database systems. This type of data has a clear and well-defined structure, with a fixed schema that defines the types and relationships of the data elements. Key characteristics of structured data:

a Tabular structure

Arrangement of the structured data typically involves organizing data into tables, rows, and columns. In this structure, each column represents a uniquely identified attribute or field, while each row corresponds to a record or specific data instance

b Schema

The schema defines the structure of the data, including the data types of each attribute, constraints, and relationships between tables. The schema provides a blueprint for how data should be organized and stored.

Examples of structured data:

- i Relational databases: structured data is commonly found in relational database management systems(RDBMS), where data is organized into tables with predefined relationships between them.
- ii Spreadsheets: data in spreadsheet applications like Microsoft Excel or Google Sheets is structured, with cells organized into rows and columns.

2 **Unstructured data**

Unstructured data pertains to information lacking a pre-established data model or is not organized in a way that is easily searchable or analyzed with traditional methods. Unlike structured data, which fits neatly into relational databases and is organized in rows and columns, unstructured data lacks a specific structure and is often in a more natural, human-readable form. Key characteristics of unstructured data:

a Lack of formal structure

Unstructured data doesn't follow a rigid and predefined data model. It doesn't conform to the organized structure typical of relational databases. It may not have a consistent format, making it more challenging to process using traditional data analysis tools.

b Human-generated content

Unstructured data often originates from human-generated content, such as text, images, audio, and video created without a specific data schema.

Examples of unstructured data:

- i Textual Data; this includes documents, emails, books, articles, and any other form of written language that doesn't follow a strict structure.
- ii Images; photographs, graphics, and other auditory data are typically unstructured.
- iii Audio; voice recordings, music, and any other auditory data are typically unstructured.
- iv Video; movies, video clips, and other visual content that include motion and sound.

3 **Semi-structured data**

Semi-structured data represents a form of information that does not fit neatly into the traditional relational database model, which is used for structured data, but also doesn't lack structure entirely like unstructured data. Semi-structured data has some level of organization, but it doesn;t conform to a rigid schema or table structure. Key characteristics of semi-structured data -

a Flexibility in schema

Unlike structured data, which adheres to a predefined and fixed schema, semi-structured data allows for more flexibility in terms of data organization. It doesn't require a strict schema, meaning that different records within the same dataset can have different fields.

b No formal schema, but some structure
Semi-structured data does not have a rigid schema, but it does have some inherent structure. This structure might come from tags, keys, or other indicators that provide a partial organization of the data.
Examples of semi-structured data:
i. Web data: HTML documents on the web are often semi-structured. While HTML has a structure, it allows for flexibility in the arrangement of elements, and not all data adheres to a strict format.
ii. Configuration files: files used for configuration settings in software applications are often semi-structured. They may have sections with key-value pairs but can also support nested structures.

9.3 TYPES OF KNOWLEDGE

Knowledge is the consciousness or state of being informed, a familiarity born out of experience or association. It embodies acquaintance with and comprehension of a subject, the state of awareness or grasping of truth. Knowledge involves information that has been grasped with an understanding of its significance, encompassing insights acquired through experience, study, or heightened awareness.

9.4 DATA-INFORMATION-KNOWLEDGE CYCLE

Data represents a potent asset available to organizations on a vast scale. When skillfully leveraged, it possesses the capability to steer decision-making, influence the crafting of strategies, and elevate the overall performance of an organization. In order to harness the potential of data to gain analytical insights and perform informed decision-making, the raw data needs to undergo a series of transformative stages. This journey of transformation is entitled as "data-information-knowledge-cycle" (see Figure 9.1).

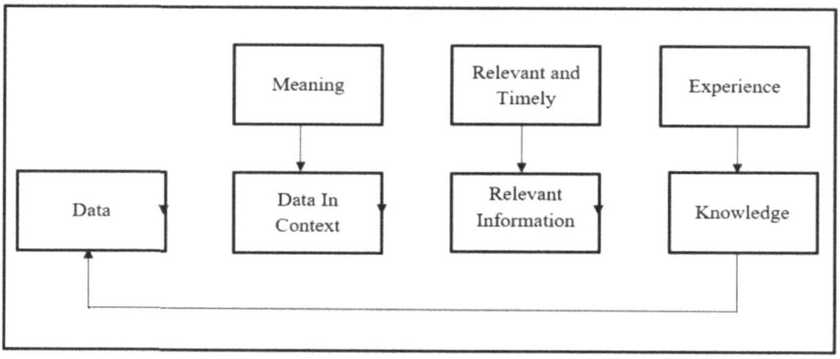

FIGURE 9.1 Data-Information-Knowledge-Cycle [5].

The cycle of data-information-knowledge involves the progression from raw data to contextualized data, then to pertinent information, further evolving into knowledge, and ultimately returning to data when that acquired knowledge is stored. Raw data consists of unstructured and contrasting facts or observations, this data undergoes transformation to gain meaning and relevance. This involves categorizing, sorting, and structuring the data based on predefined criteria. Subsequently, the organized data is interpreted by applying analytical methods, algorithms, or human reasoning to extract patterns, trends, and relationships. Ultimately, information appears as a refined and processed representation of data that contains meaning, context, and actionable insights, facilitating informed decision-making and understanding in various domains. Information adds meaning and can answer questions like "who", "what", "when", and "where". Information transforms into knowledge through a reasonable process that involves the adaptation, interpretation, and integration of data within an individual's existing understanding. When information is gained, it is initially stored in one's memory. Knowledge represents the understanding, awareness and familiarity a person has about various concepts. As the individual actively engages with this information, drawing connections and patterns, comparing it to previous knowledge, and determining its relevance and significance, knowledge begins to take shape. Critical thinking, analysis, and synthesis are critical components of this process, allowing the individual to extract important insights and build a coherent mental model. Furthermore, the ability to apply learning in real-world circumstances, as well as communicate and share that knowledge with others, strengthens and enhances the individual's understanding. As a result, knowledge is a dynamic and changing entity fashioned by continual learning, reflection, and the meaningful integration of information into one's rational thinking. Knowledge can answer questions like "how" and "why".

9.5 FOR CONVERSION OF DATA INTO DATA IN CONTEXT

Data in context means individual pieces of information that make sense and are easy to understand. Data in context refers to the interpretation and understanding of raw data within a specific framework. It involves considering the surrounding circumstances, background information, and relevant factors that give meaning and significance to data.

Analyzing raw data to glean insights, spot trends, and arrive at wise conclusions is known as data analytics. It includes a variety of methods and instruments that convert data into knowledge that can be applicable. Data collection is the first stage of data analytics, it involves gathering appropriate information gathered from diverse origins such as databases, sensors, and accessible platforms. To guarantee correctness and consistency, this raw data may be large, unstructured, and complicated, necessitating preprocessing and cleaning. Following data preparation, the next step is data analysis, which uses statistical and computational methods to find trends, patterns, correlations, and outliers in the data. With descriptive analytics, you can capture a moment in time by summarizing and characterizing the key aspects of the data. Conversely, predictive analytics forecasts future trends or results based on past data and statistical algorithms. Lastly, prescriptive analytics makes recommendations for how to improve results in light of the analysis's findings. Applications for data analytics are used in a wide range of sectors, including technology, healthcare, business, and finance, etc. Businesses employ data analytics to boost operational effectiveness, create a competitive edge, and improve decision-making procedures. Businesses can improve customer behavior, streamline operations, optimize marketing strategies, and innovate in their respective fields by utilizing data. In general, data analytics is essential for turning unstructured data into insightful knowledge that helps businesses stay ahead of the curve in today's data-driven environment and make informed decisions.

Numerous tools are available for data analysis, each tailored to meet specific requirements and preferences. Widely utilized are programming languages such as Python, which includes libraries like Pandas, NumPy, and SciPy. These programming languages offer strong and adaptable platforms for demographic analysis, data visualization, and data manipulation. Their vast ecosystems provide a multitude of community-developed packages and modules, making them adaptable options for data scientists and analysts. Owing to its versatility, data analysis and visualization software, like Microsoft Excel, Tableau, and Power BI, is another widely used tool category. These tools provide comprehensible interfaces that can be used by people with different levels of technical expertise. For example, Microsoft Excel's spreadsheet features and integrated functions for fundamental data analysis make it a popular tool. Users can more easily explore and share insights with the help of Tableau and Power BI's exceptional ability to create visually appealing and interactive dashboards. Because of their user-friendly interfaces and powerful visualization features, these tools are especially beneficial to professionals who may think data analytics is a collaborative process, where individuals with expertise in various different fields can contribute to data analytics activities. Key roles involved in data analytics include:

1 **Data analyst**

Responsibilities: data analysts are professionals who decipher and analyze data to help organizations in making informed decisions. These individuals often employ statistical techniques, data visualization tools, and SQL for data querying.

Skills: expertise in statistics, data visualization, and data manipulation tools. Understanding of programming languages like Python and R is often beneficial.

2 **Data scientist**

Responsibilities: data scientists play a crucial role in complex data analytics, predictive modeling and machine learning. These professionals are responsible for developing algorithms and training models to extract insights and patterns from datasets.

Skills: proficient programming skills, expertise in machine learning, statistical modeling, and domain-specific knowledge.

3 **Data engineers**

Responsibilities: a data engineer is an individual responsible for building, orchestrating and maintaining data pipelines; modeling and architecting data warehouses.

Skills: database management, ETL (extract, transform, load) processes and understanding of big data technologies like Hadoop.

4 **Business analysts**

Responsibilities: business analysts are skillful individuals who focus on interpreting data findings into actionable insights. They point out business areas that can be enhanced to expand effectiveness of the business.

Data analytics comprises a diverse array of methodologies that allow organizations to derive practical and usable understandings from their data, driving informed decision-making. The composite nature of data analytics is depicted in its various forms such as:

1 Descriptive analytics

Every other form of analytics is built upon the most basic form of analytics, known as descriptive analytics. It enables you to identify patterns in unprocessed data and provide a brief explanation of past or present events. "What happened?" is addressed by descriptive analytics. Consider the following scenario: you are examining the data of your business and discover that sales of a particular product – a video game console – are increasing during a particular season. Descriptive analytics reveals that this video game console encounters a surge in sales during October, November, and early December annually. Through charts, graphs, and maps, data visualization effectively illustrates trends, as well as peaks and troughs, providing a clear representation of descriptive analysis.

2 Diagnostic analysis

The next obvious question, "Why did this happen?" is addressed by diagnostic analytics. This type of analysis takes an extra step by examining simultaneous trends, uncovering correlations between variables, and, whenever feasible, establishing cause-and-effect relationships. Expanding on the earlier example, an exploration into the demographics of video game console users reveals a prevailing age range of eight to 18, while purchasers typically fall between 35 and 55. A closer examination of data from customer surveys indicates that a significant factor driving video game console purchases is parents buying it for their children. The increase in sales during the fall and early winter, aligning with gift-giving holidays, can be linked to this specific purchasing behavior.

3 Predictive analysis

Predictive analysis as the name suggests is used to make predictions about the future outcomes and primarily answers the question "What might happen in the future". You may predict potential future developments for your company by examining past data in conjunction with current market trends. For example, you have plenty of information to forecast that, as in the previous ten years, sales of video game consoles have peaked in October, November, and early December. Supported by positive tendencies in the video game sector overall, this forecast makes sense. Predicting upcoming events can aid your company in formulating strategies grounded in probable outcomes.

4 Prescriptive analysis
Prescriptive analytics serves as the conclusive response to the question, "What should we do next?" It comprehensively evaluates all facets of a situation, offering recommendations for practical solutions. This form of analytics proves invaluable in decision-making processes based on data. To illustrate with the video game example: considering the projected seasonal trend driven by winter gift-giving, what steps should your team take? Suppose you opt for an A/B test featuring two ads – one targeting customers (parents) and the other targeting the end-users, children. Analyzing the test data can reveal how to optimize profits during the seasonal spike and its presumed cause. Alternatively, you may decide to intensify marketing efforts in September, incorporating holiday-themed messaging to extend the surge for an additional month. While manual prescriptive analysis is an option, machine learning algorithms are commonly employed to sift through vast data sets and suggest optimal courses of action. Algorithms utilize "if" and "else" statements as rules for data parsing, suggesting a specific action if certain conditions are met. Though this description covers only a fraction of machine learning algorithms, it underscores their significance in algorithm training, alongside mathematical equations.

A standard data platform comprises four primary layers (see Figure 9.2):

1 Data source layer
This layer is responsible for capturing diverse types of data from various business segments, encompassing structured, unstructured, semi-structured, and other data types mentioned earlier.
2 Data warehouse layer:
Predominantly managed by data engineers, this layer involves the extraction of data from the data source layers. The data is then processed, cleaned, and transformed before being stored in the central data warehouse.
3 Data analytics layer:
In the data analytics layer, data analysts have the responsibility of retrieving, interpreting, and presenting data visually. Additionally, they perform advanced analyses encompassing predictive, prescriptive, diagnostic, and descriptive analytics to discern patterns. This layer assumes a vital role in developing sophisticated decision support systems (DSS) and generating business intelligence reports for the business. These DSS contribute to strategic decision-making by utilizing insights derived from both historical and real-time data.
4 Data science layer:
In the data science layer, data scientists are involved in the initial phase called feature engineering, where data is prepared. The processed data is subsequently employed to train and build machine-learning models. The decision support system (DSS) within this layer adopts a more specialized and predictive orientation, aligning with the distinct demands of machine learning applications.

Every element in our environment carries both advantages and disadvantages, and while data analytics empowers professionals to offer valuable insights and inform

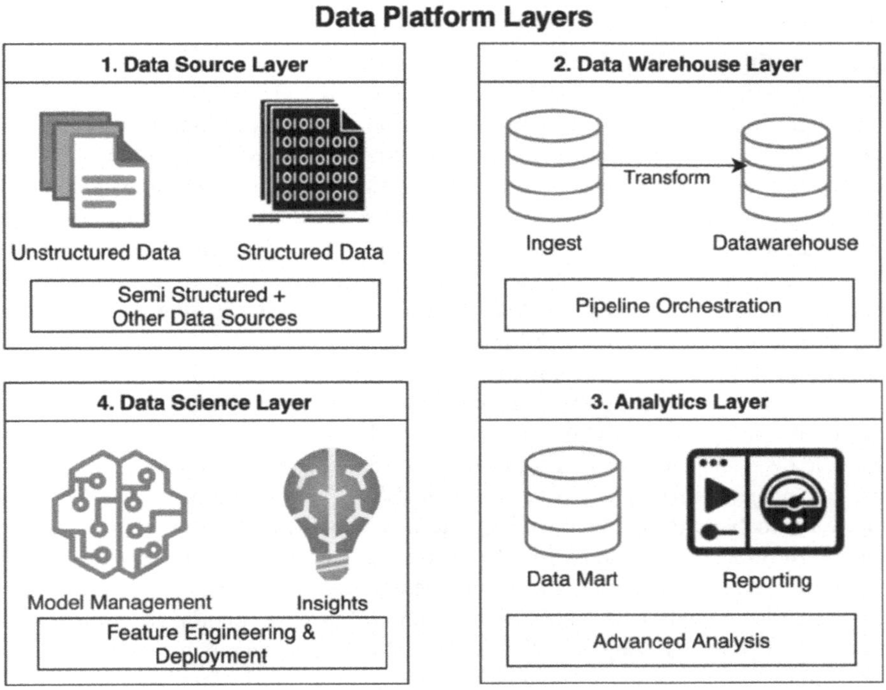

FIGURE 9.2 Data Platform Layers.

decision-making, certain drawbacks have been associated with it. Some of these disadvantages comprise:

1 Privacy concerns
Data analytics often involves collection and analysis of large amounts of personal and sensitive information which gives rise to privacy concerns.
2 Bias and discrimination
Data analytics models may come with its own set of biases in the training data, leading to potential discriminatory outcomes.
3 Cost of implementation
Implementing and maintaining robust data analytics infrastructure can turn out to be expensive. The reason behind increased implementation cost may include acquiring specialized software, hiring professionals, and ensuring top-notch security.
4 Dependency on technology
Organizations that are heavily dependent on data analytics for their decision-making process may become subject to disruptions in technology or power outages. A failure in the analytics infrastructure could result in the termination of the decision-making process.

9.6 Literature Survey

The paper "Impact of big data analytics on decision making and performance" states that big data, defined as data quantities greater than 1018 and requiring data processing tools to manage and process, is a key component of the fourth industrial revolution. Because of the volume of data involved, traditional data analysis techniques cannot be used to analyze big data. Instead, data scientists utilize the right tools to extract meaningful insights from the data. To attain performance excellence, top management must show that it is committed to using data science by providing process measurement systems, collecting and analyzing data, and giving access to a team of data scientists with varying areas of expertise. With the use of data analytics' theories, methods, and procedures, data scientists can process and derive knowledge and insights from data using data science methodologies. Leveraging big data analytics enhances the process of decision-making driven by data, while data science provides the means to derive knowledge and insights from extensive datasets. This combination supports strategic decision-making, fosters performance excellence, and provides a competitive advantage. Utilizing big data analytics (BDA), one can analyze vast datasets, uncover valuable information, identify patterns within the data, and use these patterns to forecast future performance. To fully utilize big data analytics for decision-making and performance enhancement, one must be able to build robust big data analytics skills [6]. The utilization of data science within the business domain, according to this paper, CRISP-DM has long been the most widely used data analytics technique. This methodology breaks down the broad task of finding patterns in data into several well-defined subtasks, such as deployment, modeling, assessment, and business and data understanding. The measure metrics and data stage includes comprehending various data types and formats, gaining access to relevant data, and elucidating metrics to best address previously defined questions. Data science teams can collaborate more effectively to solve big data problems by using the data analytics lifecycle. One of the core components of data science is the framework for process modelling, and data analytics is the method by which a collection of people use data analysis to assist an organization in making better decisions [7]. A research paper titled "Data-driven or Problem-driven approach: Data science for decisional problem solving" states in the field of data science, the competitive intelligence approach to problem-solving represents a paradigm shift for the decision-making process. Find data for a purpose, not a purpose for the data that already exists. Decisional problem-solving frequently begins with the information required, often without defining the stakes, especially the hypotheses related to the information required. This is comparable to applying pre-existing datasets to data science data processing methods. We suggest using a decisional-problem-driven approach where the stakes are always clearly defined. This methodology has been used to assist in decisional-problem solving in enterprises by students in our school taking courses in competitive intelligence study. Research works also continue on the various components of the extended architecture [8]. The paper "The Roles of Big Data in the Decision-Support Process: An Empirical Investigation" emphasizes on the fact that the volume and variety of data that organizations receive through electronic communication has increased dramatically in recent years, ranging from clicks on the internet to unstructured social media posts. It's astounding how much more data these organizations receive. Moreover, companies are becoming more adept at gathering, evaluating, and reacting to data in a variety of ways. Although the term "big data" has gotten overused, there are two main issues that enterprises must face with it. Business executives need to first put new technology into place before being ready for what might be a revolution in the gathering and analysis of data. The second, and most crucial, step is for the organization as a whole to embrace this new way of thinking about decision-making by realizing the potential of big data. We live in a connected world where customer preferences are always changing, so organizations need to understand the role that big data plays in decision-making, with a focus on generating possibilities from these choices. As a result, analysts can track certain profiles or decision-making habits while concurrently monitoring various communication channels [9]. The research paper "Use of Big Data and Knowledge Discovery to Create Data Backbones for Decision Support Systems" aims to offer decision support to assembly line planners in estimating assembly time. This research led to the development of a decision assistance system based on mapping work instructions for controlled language assembly to methods-time measurement (MTM) tables. Processing of vast amounts of historical process sheet data requires the automated analysis approach. Raw process sheet data as an input and mapping rules between work instruction items and MTM tables as an output are essential components of an automated analysis system. Preparing the raw input data so that knowledge discovery and data mining techniques can be applied to it is the first and most important phase in the process [10]. The authors of the research paper "Integration of decision support systems to improve decision support performance" focus on the fact that while enhancing an individual decision maker's performance that is, raising the caliber of their decision-making through increased efficacy and efficiency is the main goal of a stand-alone DSS, IDSS have proven their value in offering multiple users in an organization proactive, coordinated, global support on a range of decisions. Modern techniques and technologies, such as data mining, intelligent agents, KBSs, and Web technology, can improve the integration [11].

The research paper "Learning With Imbalanced Data in Smart Manufacturing: A Comparative Analysis?" exemplifies in the realm of smart manufacturing, often referred to as Industry 4.0, the key emphasis lies in harnessing IoT data and deploying machine learning (ML) to automate the prediction of faults. This not only reduces maintenance time and costs but also elevates overall product quality [12]. The authors of the research paper entitled "The enabling technologies of industry 4.0: examining the seeds of the fourth industrial revolution" state that the seeds of the fourth industrial revolution are found in the complex and heterogeneous cluster of emerging technologies known as Industry 4.0. When considering Industry 4.0 systems from a broad technological perspective, semiconductor and internet technologies which are becoming more and more rich in AI content, make up the majority of the system. It can be argued that, rather than the unmistakable beginning of a fourth industrial revolution, we have up to now been witnessing the continuation, or perhaps amplification, of the third industrial revolution given the information technology origins of these fields. The present study has delineated and scrutinized the six principal constituents of the novel digital economy, which has emerged from the well-established semiconductor-cum-internet paradigm. Big data and artificial intelligence (AI) are the two technological categories that, as of this writing, show the greatest promise of evolving into unique, all-purpose technologies, even though it is impossible to forecast with certainty what the future will bring. It's crucial to keep in mind that many of the components of Industry 4.0 have long histories. For example, the mechatronic industry is the foundation for robotics and human-machine interfaces. The use of sensors in machines and computers-connected machines is not new; 3D printing has been around for more than 30 years. Even artificial intelligence (AI) has been around for a while, but its effects on businesses are only now becoming apparent [13]. The authors of "Digital Twin and Big Data Towards Smart Manufacturing and Industry 4.0: 360 Degree Comparison" emphasize that big data and digital twins are two key components that support smart manufacturing. Cyber-physical integration is made possible by digital twins, which give manufacturers the ability to control the two-way and real-time mappings between physical objects and digital representations. When combined with big data's precise analysis and prediction capabilities, digital twin-driven smart manufacturing will become more predictive and responsive, which will improve factory management in many ways. They work well together to support the advancement of smart manufacturing. However, the practical applications of smart manufacturing, along with its dynamic evolution, are highly difficult. The functions of big data and digital twins in smart manufacturing were briefly examined in this research paper. More study is still required to enhance and enrich related studies by taking into account more variables and real-world scenarios [14].

9.7 Description on Various Tools for Business Analytics

Data analysis is a crucial practice for modern enterprise, but selecting the appropriate tool can be difficult. To determine the best tool, consider the types of data to analyze and data integration requirements. Additionally, select data sources, tables, and columns, and replicate them into a data warehouse to establish a unified and authoritative source for analytics. Popular options on the market include SQL, BI, and XML.

9.7.1 RapidMiner

RapidMiner is a versatile analytics platform that caters to various user needs with both free and open-source options, as well as an economical version. Recognized as a popular predictive analytics tool, it serves as a comprehensive data science platform. RapidMiner empowers users, even those without extensive programming skills, to handle data preparation, machine learning, deep learning, text mining, and predictive analytics. This facilitates the easy creation, evaluation, and deployment of predictive models. With its user-friendly interface and a robust library of data science and machine learning algorithms, RapidMiner Studio provides an all-in-one programming environment. This enables users to seamlessly incorporate predictive analysis into their workflows. Beyond fundamental data mining tasks like cleansing and clustering, the tool supports scripting for more intricate operations. Its versatility extends to various sectors, including business, research, and education. Operating on a client/server model, RapidMiner efficiently performs tasks such as extraction, transformation, and data processing. [15].

RapidMiner is a user-friendly data analysis tool with a visual interface, drag-and-drop functionality, data integration, data preprocessing, machine learning and predictive analytics, deep learning, and text mining capabilities. It supports data from various sources, allows users to perform tasks like cleaning, transforming, and aggregating data, and supports deep learning techniques for advanced methods.

RapidMiner is a popular tool in academia and industry for business analytics and scientific research, offering versatility and a user-friendly interface for leveraging machine learning and data science [16].

9.7.2 Orange Tool

Orange is a data visualization and analysis tool available as open-source software, employing a visual programming interface tailored for data science and machine learning. It provides a user-friendly platform encompassing interactive visualization tools, data preprocessing capabilities, support for machine learning, text mining components, and tools for data exploration and analysis. Notable features include a visual programming interface, interactive data exploration, and model evaluation functionalities. Widely utilized in educational settings, Orange serves as a valuable resource for teaching data science and machine learning concepts. Its adaptability is further enhanced by a diverse ecosystem of add-ons and extensions. With an engaged user community, Orange offers resources, tutorials, and support to its users [17].

9.7.3 Tableau

Tableau stands out as a robust and extensively employed tool for data visualization and business intelligence, empowering users to convert raw data into a comprehensible and visually appealing format. This software facilitates the creation of interactive and dynamic dashboards, reports, and charts, simplifying the analysis and comprehension of intricate datasets for both individuals and organizations. Tableau offers compatibility with a range of data sources, including spreadsheets, databases, cloud services, and big data platforms, enabling seamless connection and visualization of data from diverse origins. A notable strength of Tableau lies in its intuitive and user-friendly interface, eliminating the need for extensive coding or technical expertise. Users can effortlessly drag and drop elements to craft visually striking and insightful visualizations [18].

9.7.4 PowerBi

Power BI, a robust business analytics tool developed by Microsoft, offers users a platform to visualize and analyze data effectively. It serves as a centralized hub, connecting to diverse data sources and transforming raw data into actionable insights, fostering seamless sharing across an organization. The tool's versatility is highlighted by its support for various data connectors, including Excel spreadsheets, cloud-based and on-premises sources, databases, and online services, providing accessibility to a wide array of datasets. What sets Power BI apart is its user-friendly interface and powerful visualization capabilities. Users can effortlessly craft interactive reports and dashboards by intuitively dragging and dropping elements onto the canvas. The tool offers a spectrum of visualization options, ranging from simple charts to sophisticated visuals, enabling users to convey data in a compelling and easily comprehensible manner. Furthermore, Power BI allows the creation of real-time dashboards, accessible and shareable across different devices. Its seamless integration with other Microsoft tools and services positions Power BI as a potent solution for businesses seeking to unlock the full potential of their data, fostering informed decision-making and data-driven insights [19].

9.7.5 Python

Python offers a versatile and expressive environment for managing, cleaning, and analyzing data. Key libraries such as NumPy, Pandas, and Matplotlib provide robust tools for numerical computations, data manipulation, and visualization. The inclusion of Scikit-learn extends the capabilities by offering a comprehensive set of machines learning algorithms, simplifying the process for data scientists to construct predictive models and inform data-driven decisions. Python's popularity in decision-making stems from its accessibility and support for frameworks like Jupyter Notebooks, facilitating interactive and collaborative data analysis. Decision-makers can leverage Python for prototyping, experimenting with diverse models, visually exploring data, and effectively communicating findings. Moreover, Python's seamless integration with web frameworks and business intelligence tools enables the efficient deployment of data science solutions, facilitating organizations in implementing streamlined data-driven decision-making processes. Overall, Python plays a pivotal role in the complete data science pipeline, from initial data exploration to the deployment of models, contributing significantly to informed decision-making across various industries [20].

9.8 Implementation and Result

In today's competitive banking landscape, understanding and mitigating customer churn is paramount for sustained success. Through meticulous churn analysis, financial institutions gain valuable insights into customer behaviors, enabling them to proactively address attrition risks and implement targeted retention strategies.

9.8.1 Problem Statement – Bank Customer Churn Analysis

Every business has customers and not all customers remain with the business forever. Customers come and go, the rate at which customers leave a business against the total number of customers they have is called loss of customers or churn rate which is also sometimes referred to as customer attrition. Churn analysis for bank customers involves identifying and analyzing the elements that lead to customers ending their banking connection. Financial companies need to do churn research since it is usually less expensive to keep current clients than to find new ones. The bank loses potential future business as well as existing business when consumers depart. Key points for churn analysis:

1. Data Collection
   Customer information – the process begins with gathering data on customer demographics, transaction history, account details, etc.
   Behavioral data – collect data on customer behavior such as frequency of transactions, mode of payment, etc.
2. Define churn
   Determine the parameters that will determine whether a consumer is deemed to have churned. This could be determined by closing an account, not using it for a predetermined amount of time, or any other pertinent metric.
3. Data cleaning and processing
   The gathered data should be cleaned and preprocessed to address missing values, outliers, and guarantee consistency. This stage is essential to a precise analysis.
4. Feature selection
   Determine which essential elements could affect client attrition. Account balance, transaction frequency, customer service exchanges, and other attributes might be included in this list.

Steps to perform detailed analysis:

Step 1: Understanding the dataset
The raw data in the dataset is available as a CSV file (see Figure 9.3). The dataset, which consists of 10,000 rows of data, gives a general idea of a bank's customer base. It contains information on the customer's gender, country of residence, credit score which is a prediction of the likelihood that a customer will repay debt, and tenure which indicates how long the customer has been a bank customer. Balance represents cash amount in the bank. Product number represents the number of products that are owned by the customer. The credit card attribute signifies whether the customer

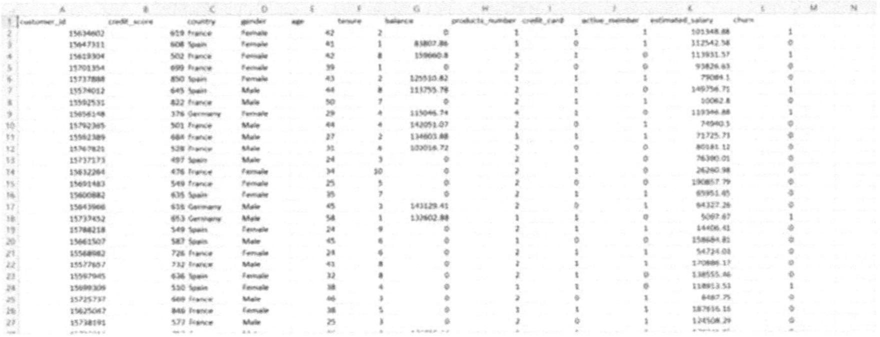

	customer_id	credit_score	country	gender	age	tenure	balance	products_number	credit_card	active_member	estimated_salary	churn
2	15634602	619	France	Female	42	2	0	1	1	1	101348.88	1
3	15647311	608	Spain	Female	41	1	83807.86	1	0	1	112542.58	0
4	15619304	502	France	Female	42	8	159660.8	3	1	0	113931.57	1
5	15701354	699	France	Female	39	1	0	2	0	0	93826.63	0
6	15737888	850	Spain	Female	43	2	125510.82	1	1	1	79084.1	0
7	15574012	645	Spain	Male	44	8	113755.78	2	1	0	149756.71	1
8	15592531	822	France	Male	50	7	0	2	1	1	10062.8	0
9	15656148	376	Germany	Female	29	4	115046.74	4	1	0	119346.88	1
10	15792365	501	France	Male	44	4	142051.07	2	0	1	74940.5	0
11	15592389	684	France	Male	27	2	134603.88	1	1	1	71725.73	0
12	15767821	528	France	Male	31	6	102016.72	2	0	0	80181.12	0
13	15737173	497	Spain	Male	24	3	0	2	1	0	76390.01	0
14	15632264	476	France	Female	34	10	0	2	1	0	26260.98	0
15	15691483	549	France	Female	25	5	0	2	0	0	190857.79	0
16	15600882	635	Spain	Female	35	7	0	2	1	1	65951.65	0
17	15643966	616	Germany	Male	45	3	143129.41	2	0	1	64327.26	0
18	15737452	653	Germany	Male	58	1	132602.88	1	1	0	5097.67	1
19	15788218	549	Spain	Female	24	9	0	2	1	1	14406.41	0
20	15661507	587	Spain	Male	45	6	0	1	0	0	158684.81	0
21	15568982	726	France	Female	24	6	0	2	1	1	54724.03	0
22	15577657	732	France	Male	41	8	0	2	1	1	170886.17	0
23	15597945	636	Spain	Female	32	8	0	2	1	0	138555.46	0
24	15699309	510	Spain	Female	38	4	0	1	1	0	118913.53	1
25	15725737	669	France	Male	46	3	0	2	0	1	8487.75	0
26	15625047	846	France	Female	38	5	0	1	1	1	187616.16	0
27	15738191	577	France	Male	25	3	0	2	0	1	124508.29	0

FIGURE 9.3 Understanding the Dataset.

possesses a credit card or not. The active member attribute indicates if the customer has engaged in transactions and is currently an active member of the bank. The estimated salary attribute provides an estimate of the customer's annual salary and churn indicates whether the customer is still connected to the bank.

Step 2: Importing the dataset into power Bi
Home > Get Data > Text/CSV > Select csv file > Transform Data (see Figure 9.4).

Step 3: Data Cleaning / Data Transformation
In this instance, renaming the columns and altering their data types is the first step in the transformation process (see Figure 9.5). For example, the data types balance

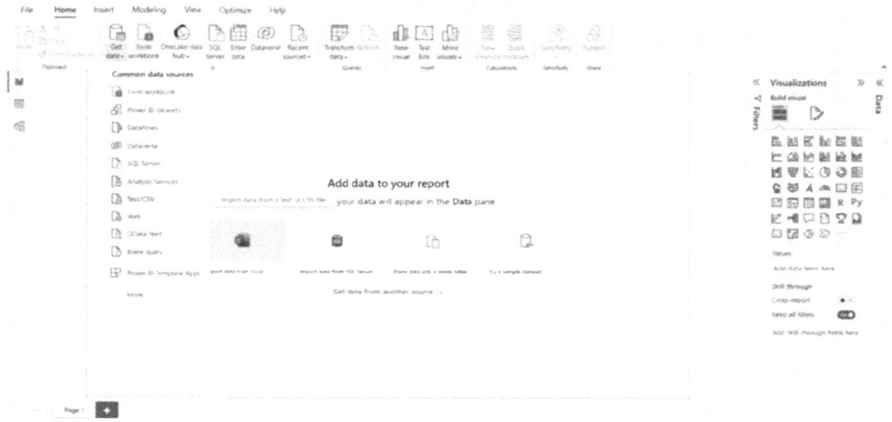

FIGURE 9.4 Importing the Dataset.

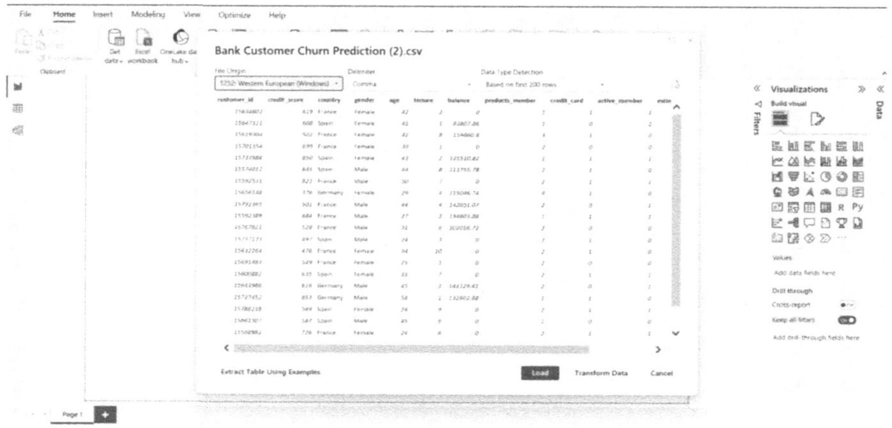

FIGURE 9.5 Loading the Data.

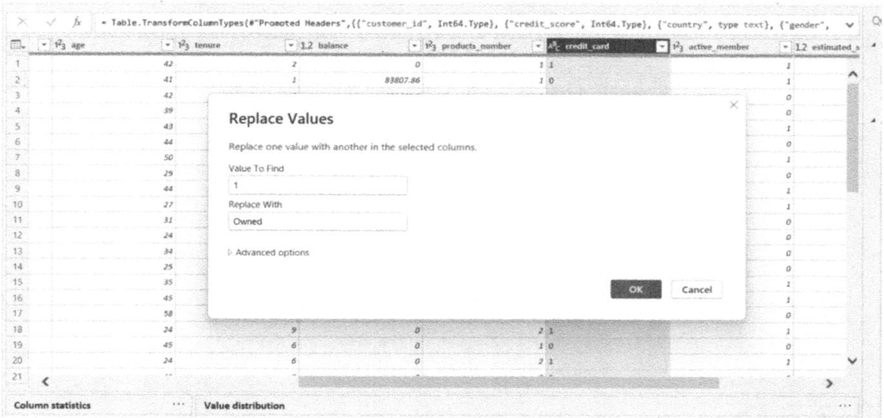

FIGURE 9.6 Replacing the Values to Clean Data.

and estimated_salary were converted into fixed decimal values, while the data types customer id, credit_score, age, tenure, product_number, credit_card, active member, and churn were changed from fixed decimal numbers to whole numbers. In contrast, the gender and country data types were converted to text data types. The next critical stage in data transformation is to eliminate any unnecessary columns, rename the columns to make them easier to comprehend, and change any categorical columns to have distinct values. The credit card status column will first be transformed by changing its data type to text. Next, we will replace the values 1 and 0 with owned and not owned, respectively (see Figure 9.6). To change the values, simply right-click on the column > Replace Values.

The active_member column needs to be renamed as Activity status, and its values should be changed from 0 and 1 to active and inactive, accordingly. Similarly, for the churn status column, change the value of 0 as not churned and 1 as churned. Next, we need to categorize the age column so as to understand which age group has the highest percentage of churn and what is the reason behind the churn. So for this, we need to create a conditional column called Age Groups . To create a conditional column - Add column > Conditional Column (see Figure 9.7).

Make conditional columns for balance and credit score in the same manner.

Step 4: Visualization

Figures 9.8 and 9.9 show that we can conclude that a total of 2037 customers have churned or are no longer associated with the bank. The highest churn rate is for customers who age between 41–50 years, and credit scores between 601–700 with account balance between 100k–200k.

Step 5: Practicality

Bank churn analysis through data science assists banks in making informed decisions by leveraging historical customer data and predictive modeling techniques. By identifying patterns and factors associated with churn, banks can predict the likelihood of a

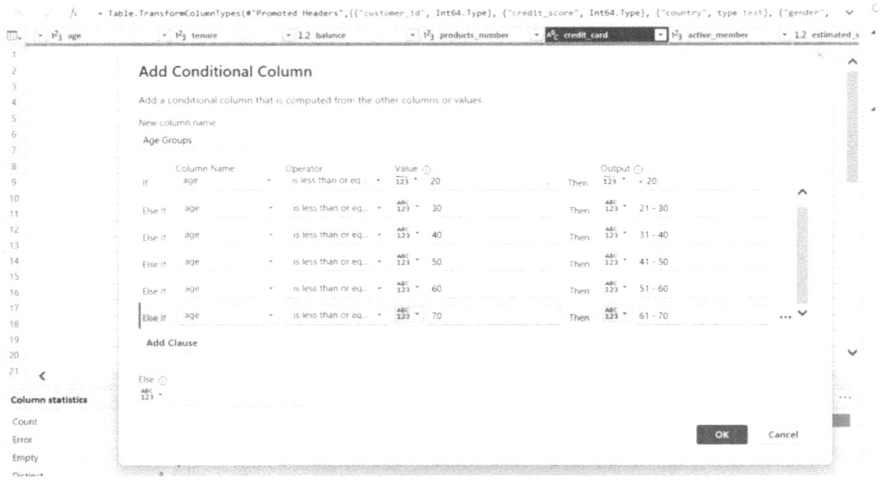

FIGURE 9.7 Adding Conditional Column.

customer leaving, enabling proactive measures. This analysis also facilitates customer segmentation, allowing for targeted retention strategies based on unique characteristics and behaviors. The assessment of risk associated with individual customers helps in resource allocation, directing efforts towards those most at risk. Moreover, understanding the reasons behind customer dissatisfaction from feedback analysis guides product and service improvements, ultimately enhancing customer satisfaction. Through real-time monitoring, banks can stay vigilant, adapting strategies as needed to reduce churn rates and optimize resource allocation, while competitive benchmarking provides insights into industry performance for setting realistic goals. Overall, bank churn analysis empowers decision-makers to implement effective and efficient retention strategies, improving customer loyalty and business profitability.

9.9 CONCLUSION

The research undertaken in this paper shows the pivotal role of data science in enhancing the decision-making processes. The implementation of various tools for data analytics and visualization has proved to be instrumental in extracting meaningful insights from diverse types of data. The research highlights the significance of adopting a data-driven mindset, wherein decision-makers leverage empirical evidence and quantitative analysis to improve their strategies. This research illuminates the transformative potential of data science in the realm of decision- making, particularly through the lens of bank churn analysis.

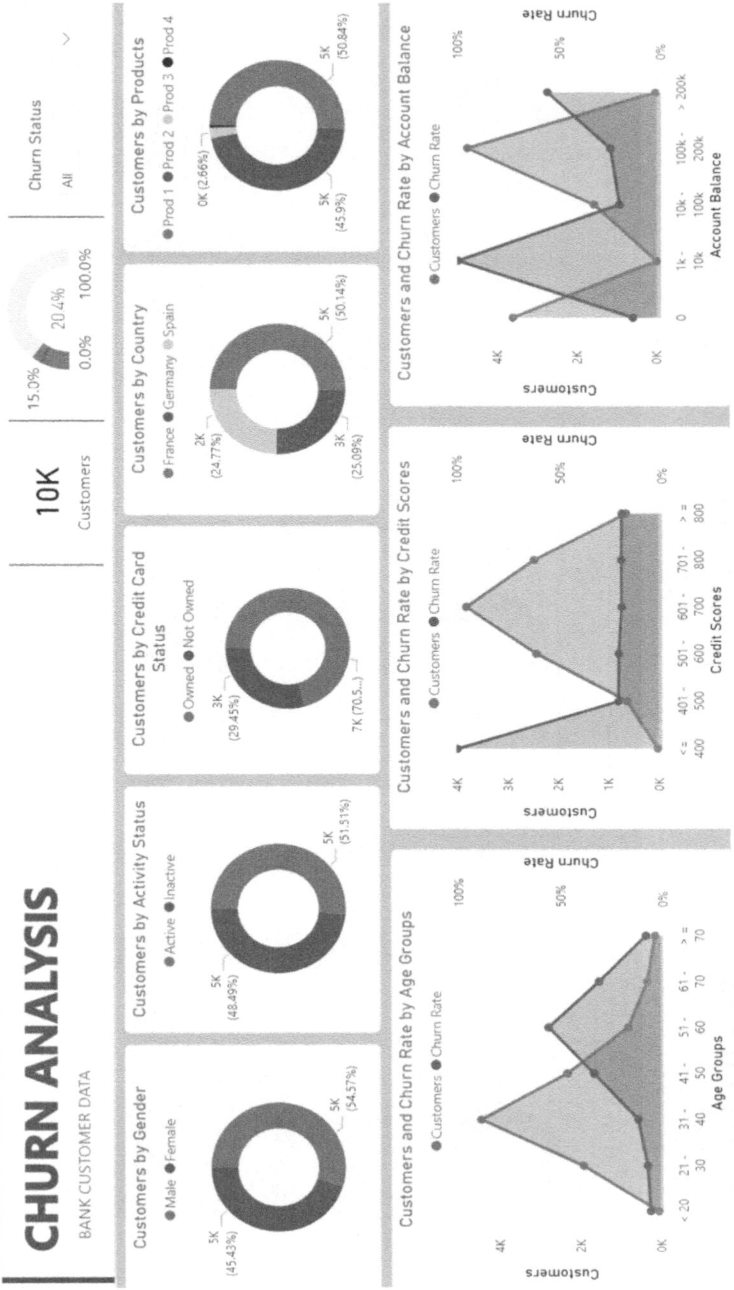

FIGURE 9.8 Visualization in PowerBi for All Customers.

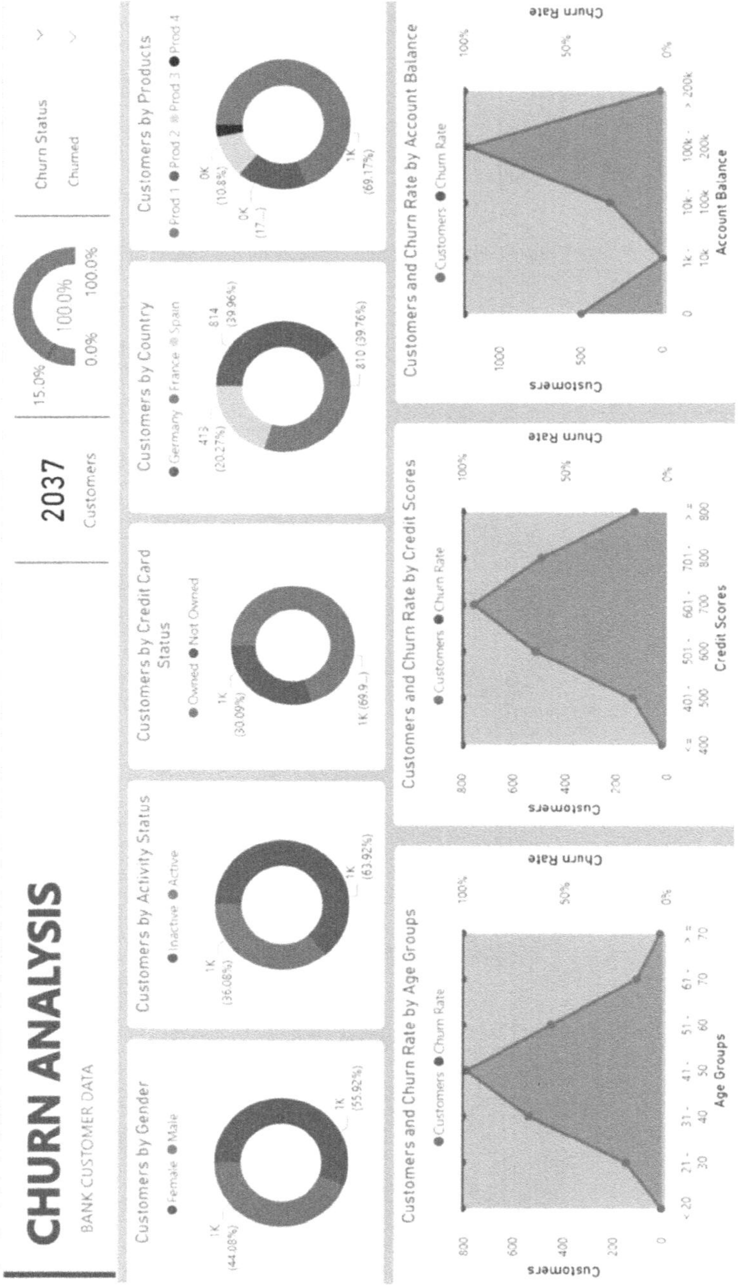

FIGURE 9.9 Visualization in PowerBi for Churned Customers.

REFERENCES

[1] Taylor, Petroc. "Data Growth Worldwide 2010–2025." Statista, 16 Nov. 2023, www.statista.com/statistics/871513/worldwide-data-created/.

[2] Balakrishnan, Tara, et al. "The State of AI in 2020." *McKinsey & Company*, McKinsey & Company, 18 Nov. 2020, www.mckinsey.com/capabilities/quantumblack/our-insights/global-survey-the-state-of-ai-in-2020.

[3] Cao, Longbing. "Data science: a comprehensive overview." ACM Computing Surveys (CSUR) 50.3 (2017): 1–42.

[4] Sarker, Iqbal & Hoque, Moshiul & Uddin, Kafil & Alsanoosy, Tawfeeq. (2021). Mobile Data Science and Intelligent Apps: Concepts, AI-Based Modeling and Research Directions. Mobile Networks and Applications. 26. 10.1007/s11036-020-01650-z.

[5] Kempe, Shannon. "The Data – Information – Knowledge Cycle." *DATAVERSITY*, 14 Nov. 2013, www.dataversity.net/the-data-information-knowledge-cycle/.

[6] Thirathon, U., Wieder, B., Matolcsy, Z., & Ossimitz, M. L. (2017, November). Impact of big data analytics on decision making and performance. In *International conference on enterprise systems, accounting and logistics*.

[7] Lu, Jing, et al. *Data Science in the Business Environment: Customer Analytics Case Studies in SMEs*. https://doi.org/10.1108/JM2-11-2019-0274/full/html.

[8] David, A. (2023). Data science for decisional problem solving: Data-driven or Problem-driven approach. Amos DAVID.

[9] Poleto, Thiago & Carvalho, Victor & Costa, Ana. (2015). The Roles of Big Data in the Decision-Support Process: An Empirical Investigation. 10.1007/978-3-319-18533-0_2.

[10] Renu, Rahul & Mocko, Gregory & Koneru, Abhiram. (2013). Use of Big Data and Knowledge Discovery to Create Data Backbones for Decision Support Systems. Procedia Computer Science. 20. 446–453. 10.1016/j.procs.2013.09.301.

[11] Liu, Shaofeng & Duffy, Alex & Whitfield, Robert & Boyle, Iain. (2010). Integration of decision support systems to improve decision support performance. Knowl. Inf. Syst.. 22. 261–286. 10.1007/s10115-009-0192-4.

[12] Y. Fathy, M. Jaber and A. Brintrup, "Learning With Imbalanced Data in Smart Manufacturing: A Comparative Analysis," in IEEE Access, vol. 9, pp. 2734–2757, 2021, doi: 10.1109/ACCESS.2020.3047838.

[13] Martinelli, Arianna & Mina, Andrea & Moggi, Massimo. (2021). The enabling technologies of industry 4.0: examining the seeds of the fourth industrial revolution. Industrial and Corporate Change. 30. 10.1093/icc/dtaa060.

[14] Q. Qi and F. Tao, "Digital Twin and Big Data Towards Smart Manufacturing and Industry 4.0: 360 Degree Comparison," in IEEE Access, vol. 6, pp. 3585–3593, 2018, doi: 10.1109/ACCESS.2018.2793265.

[15] https://docs.rapidminer.com/9.9/studio/installation/

[16] www.analyticsvidhya.com/blog/2021/10/intro-to-rapidminer-a-no-code-development-platform-for-data-mining-with-case-study/

[17] https://orangedatamining.com/

[18] www.tableau.com/why-tableau/what-is-tableau

[19] www.techtarget.com/searchcontentmanagement/definition/Microsoft-Power-BI

[20] www.simplilearn.com/why-python-is-essential-for-data-analysis-article

10 Using Ensemble Model to Reduce Downtime in Manufacturing Industry

An Advanced Diagnostic Framework for Early Failure Detection

S. Thenmozhi, Kumudavalli M.V., P. Karthikeyan, Kumar Chandar S, and Selva Sharmila

10.1 INTRODUCTION

In modern business landscapes, companies are facing global pressure to adopt various strategies and meet different requirements. These demands encompass production cost reduction, satisfying customer quality expectations, fostering innovations, and ensuring operational safety [1]. The reliable operation of production machinery along with effective maintenance practices is crucial in addressing these constraints. Maintenance is of prime importance for modern companies' profitability and competitiveness since it ensures that the standards regarding the system life, availability, and efficiency are respected. Maintenance activities are integral to the manufacturing lifecycle, constituting approximately 60–70% of total production costs [2]. Therefore, efficient prediction and maintenance tasks can expedite issues resolution, improve machine tool availability [3], and optimize troubleshooting processes. Conversely, inadequate maintenance techniques can lead to a decline in overall plant productivity by 5–20% [4]. This highlights the importance of effective maintenance in ensuring the reliable performance of production machinery and equipment.

Recent years have witnessed a growing emphasis on the maintenance of systems to enhance product efficiency and continuity. Maintenance approaches can be categorized into four types: reactive, planned, proactive, and predictive [5]. Reactive maintenance solves issues only after a breakdown. Planned maintenance involves scheduled inspections and tasks to extend system life. Predictive maintenance

DOI: 10.1201/9781003480860-10

(PdM) uses sensor data and data analytics to predict failure and optimize maintenance intervals, reduce downtime and enhance reliability. In the realm of continuous health monitoring, predictive analytics is used to record real-time equipment data and assess past data to make estimations about the equipment life cycle [6]. This approach minimizes expenses associated with machine maintenance and optimizes the duration of machine operation, resulting in heightened production.

The advancements of low-cost sensors, real-time monitoring systems, and expert algorithms have significantly improved PdM in recent times. Artificial intelligence (AI) based algorithms and models are also gaining attention for their ability to enhance autonomy, flexibility, efficiency, and safety in robotic systems within complex industrial environment. Researchers are actively developing AI based algorithms to further enhance prediction accuracy and reduce labor costs in maintenance practices [8] [9] [10] [11]. The prime target of this paper is to build an automated method for PdM using AI techniques. The following is a summary of this study's main contributions:

- Extensive efforts are done to analyse predictive maintenance data and develop a suitable model for predictive maintenance in manufacturing industry.
- Three fundamental classifiers multilayer perceptron (MLP), K-nearest neighbors (KNN), and support vector machine (SVM) are used to create an inventive ensemble model.
- The scalability of the developed model as well as standard classifiers is validated using AI4I2020 predictive maintenance database.
- This is the first ensemble model for predictive manufacturing process maintenance, to the authors' knowledge.
- A comparative analysis is done between the proposed model and past approaches to prove its superiority.

The subsequent parts will be delineated in the following: an overview of earlier studies in this field is included in Section 10.2. The dataset's details and the suggested technique are outlined in Section 10.3. The model's experimental findings are detailed in Section 10.4. The empirical results and future scope are summed up in Section 10.5.

10.2 REVIEW OF PAST APPROACHES

PdM applications use AI methods, including classification or regression problem type of analysing the large amount of data gathered from real-time monitoring systems using supervised learning. PdM data has been effectively classified using a variety of AI models, including MLP, random forest (RF), KNN, and SVM. Mourtzis et al. [3] implemented a monitoring system based on machine learning tools. This system gathered data from machine tools and analysed it using a fusion technique to aid in preventive maintenance. Dalochio et al. [5] presented comparative survey of existing approaches for PdM. Authors also discussed the merits and demerits of AI based PdM approaches.

In [6], Ahn et al. introduced a hybrid model for PdM of manufacturing processes that combines Bi Long Short-Term Memory (BiLSTM) and convolutional neural network (CNN). In this approach, CNN was used for feature extraction whereas BiLSTM was utilized for anomaly detection. Performance of the model was validated using pump dataset and showed better accuracy. Matzka [7] applied bagged DT (BDT) to predict machine failures and showed high classification rate. Torcianti and Matzka [8] used random undersampling boosting (RUSBoost) tress for predicting machine failures. They also employed a local surrogate model to improve the explainability of the RUSBoost model. Pastorino and Biswas [9] developed a data blind machine learning (DBML) model for PdM. In Sharma et al. [10] a comparison analysis was conducted to examine machine learning models in the context of diagnosing machine problems. In this approach, data was preprocessed using min-max method. Diverse AI algorithms including RF, decision tree (DT), KNN, SVM, and logistic regression (LR) were employed to predict machine fault based on the normalized data. Experiments results revealed that the SVM model gave better outcome compared to other classifiers.

An interesting conventional machine learning (CML) model for machine activity detection was presented by Harichandran et al. [11]. The CML model was designed by fusing supervised and unsupervised machine learning model. Results showed that the fusion of different machine learning models achieved more accurate results in equipment monitoring and maintenance. Kamel [12] developed a PdM model using artificial neural network (ANN). The model was tested using PdM dataset. Results demonstrated the effectiveness of using AI models for PdM. Vittipittayamongkol and Arreeras [13] explored various AI models in predictive maintenance such as DT, SVM, KNN, and RF. RF model displayed better performance than other AI models.

Equipment health monitoring system based on AI models such as SVM and gradient boosting (GB) was proposed by Mota et al. [14]. Initially, data was preprocessed and then served as inputs for GB and SVM. Experimental results showed GB model displayed better outcome than SVM model. Souca and Lughofer [15] designed a fuzzy neural classifier (FNC) for predictive maintenance in manufacturing process. Ghasemkhani et al. [16] introduced a balanced K-star model for PdM in manufacturing industry. Shahin et al. [17] analysed the potential of machine learning models in predicting machine failures in manufacturing industry. Results showed light gradient boosted machine (LightGBM) showed better accuracy in PdM.

10.3 MATERIAL AND METHODS

10.3.1 Research Dataset

In this investigation, AI4I2020 predictive maintenance dataset is used for experimentation. Information about the dataset is given in Table 10.1. The dataset is publicly accessible at the UCI machine laboratory [7]. The dataset is complete without any missing values. The dataset contained 10,000 samples and 14 variables. The variables of the data are listed in Table 10.2. Three columns such as UID, product ID, and type are removed. The input attributes consist of five variables namely AT, PT, RS, T, and TW. Machine failure is considered the objective. Class label 1 indicates machine failure and 0 signifies the non-failure. Statistical properties of the input features are reported in Table 10.3. Further to this, machine failure is reported as true, if any one of the failure modes (TWF, PWF, OSF, HDF, or RNF) are detected.

TABLE 10.1
Details of AI4I2020 Dataset

Tasks	Data type	No. of features	No. of data	Missing values	Field	Web hits
Classification, regression	Numeric, Boolean	6	10000	N/A	Manufacturing	36510

TABLE 10.2
Details of the Variables in the Dataset

Variable	Description
UID	Identifier
Product ID	Serial numbers
Type	Quality of products
AT	Air Temperature (kelvin)
PT	Process Temperature (kelvin)
RS	Rotational Speed (rpm)
T	Torque (Newton meters)
TW	Tool Wear (minutes)
Machine failure	Failure=1, non-failure=0
TWF	Torque Wear Failure
PWF	Power Failure
HDF	Heat Dissipation Failure
RNF	Random Failure
OSF	Overstrain Failure

TABLE 10.3
Statistical Properties of Input Attributes

Features	Mean	Min. value	Max. value	Standard deviation
AT	300	295.3	304.5	2
PT	310	305.7	313.8	1.4
RS	1538.8	1168	2886	179.2
T	39.98	3.8	76.6	9.96
TW	107.9	0	253	63.6

10.3.2 Proposed Methodology

Equipment breakdowns in the industrial sector provide a substantial risk to overall productivity. Thus, process equipment PdM models must be resilient. This framework should be prioritizing maintaining production schedules and product quality while

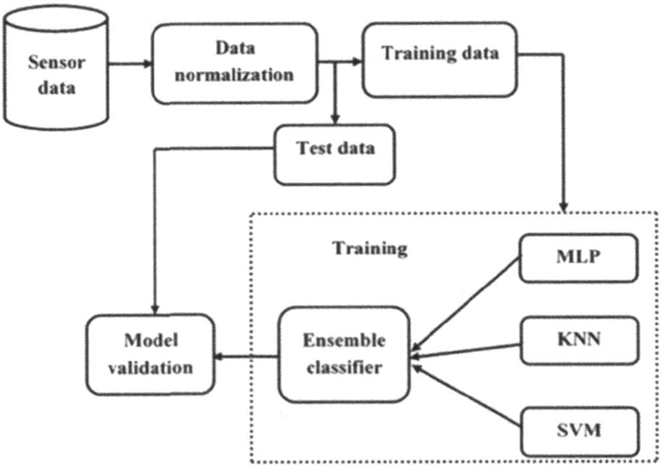

FIGURE 10.1 Schematic Architecture of the Developed Framework for Predictive Maintenance.

minimizing energy wastage and reducing accidents. To achieve this goal, an innovative framework for predictive maintenance is developed as depicted in Figure 10.1. The proposed model encompasses three phases such as:

- data collection
- data preprocessing
- classification.

10.3.2.1 Data Collection Phase

In industry, predictive maintenance begins with data collecting. Data is acquired from industrial equipment sensors utilizing IoT technologies. The analog-to-digital converter in these sensors converts real-world data to data is transported to storage devices.

10.3.2.2 Data Preprocessing Phase

In the data preprocessing stage, missing values are handled and normalization is used to transform raw data into processed data, and discording irrelevant data to make them suitable for subsequent analysis. The raw data degrade the performance of the predictions and make learning process very complex, so it is important to standardize data. Furthermore, AI models are sensitive to input variables. In this study, min-max technique is used to standardize the values. Normalization technique brings raw data into a common range between 0 and 1 and guarantees better training results. Mathematically, data normalization can be expressed as:

$$y = \frac{x - x_{min}}{x_{max} - x_{min}} \tag{10.1}$$

where, y is normalized value, x is input feature, x_{min} is minimum value, and x_{max} is maximum value.

10.3.2.3 Classification Phase

MLP: MLP is feed forward, supervised neural network. It can be used to solve non-linear problems due to their simplicity, adaptability, and efficiency [18]. The MLP has input, hidden, and output layers. A mathematical synaptic summation of weight and bias connects layers:

$$h_j = \sum_{p=1}^{N} \left(X_p . W_{pj} \right) + b_j \tag{10.2}$$

where, x is input, w_{pj} is weight between input and hidden layer, b is bias, and p is number of input features. Hidden layer outputs are calculated via the activation function. This research uses sigmoid activation. The concealed layer's output is sent to the output layer. The final output is estimated using Equation (10.3):

$$y_p = g\left(\sum_{p=1}^{N} \left(X_p . W_{pj} \right) + b_j \right) \tag{10.3}$$

where, y is output and g is activation function.

KNN: the KNN algorithm is a supervised learning technique and a basic kind of classification [19]. The underlying principle of this notion is resemblance. In this work. Euclidean distance is used as distance measure.

$$D\left(X_r , x \right) = \sqrt{ \sum_{r=1}^{N} \left(X_r - x_r \right)^2 } \tag{10.4}$$

where, X is input vector and x is test data.

SVM: in both classification and regression applications, the SVM classifier is renowned for its superior performance and resilience [20]. This study employs SVM with radial basis function (RBF) because to its exceptional performance and computational efficiency. The RBF kernel can be defined as:

$$\text{Gaussian kernel}\left(\mathrm{x}, \mathrm{y} \right) = \exp\left(-\frac{\left\| \mathrm{x} - \mathrm{y} \right\|^2}{2\sigma^2} \right) \tag{10.5}$$

where, $\|x - y\|$ is Euclidian distance between two points x and y and σ is kernel hyper parameter (Gamma).

The main objective of decision support systems is to consistently provide dependable and precise results for each categorization activity. To achieve this, this paper introduces an ensemble classifier by combining predictions through majority voting. Figure 10.2 depicts the ensemble classifier scheme.

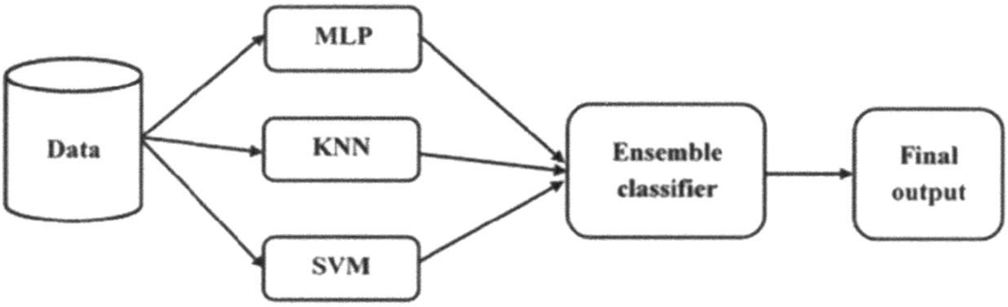

FIGURE 10.2 Developed Ensemble Classifier.

10.4 NUMERICAL RESULTS AND DISCUSSION

The model was tested extensively to prove its efficacy. All tests were executed using MATLAB2021a in a Windows 10 environment, running an Intel core i5 processor RAM.

10.4.1 Performance Evaluation Metrics

The following metrics are utilized to assess the efficacy of the developed ensemble model.

$$\text{Accuracy} = \frac{TP + TN}{TP + TN + FP + FN} \tag{10.6}$$

$$\text{Precision} = \frac{TP}{TP + FP} \tag{10.7}$$

$$\text{Recall} = \frac{TP}{TP + FN} \tag{10.8}$$

$$F1 - \text{score} = 2 * \frac{(\text{Precision} * \text{Recall})}{(\text{Precision} + \text{Recall})} \tag{10.9}$$

where TN (true negative) is the number of negative instances identified as negative, FP (false positive) is the number of negative instances incorrectly predicted as positive, and FN (false negative) is the number of positive instances incorrectly classified as negative. TP (true positive) represents the number of positive instances correctly predicted as positive.

10.4.2 Result Analysis

The effectiveness of the developed ensemble model for predictive maintenance in manufacturing industry is analysed using 10-fold cross validation approach. Ten subsets of the data are used in the K fold cross validation procedure. Training and

TABLE 10.4
Performance of the Developed Model in Predictive Maintenance

Classifiers	Accuracy (%)	Precision (%)	Recall (%)	F1-score (%)
MLP	90.81	91.29	90.2	90.74
KNN	91.53	92.26	90.64	91.42
SVM	94.3	94.92	93.61	94.26
Ensemble	98.72	98.99	98.43	98.69

testing employ nine and one subsets, respectively. The method is performed ten times. The performance of the introduced model is computed by taking average of these ten iterations. Table 10.4 summarizes the outcomes of the developed ensemble model. As reported in Table 10.4, the ensemble model attained the highest accuracy of 98.7%, followed by the SVM, KNN, and MLP. This shows that the ensemble model exhibited better results than other classifiers. This high accuracy can be attained by the fusion of many classifiers in the ensemble and using their collective outcomes. On observing precision, the MLP, KNN, and SVM achieved precision values of 91.29%, 92.26%, and 94.92%, respectively. However, the ensemble classifier showed the highest precision of 98.99% which indicates that it has very few FP predictions. Similarly, the ensemble classifier yielded the highest recall of 98.43%, indicating its ability to capture a high percentage of positive instances. SVM closely follows with a recall of 93.61%, followed by KNN (92.26%) and MLP (90.2%). With its maximum F1-score of 98.69%, the ensemble model demonstrated a balance between recall and accuracy. SVM possessed the second highest F1-score of 94.26%, followed by KNN and MLP. It is observed from the outcomes that the ensemble model showed excellent performance across all metrics. These findings suggest that the ensemble model is a promising tool for predictive maintenance in manufacturing due to its robust performance across all metrics. By fusing the advantages of many classifiers and minimizing their individual limitations, the ensemble classifier provides a reliable approach to predictive maintenance. A pictorial depiction of the Table 10.4 is illustrated in Figure 10.3.

To further show the performance of the developed model, a receiver operating characteristic (ROC) curve is generated, as shown in Figure 10.4. For evaluating the trade-off between true positive rate (TPR) and false positive rate (FPR), the ROC curve is an often utilized tool. As shown in Figure 10.4, it is evident that the ensemble model outperformed the other models by reporting an impressive area under curve (AUC) value of 0.987. This AUC value signifies the model's ability to accurately predict classes. Next better model is SVM by achieving AUC of 0.943, followed by KNN (0.915) and MLP (0.908). These results demonstrate the generated model's better performance when compared to other classifiers.

10.4.3 Comparison with Benchmark Models

The key objective of this paper is to design a more robust and accurate model for predictive maintenance in manufacturing industry using AI techniques. As reported in

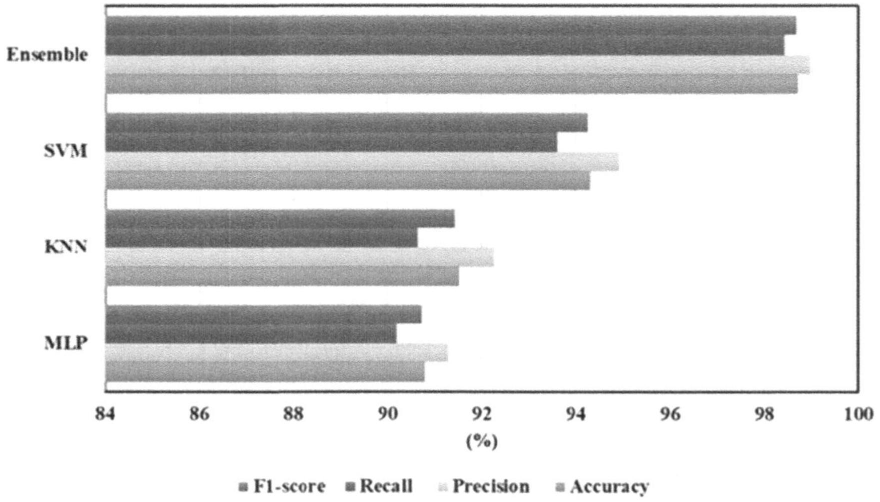

FIGURE 10.3 Efficacy of the Introduced Model for Predictive Maintenance.

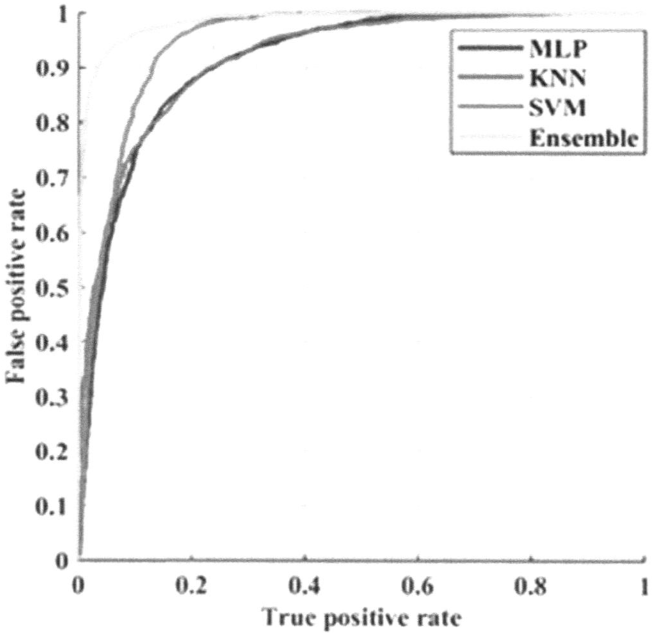

FIGURE 10.4 ROC Curve.

Table 10.4, the introduced system exhibited excellent performance across all metrics followed by SVM, KNN, and MLP. The MLP displayed lower results compared to other classifiers. Nevertheless, the accuracy is remarkably improved when ensemble classifier is used to predict the classes. The experimental outcomes of the ensemble classifier evidence its stability and ability to generalize effectively in light of all metrics in distinguishing between failure and non-failure cases.

Over the past years, numerous researchers have attempted to build a model for predictive maintenance in manufacturing using machine learning models. Each method has its own strength and weakness. To prove the superiority of the developed model, its performance is compared with the other models including BDT [7], RUSBoost [8], DBML [9], SVM [10], CML [11], ANN [12], RF [13], GB, [14] FNC [15], Balanced K-star [16], and LightGBM [17]. Only past approaches using the same data for investigation are considered to ensure a realistic and accurate comparison. Table 10.5 presents a comparative analysis between the introduced model and earlier models. BDT model performed well in recall but lower precision compared to other models, while RUSBoost model maintained good balance between precision and recall but attained lower values compared to the introduced model. DBML model showed higher accuracy than other models and balanced precision-recall trade off. Both SVM and CML exhibited high accuracy but had lower precision. ANN model excelled in precision but had lower recall compared to other models, whereas RF model showed lower performance compared to other models. Like RUSBoost, GB maintained good balance between precision and recall. Balanced k-star and LightGBM models demonstrated better performance than that of other models except the proposed model. The proposed model outperformed all models

TABLE 10.5
Comparative Analyses between Proposed and Earlier Models

Contributors	Year	Classifier	Accuracy (%)	Precision (%)	Recall (%)	F1-score (%)
Matzka [7]	2020	BDT	91.74	86.73	97.1	91.62
Torcianti et al. [8]	2021	RUSBoost	92.74	91.33	90.85	91.09
Pastorino and Biswas [9]	2021	DBML	97.30	93.45	92.56	93.01
Sharma et al. [10]	2022	SVM	97.42	91.23	93.57	92.39
Harichandran et al. [11]	2022	CML	97.99	65.1	58.5	61.62
Kamel [12]	2022	ANN	96.50	97.53	87.23	92.09
Vuttipittayamongkol and Arreeras [13]	2022	RF	82.67	82.67	61.39	70.46
Mota et al. [14]	2022	GB	92.74	91	92	91.50
Souza and Lughofer [15]	2023	FNC	96.80	95.67	94.67	95.17
Ghasemkhani et al. [16]	2023	Balanced K-star	97.55	97.76	97.75	97.75
Shahin at al. [17]	2023	LightGBM	93	94	93	93.50
Proposed		Ensemble	98.72	98.99	98.43	98.69

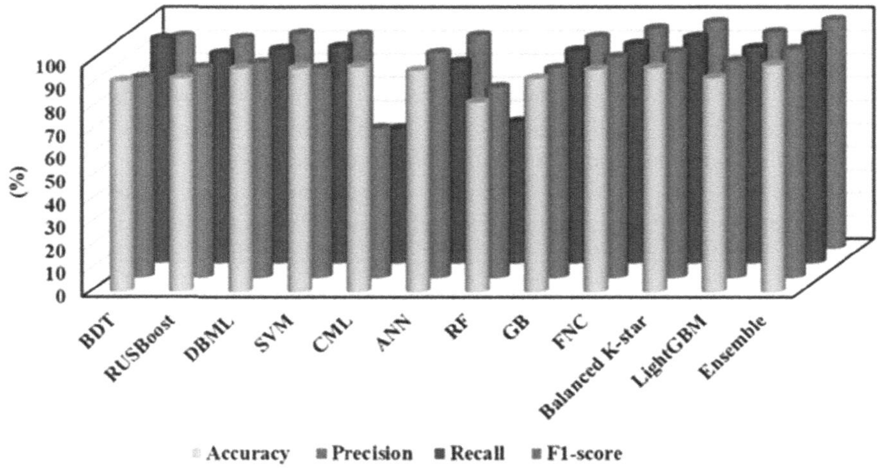

FIGURE 10.5 Comparing the Ensemble Model on the Same Dataset with the Benchmark Models.

in terms of all metrics. The comparative analysis underscores the potential of the introduced ensemble model to achieve superior performance by fusing the strength of many classifiers. Performance comparison is visually illustrated in Figure 10.5 to further highlight the power of the ensemble classifier for predictive maintenance in manufacturing industry.

10.5 CONCLUSION

Predictive maintenance using AI methodologies to forecast potential equipment failures in advance, offer substantial advantages such as lower repair costs, minimized downtime, improved safety, and enhanced overall efficiency. While numerous AI methods have been developed for predictive maintenance, they often yield better results but suffer from lower accuracy. To tackle such an issue, an innovative ensemble model is introduced. The proposed ensemble classifier is developed by combining three base classifiers such as SVM, KNN, and MLP. Effectiveness of the developed ensemble classifier is validated using PdM dataset. This study evaluated the efficacy of KNN, MLP, SVM, and ensemble classifier in early detection of machine failures. The ensemble classifier outperformed other classifiers in all measures.

Future research will focus on exploring additional features to better represent input and capture underlying patters, potentially leading to enhanced performance. In addition to this, explainable artificial intelligence models will be investigated to gain a deeper understanding of the input variables affecting predictions and facilitate more informed and effective decision-making processes in predictive maintenance in manufacturing industry.

REFERENCES

1 P. Vrignat, F. Kratz, and M. Avila, Sustainable manufacturing, maintenance policies, prognostics and health management: A literature review, *Reliab. Eng. Syst. Saf.*, 218, 2022, 108140.

2 D.K. Abideen, A. Yunusa-Kaltungo, P. Manu, and C. Cheung, A systematic review of the extent to which BIM is integrated into operation and maintenance, *Sustainability*, 14, 2022, 8692.

3 D. Mourtzis, E. Vlachou, N. Milas, and N. Xanthopoulos, A cloud-based approach for maintenance of machine tools and equipment based on shop-floor monitoring. *Procedia CIRP*, 41, 2016, 655–660.

4 D. Mourtzis, J. Angelopoulos, and N. Panopoulos, Intelligent predictive maintenance and remote monitoring framework for industrial equipment based on mixed reality, *Front. Mech. Eng.*, 6, 2020, 578379.

5 J. Dalzochio, R. Kunst, E. Pignaton, A. Binotto, S. Sanyal, J. Favilla, and J. Barbosa, Machine learning and reasoning for predictive maintenance in industry 4.0: Current status and challenges. *Comput. Ind.*, 123, 2020, 103298.

6 J. Ahn, Y. Lee, N. Kim, C. Park, and J. Jeong, Federated learning for predictive maintenance and anomaly detection using time series data distribution shifts in manufacturing processes, *Sensors*, 23, 2023, 7331.

7 S. Matzka, Explainable artificial intelligence for predictive maintenance applications. In Proceedings of the Third International Conference on Artificial Intelligence for Industries, Irvine, 2020, 69–74.

8 A. Torcianti, and S. Matzka, Explainable artificial intelligence for predictive maintenance applications using a local surrogate model. In Proceedings of the 4th International Conference on Artificial Intelligence for Industries, 2021, 86–88. DOI: 10.1109/AI4I49448.2020.00023

9 J. Pastorino, and A.K. Biswas, Data-blind ML: Building privacy-aware machine learning models without direct data access. In Proceedings of the IEEE Fourth International Conference on Artificial Intelligence and Knowledge Engineering, 2021, 95–98. DOI: 10.1109/AIKE52691.2021.00020

10 N. Sharma, T. Sidana, S. Singhal, and S. Jindal, Predictive maintenance: comparative study of machine learning algorithms for fault diagnosis, *Social Sci. Res. Network*, 2022, 21–28. DOI: 10.2139/ssrn.4143868

11 A. Harichandran, B. Raphael, and A. Mukherjee, Equipment activity recognition and early fault detection in automated construction through a hybrid machine learning framework. *Computer-Aided Civ. Infrastruct. Eng.* 38, 2022, 253–268.

12 H. Kamel, Artificial intelligence for predictive maintenance. *J. Physics: Conf. Ser.*, 2299, 2022, 012001.

13 P. Vuttipittayamongkol, and T. Arreeras, Data-driven industrial machine failure detection in imbalanced environments. In Proceedings of the IEEE International Conference on Industrial Engineering and Engineering Management, 2022, 1224–1227.

14 B. Mota, P. Faria, and C. Ramos, C. Predictive maintenance for maintenance-effective manufacturing using machine learning approaches. In Proceedings of the 17th International Conference on Soft Computing Models in Industrial and Environmental Applications, Lecture Notes in Networks and Systems. 531, 2022, 13–22. DOI: 10.1007/978-3-031-18050-7_2

15 P.V.C. Souza, and E. Lughofer, E. EFNC-Exp: An evolving fuzzy neural classifier integrating expert rules and uncertainty. *Fuzzy Sets Syst.*, 466, 2023, 108438.

16 B. Ghasemkhani, O. Aktas, and D. Birant, Balanced K-star: An explainable machine learning method for internet-of-things-enabled predictive maintenance in manufacturing. *Machines*, 11, 2023, 322.
17 M. Shahin, F.F. Chen, and A. Hosseinzadeh, A. et al. Using machine learning and deep learning algorithms for downtime minimization in manufacturing systems: an early failure detection diagnostic service, *Int. J. Adv. Manuf. Technol.*, 128, 2023, 3857-3883.
18 H. Kaur and M. Kaur, Fault classification in a transmission line using levenberg-marquardt algorithm based artificial neural network, *Data Commun. Networks*, 2020, 119–135. DOI: 10.1007/978-981-15-0132-6_9
19 Z. M. Çınar, A. Abdussalam Nuhu, Q. Zeeshan, O. Korhan, M. Asmael, and B. Safaei, Machine learning in predictive maintenance towards sustainable smart manufacturing in industry 4.0, *Sustainability*, 12 (19), 2020, 1–42.
20 A. Buabeng, A.Simons, N.K. Frempong, and Y.Y. Ziggah, Predictive maintenance model based on multi sensor data fusion of hybrid fuzzy rough set theory feature selection and stacked ensemble for fault classification, *Mathematical Problemms in Engineering*, 2022, 4372567. DOI: 10.1155/2022/4372567

11 Use Cases of Digital Twin in Smart Manufacturing

Vineet Bhatia, Bhavesh Bhatia, Nilesh Ware, and Sanjeev Kumar Khare

11.1 INTRODUCTION

In the evolving field of manufacturing, digital twins (DT) fulfill a vital function in reshaping traditional norms and unlocking transformative opportunities. Consider envisioning the construction of a representation for an object or system, an interactive counterpart that mirrors its real-world existence in real time. This groundbreaking technology goes beyond imitation; it is a game changer with significant implications across the entire manufacturing spectrum.

With applications ranging from manufacturing and aerospace to healthcare and energy, digital twins are employed across various industries to harmonize physical and digital processes, unlocking new value.

At the forefront digital twins revolutionize the product design and development cycle. Engineers can now construct prototypes that faithfully replicate every detail of a product even before it is physically manufactured. This not only speeds up the design phase, it also enables extensive testing in a risk-free virtual environment. As a result, the overall design process becomes more efficient and streamlined, significantly reducing the need for physical prototypes.

Another compelling application of DT in manufacturing is maintenance. By utilizing real-time data gathered from sensors embedded in assets, manufacturers can accurately anticipate when equipment maintenance will be required. This proactive approach prevents downtime maximizes the lifespan of equipment and ultimately contributes to an operation that's both efficient and cost effective.

Furthermore, digital twins offer manufacturers a view of their production ecosystem making them invaluable, for process optimization purposes. With this technology, manufacturers can gain insights into tuning their production processes for efficiency.

By utilizing models that simulate the manufacturing process companies can address bottlenecks, inefficiencies, and areas that can be improved. Adjusting parameters within this environment facilitates the enhancement of production processes, resulting in increased efficiency and reduced costs.

DTs also bring changes to supply chain management. Integrating these replicas into the supply chain provides manufacturers with real-time insights, into the flow of materials, inventory levels, and logistics. This level of visibility allows for responses to changes in demand and disruptions fostering an adaptable and responsive supply chain.

DOI: 10.1201/9781003480860-11

With digital twins quality control takes a leap forward as it enables real-time comparisons between specifications and actual production data. This meticulous monitoring ensures that products consistently meet the standards of quality reducing defects and enhancing product reliability.

Sustainability is an aspect in manufacturing practices, where digital twins play an essential role in achieving energy efficiency goals. By modeling and analyzing energy consumption patterns manufacturers can pinpoint areas for enhancement, implement energy-saving measures, and contribute to an environmentally friendly production process.

A wind turbine serves as a perfect example of how DT has been utilized. It incorporates a network of sensors placed in vital functional zones [1]. These sensors collect data on the real-world performance of the turbine, encompassing aspects such as energy production, temperature, and environmental conditions.

11.2 BACKGROUND AND DEFINITIONS

11.2.1 Smart Manufacturing

The factory floor is no longer a symphony of clanging machines and grimy overalls. Smart manufacturing, fueled by the digital revolution, is rewriting the industrial playbook. Imagine a world where sensors dance across production lines, feeding real-time data to virtual twins that mirror every bolt and gear. This digital ensemble, conducted by artificial intelligence, tunes processes to perfection, predicting glitches before they occur and optimizing production with laser precision. Gone are the days of guesswork and inefficiencies; smart manufacturing empowers factories to become agile organisms, adapting to changing demands and crafting products with unparalleled quality and speed. The backbone of smart manufacturing is technologies like internet of things (IOT), big data analytics (BDA), digital twin (DT), etc.

11.2.1.1 Smart vs Traditional Manufacturing

Traditional manufacturing and smart manufacturing represent two different approaches to the production of goods, with smart manufacturing incorporating advanced technologies to improve efficiency, productivity, and responsiveness. Smart manufacturing represents a shift towards more technologically advanced and interconnected systems, enabling greater efficiency, flexibility, and responsiveness compared to traditional manufacturing methods. Traditional manufacturing, while still a vital contributor to the global economy, often relies on manual processes, limited data capture, and fixed production lines. This can lead to slower response times, higher waste, and susceptibility to unexpected disruptions. Table 11.1 offers a detailed comparison of these two approaches, highlighting the specific technologies and features that differentiate them.

11. 2.1.2 Enabling Technologies for Smart Manufacturing

The fundamental components of smart manufacturing include data analytics, robotics, 3D printing, and additive manufacturing, which work seamlessly together to increase

TABLE 11.1
Traditional vs Smart Manufacturing

Features	Traditional Manufacturing Technologies	Smart Manufacturing Technologies	Traditional Manufacturing	Smart Manufacturing
Data & Connectivity	• Limited sensor usage, mainly for basic monitoring. • Paper-based record keeping and manual data transfers.	• Industrial IoT (IIoT) sensors for real-time data collection (temperature, pressure, vibration). • Machine vision for automated defect detection and process monitoring. • RFID tags for tracking materials and products. • Cloud platforms for data storage, analysis, and sharing.	Data collection and analysis are restricted, often relying on manual procedures or disparate systems.	Real-time data capture and analysis through sensors IoT devices, and interconnected systems, enabling continuous data flow and insights.
Processes & Operations	• Fixed production lines with limited flexibility. • Manual operation of machines and processes. • Scheduled maintenance based on time or usage intervals.	• Programmable logic controllers (PLCs) for basic automation. • Industrial robots for repetitive tasks. - Digital twins for virtual simulations and process optimization. • Automated guided vehicles (AGVs) for material handling. • Predictive maintenance using machine learning algorithms to anticipate equipment failures.	Manual, static processes with limited flexibility, often relying on fixed production lines and manual adjustments.	Automated, dynamic processes with continuous optimization, leveraging data-driven insights and flexible automation to adapt to changing conditions and requirements.

(continued)

TABLE 11.1 (Continued)
Traditional vs Smart Manufacturing

Features	Traditional Manufacturing Technologies	Smart Manufacturing Technologies	Traditional Manufacturing	Smart Manufacturing
Decision-Making	• Reliant on human expertise and intuition based on experience. - Limited data availability for informed decision-making.	• Artificial intelligence (AI) for pattern recognition and anomaly detection. • Algorithms based on machine learning for predictive maintenance and optimization of processes. • BDA for identifying trends and insights from large datasets. • Real-time dashboards and visualizations for data-driven decision-making.	Reliant on human expertise and intuition, based on experience and limited data availability.	Data-driven decision-making powered by AI and machine learning, enabling proactive optimization, predictive maintenance, and informed adjustments based on real-time insights.
Maintenance & Quality Control	• Reactive maintenance based on equipment failure or downtime. - Manual quality checks with potential for human error.	• Advanced sensors for continuous monitoring of equipment health. • Predictive maintenance software to anticipate problems before they occur. • Automated inspection systems for fast and accurate quality control. - Statistical process control (SPC) for monitoring and correcting production variations.	Reactive maintenance based on equipment failure or scheduled downtime, with manual quality checks that are prone to consuming significant time and prone to errors.	Predictive maintenance, which reduces costs and downtime by anticipating any problems before they arise, in conjunction with automated quality control systems to guarantee constant product quality.

TABLE 11.1 (Continued)
Traditional vs Smart Manufacturing

Features	Traditional Manufacturing Technologies	Smart Manufacturing Technologies	Traditional Manufacturing	Smart Manufacturing
Flexibility & Customization	• Limited ability to adjust production lines quickly. • Time-consuming and costly changeovers for new products.	• Modular production systems with reconfigurable components. • Virtual prototyping for design iteration and optimization before physical production. • Additive manufacturing (3D printing) for on-demand creation of customized parts. • Configurable product platforms with interchangeable components.	Limited ability to quickly adjust production lines or accommodate new product designs, often requiring significant setup time and manual adjustments.	Agile production lines that can readily adapt to variations in demand, product customization, and new product introductions, enabling a more responsive and customer-centric approach.

productivity and empower the human workforce. The essential technologies facilitating this process are briefly outlined in the following paragraph and illustrated in Figure 11.1:

a **Internet of things (IoT):** the internet of things (IoT) facilitates the connection of physical objects and sensors to the internet, enabling communication and data gathering capabilities. In smart manufacturing, IoT devices can be embedded in machines, tools, and products for tracking and managing several areas of the production process.
b **Big data and analytics:** beyond mere tracking and management, smart manufacturing empowers producers to harness the power of real-time data from an interconnected web of sensors, machines, and devices. This continuous flow of information, fueled by cutting-edge analytics, becomes the lifeblood of optimized production processes. Predictive maintenance capabilities minimize downtime, ensuring consistent output.
c **Artificial intelligence (AI) and machine learning (ML):** AI and ML algorithms are capable of pattern recognition, result prediction, and process

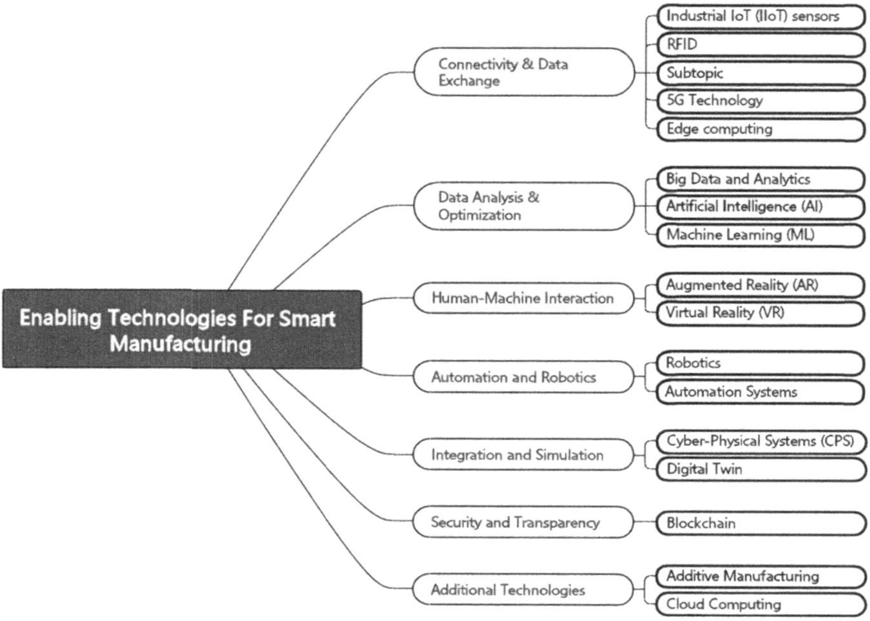

FIGURE 11.1 Key Enabling Technologies for Smart Manufacturing.

optimization through data analysis. In smart manufacturing, they are utilized for process optimization, demand forecasting, quality control, and predictive maintenance.

d **Augmented reality (AR) and virtual reality (VR):** augmented reality and virtual reality technologies enhance human-machine interaction by providing real-time data, guidance, and visualizations. They are used for training, maintenance, and troubleshooting tasks in smart manufacturing.

e **Robotics and automation:** the manufacturing landscape is evolving, with robots no longer confined to cages. Cobots, robots designed to safely interact with humans, are now common teammates on assembly lines, assisting with tasks and taking on the most physically demanding or hazardous jobs.

f **Cyber-physical systems (CPS):** cyber-physical systems (CPS) integrate computational and physical processes, allowing for real-time monitoring, control, and optimization of physical processes through digital technologies. This integration forms the backbone of smart manufacturing.

g **Digital twin:** a digital twin is "a virtual replica of a physical system or process". Digital twins are used in smart manufacturing to model and examine production processes, facilitating improved comprehension, optimization, and preventative maintenance.

h **Blockchain:** supply chains may become more transparent and secure with the use of blockchain technology. It ensures traceability and authenticity of products, prevents counterfeiting, and facilitates secure transactions between different entities in the manufacturing ecosystem.

i **5G technology:** 5G networks' high speed and low latency allow devices and systems in smart manufacturing to communicate with each other seamlessly. This is essential for remote monitoring, real-time data interchange, and industrial process management.

j **Edge computing:** rather than relying solely on centralized cloud servers, edge computing involves processing data in proximity to its source, or at the edge of the network. For time-sensitive applications in smart manufacturing, this lowers latency and improves real-time processing capabilities.

k **Additive manufacturing:** 3D printing and additive manufacturing are also essential components of smart manufacturing, as they enable manufacturers to produce complex and customized parts quickly and efficiently. As manufacturers can now build parts with more accuracy and precision, this technology can help reduce waste and enhance the quality of the end product.

11.2.2 Digital Twin

A digital twin (DT) is a "Virtual representation of a physical object, system, or process that spans its entire lifecycle". It involves creating a "digital counterpart that mirrors the real-world entity in a virtual environment". Advancements in technology, particularly in fields such as data analytics (DA), artificial intelligence (AI), simulation, and the internet of things (IoT), have made this concept increasingly prevalent across diverse industries.

11.2.2.1 The Development of Digital Twin Technology

The concept of DT technology has evolved over several decades. The evolution of DT technology has seen it transition from a concept into increasingly sophisticated and transformative systems [2]. Various stages of the evolution of the digital twin concept to maturing of technology is covered in subsequent sub-sections [3]:

11.2.2.1.1 Early Beginnings (1960s)

- The concept of creating a digital replica of a physical entity finds its origins in NASA's Apollo missions during the 1960s. Sparked by the Apollo 13's near-disaster in 1970, NASA pioneered a revolutionary concept: a "living model" of the spacecraft. This wasn't just a blueprint or a static replica, but a dynamic system that mirrored the real Apollo 13 in real-time. [4]
- When disaster struck, this "living model" became their lifeline. By feeding in data from the crippled spacecraft and tweaking the digital version, NASA could test solutions in a virtual world before putting the crew at further risk. It was like having a crystal ball reflecting the future, showing them how to navigate the unforgiving vacuum of space.
- This idea, though not explicitly termed "digital twin", laid the foundation for future developments.

11.2.2.1.2 Formalization and Expansion (2000s)

- In 2002, John Vickers of NASA officially formulated the phrase "digital twin" while working on product lifecycle management in manufacturing [5].
- Michael Grieves further refined the concept, highlighting its potential for optimizing entire production processes [6], [7].
- This era saw the development of digital twins primarily focused on industrial applications, particularly simulations for performance prediction and anomaly detection.

11.2.2.1.3 The Rise of Connectivity and Intelligence (2010s onwards) [8]

- The evolution of sensor technology, artificial intelligence (AI), and the internet of things (IoT) has revolutionized digital twins.
- Data captured in real-time by sensors connected to physical assets allowed for continuous updates and dynamic virtual representations.
- AI and machine learning empowered digital twins to perform predictive analytics, optimize operations, and even suggest maintenance interventions before failures occur.

11.2.2.2 Level of Technological Evolution

The "Institute for Information and Communication Technology Planning and Evaluation" in Korea has developed a framework to define the level of technological evolution of DTs. This framework consists of five stages or levels, each representing a different level of advancement and capability in terms of DT technology [8]. The five stages are as follows:

1. Phase I **Mirroring** (replication): duplicating a physical object into a digital twin.
2. Phase II **Monitoring** (oversight): using the digital twin analysis to monitor and manage the physical thing.
3. Phase III **Modeling and simulation**: improving the physical object based on insights gained from simulations conducted on its digital twin.
4. Phase IV **Federation** (collaboration): setting up digital twins that are federated, optimizing complicated physical things, and promoting federated digital twin interoperability with complex physical objects.
5. Phase V **Autonomous operation**: independently identifying and solving challenges inside "federated digital twins" and enhancing physical things according to solutions generated from "federated digital twins".

During the initial three phases, a singular physical object is replicated as a digital twin, and optimization takes place using simulation results exclusively from this individual digital twin. To reduce the workload during the mirroring step, it is essential to carefully pick the key replicated elements. Depending on the key objective of the DT, these key replicated elements are chosen; examples include traffic volumes and roadmaps in digital twins for transportation, or varieties of vegetables, sunshine, and water in digital twins for smart farms.

Moving to the third phase, a digital twin model is constructed by combining objects replicated during the initial stage. This model conducts various simulations aimed at optimizing real-world scenarios or addressing real-world challenges. The procedures of digital twin construction, simulation, and optimization are covered in the third phase. The completion of stage one and two is essential to establishing a comprehensive three-step digital twin model.

The fourth phase involves the collaboration of each individually optimized digital twin to optimize the intricate real world. Various types of digital twins must be carefully considered and systematically interconnected to construct a large-scale digital twin, such as a city digital twin. Complex real-world scenarios, particularly, are ideally suited for replication and optimization using the federated digital twin structure. In this framework, individually optimized digital twins contribute to enhancing the overall system.

In the fifth phase, provided that DT models exhibit stability, reliability, and dependability for seamless integration with the real world, DT technology advances to a level of autonomous operation and optimization. In this stage, federated digital twins operate independently to identify real-world challenges and offer solutions, essentially functioning as an independent digital twin ecosystem.

11.2.2.3 The Five-Dimensional Digital Twin Model

The formulation of the five-dimensional digital twin model is expressed through the following formula (1) [9].

$$M_{DT} = (PE, VM, Ss, DD, CN) \tag{11.1}$$

Here, M_{DT} = digital twin model, PE = physical entities, VM = virtual models, Ss = services, DD = digital twin data, and CN = connections. The visual representation of the five-dimensional model of DT is depicted in Figure 11.2, adhering to formula (1).

- **Representation of physical entities in the DT**: essence of the digital twin (DT) lies in creating digital models that simulate the behaviors of physical entities [10]. The foundation of DT is rooted in the physical world, encompassing devices, products, physical systems, activity processes, and even entire organizations. According to their structure and purpose, they can be categorized into three levels: system, unit, and system of systems (SoS) [11].
- **Modeling virtual entities in the DT**: the DT virtual models have to be precise duplicates of their real-world counterparts, encompassing the real-world geometry, characteristics, actions, and regulations [12].
    - Three-dimensional geometric models represent the real object in terms of size, shape, tolerance, and structural relationships.
    - Physics models, which rely on characteristics such as velocity, abrasion, and force, mimic the physical occurrences of entities, encompassing deformation, delamination, fracture, and corrosion.

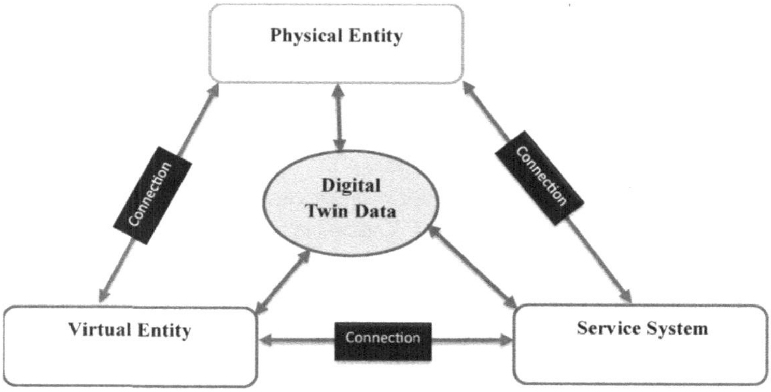

FIGURE 11.2 Dimensions of Digital Twin.

- Behavior models include state transitions, performance degradation, and coordination. They also describe how entities behave and adapt to alterations in their external surroundings.
- Rule models equip digital twins with logical capabilities, enabling judgment, reasoning, assessment, and independent decision-making by adhering to rules derived from historical data or provided by subject matter experts.

- **Data dynamics in the DT**: data is the primary force behind DT, tackling problems with heterogeneous, multi-temporal, multi-dimensional, and multi-source data [13]. Data, which includes both dynamic condition data and static attribute data, is sourced from physical entities. Virtual models generate data reflecting simulation results, while services contribute data describing service invocation and execution. The combination of all these data sources yields fusion data.
- **Role of services in the DT**: in the context of product-service integration, the importance of service is emphasized in digital twins (DT). DT offers consumers application services including simulation, verification, monitoring, optimization, diagnosis, and prognosis and health management (PHM) as part of the "everything-as-a-service" (XaaS) paradigm [11].

  Furthermore, third-party services become essential to building a working DT, including data, knowledge, and algorithm services. DT requires ongoing platform services maintenance in order to provide alterations in software development, model formulation, and service provision.
- **Dynamic connections in the DT**: for sophisticated simulation, operation, and analysis in DT, twin dynamic linkages between digital representations and their real counterparts are essential. Six distinct connections are identified for DT: the connection between "virtual models and physical entities (CN-PV)", the connection between "physical entities and data (CN-PD)", the connection

between "physical entities and services (CN-PS)", the connection between "virtual models and data (CN-VD)", the connection between "virtual models and services (CN-VS)", and the connection between "services and data (CN-SD)".

11.2.2.4 Definitions

A digital twin (DT) is a "virtual representation of a real-world entity or process, composed of three elements: a physical entity in real space, the digital twin in software form, and data that links the first two elements together".

Different authors and organizations provide varied definitions of the digital twin, and these diverse perspectives are presented in Table 11.2. While various definitions of digital twin (DT) may emphasize particular facets or elements of DT systems, a comprehensive definition of DT encompasses a digital illustration of physical assets, processes, individuals, locations, and systems. This representation encapsulates both the components and the dynamic aspects, illustrating how the intricate system functions and transforms over its lifespan. Table 11.2 covers most prevalent definitions of digital twin.

11.2.2.5 Types of Digital Twin (DT)

Diverse types of digital twins emerge based on the extent of product magnification, with each type serving a unique role and contributing to a comprehensive understanding of complex systems. These variations in digital twins primarily hinge on their application areas, often coexisting within a single system or process [20], [21]. These are covered in brief in succeeding paragraph's

- **Parts twin / component twins**: component twins are the microscopic view of a DT, zooming in on the tiniest elements like sensors or valves. Parts twins, while still small, take a slightly wider lens, focusing on components with less critical roles in the overall system.

  Though this category of digital twin replicates the manufacturing layout on a smaller scale, it facilitates advanced surveillance of individual equipment parts, ensuring timely maintenance. As a result, this ensures the stability of the production process and maintains the quality of the end product.

  Example includes a digital twin of a jet engine turbine blade.
- **Product twins /asset twins:** when multiple components collaborate, they collectively constitute what is referred to as an asset. Asset twins facilitate the analysis of interactions among these components, generating a wealth of performance data that can be analyzed and then translated into actionable insights.

  An asset twin scrutinizes the seamless interaction and overall performance efficiency of distinct components when combined into a holistic solution.

  Employing DT for assets or products helps in productivity enhancement, optimization of key metrics such as "mean time between failures (MTBF)" and "mean time to repair (MTTR)", and a reduction in resource consumption. Example includes a DT of a wind turbine.
- **System twins:** system twins afford a clear view into the intricate interactions among assets, providing insights into their collaborative dynamics. Additionally,

TABLE 11.2
Definitions of DT

Source	Definition	Key Points
Digital Twin Consortium (2020) [14]	"A virtual representation of real-world entities and processes, synchronized at a specified frequency and fidelity"	• Emphasis on synchronization and fidelity. • Acknowledges the utilization of both real-time and historical data. • Highlights potential for simulation and prediction.
NASA (2012) [15]	" An integrated Multiphysics, multiscale, probabilistic simulation of an as-built vehicle or system that uses the best available physical models, sensor updates, fleet history, etc. to mirror the life of its corresponding flying twin"	• Focuses on mirroring throughout the lifecycle. • Emphasizes information integration and use for optimization.
Dr. Michael Grieves (2014) [5]	" The virtual digital representation equivalent to physical products"	• Highlights real-time data and model updates. • Includes advanced capabilities like simulation, ML, and reasoning. • Explicitly mentions decision-making support.
Gartner (2017) [16]	"A software model of a physical thing or system that relies on sensor data to understand its state, respond to changes, improve operations, and add value"	• Focuses on software models and sensor data. • Highlights enhances operational efficiency and decision-making capabilities
IBM (2022) [17]	"A dynamic digital replica of a physical asset or system that combines data from sensors, models, and AI to learn, reason, and predict behavior"	• Emphasizes dynamic nature and learning capabilities. • Highlights the role of AI and prediction. • Positions Digital Twins as a key component of their hybrid cloud strategy.
Siemens (2023) [18]	"A living, digital twin that is a virtual copy of a real-world object or process, continuously updated with real-time data and used to monitor, analyze, and improve performance"	• Uses the term "living" to emphasize active monitoring and improvement. • Focuses on continuous updates and performance optimization. • Links Digital Twins to their Mind Sphere IoT platform.

TABLE 11.2 (Continued)
Definitions of DT

Source	Definition	Key Points
GE Digital (2021)	"A Digital Twin is a digital model of an industrial asset – like a jet engine or a wind turbine. They are built by continuously collecting data off physical and virtual sensors on the asset and analyzing the data to gain unique insights about performance and operation that drive business value and outcomes"	• Data-driven mirror: a digital model continuously reflecting a real-world industrial asset. • Sensor symphony: constructed by gathering data from sensors, both physical and virtual, attached to the asset. • Insights engine: analyses the information gathered to reveal novel perspectives regarding operation and performance. • Business Booster: drives valuable business outcomes through improved efficiency, prediction, and optimization.
Microsoft Azure (2023) [19]	"A digital twin is a virtual model of a real-world environment that is driven with data from business systems and IoT devices. It is used to enable insights and actions for a business or organization. Developers and architects are looking to digital twins as the solution that enables intelligent and connected environments"	• Positions cloud computing, AI, and IoT as key enablers for Digital Twins. • Focuses on operational efficiency and predictive insights. • Connects Digital Twins to their Azure cloud platform and IoT services.

these twins may propose recommendations for enhancing overall performance based on the observed interdependencies.

Further system twins provide an extensive overview of the factory or plant, enabling the exploration of various system configurations to attain optimal effectiveness or uncover potential business opportunities for introducing additional sources of profit.

This comprehensive system twin coverage encompasses a set of assets engaged in specific operations, such as energy supply or the production of a particular base product within the plant. It serves as a pathway to acquiring valuable data for strategic decision-making and achieving full visibility into the entire process.

Example includes a DT of a manufacturing plant, a supply chain network, or a healthcare system.

- **Process twins:** zooming in at the macro level, process twins unveil the collaborative workings of systems in generating a complete production facility. The critical inquiry involves assessing whether these systems operate in synchrony to achieve peak efficiency or if delays in one system have cascading effects on others. Process twins play a crucial role in identifying the precise timing schemes that impact the general effectiveness of the production process.

  Process twinning allows the collection of data on outputs without interrupting the actual manufacturing process or compromising production quality. This approach enables efficient tracking of essential business metrics, fostering data-driven decision-making rather than relying on intuition or speculations.

  Example includes a DT of a drug production process, a customer order fulfillment process, or a financial forecasting process.
- **Integrated Product -Process Twins**: the term "integrated product-process digital twin" refers to a comprehensive digital model that encompasses not only a product's design and specifications but also the manufacturing or production processes required to create it [3]. The idea of this concept involves the combination of digital twins representing both the physical product and the manufacturing processes into a unified, interconnected model. This approach results in a single, integrated model that brings together the virtual representations of the product and the production processes. By interconnecting these digital twins, the model provides a holistic view of the entire product lifecycle, from design and production to operation and maintenance. This integration enables real-time monitoring, analysis, and optimization of both the product and the production processes, leading to improved efficiency, reduced costs, and enhanced performance.

  The series of steps that lead to the configuration of the actual system, initiated by the virtual product's chosen selection, is depicted in Figure 11.3.

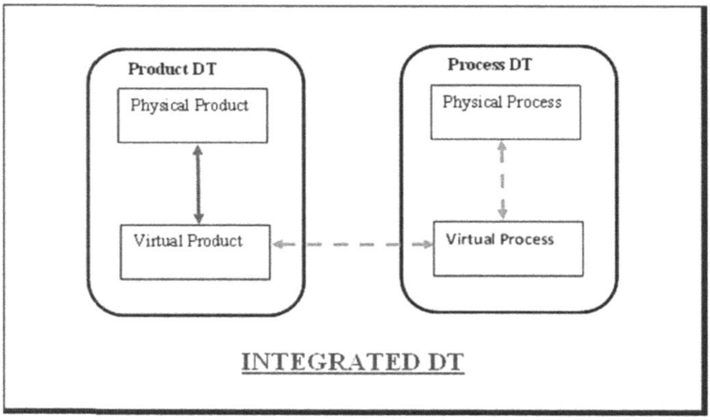

FIGURE 11.3 Integrated Product Process Twin.

This process consists of a sequence where the chosen product triggers a chain of events, ultimately resulting in the physical system's configuration. In other words, the figure illustrates the journey of the virtual product, which sets off a succession of actions that culminate in the preparation and setup of the corresponding physical system. The specifications associated with the chosen product play a pivotal role in determining the configuration of the virtual process model. Subsequently, the configured virtual process model serves as the catalyst, setting in motion the configuration of the corresponding physical system. The diagrammatic representation of integrated twin is presented in Figure 11.3.

11.3 USE CASES: DIGITAL TWIN IN MANUFACTURING

The digital twin concept is used in the manufacturing industry at three main levels: the product level, the unit or system level, and the system of systems/shop-floor level. At these levels, digital twins act as virtual replicas that integrate various components such as processes, raw materials, tools and equipment, finished products, and services. This comprehensive depiction captures the production system's static and dynamic compositions. [13].

- **Product level**: digital twins can be used to design and test virtual prototypes of products before they are physically manufactured. This can help identify and resolve potential issues early in the development process, reducing costs and time-to-market. Digital twins can also be used to optimize product performance and customize products to individual customer needs.
- **Unit/system level**: digital twins can be used to monitor and optimize the performance of individual machines or systems on the factory floor. This can help improve efficiency, reduce downtime, and prolong the lifespan of equipment. Digital twins can also be used to simulate and test different maintenance scenarios, enabling predictive maintenance and reducing the risk of unplanned downtime.
- **System of systems/shop-floor level**: digital twins can be used to optimize the performance of the entire manufacturing system, including the coordination of multiple machines, systems, and processes. This can help improve overall efficiency, reduce waste, and increase flexibility in production. Digital twins can also be used to simulate and test different production scenarios, enabling what-if analysis and optimizing the use of resources.

The utilization of digital twins in diverse manufacturing domains is elucidated in the subsequent sub-sections.

11.3.1 Product Design

The entire process of designing and developing a new product is shaped by a number of crucial phases, such as market research, conceptual design, design verification, and more [22]. Digital twin (DT) can be applied to following:

- Conceptual design
- Detailed design
- Design verification
- Redesign.

11.3.1.1 Conceptual Design

The stages within the conceptual design subprocess comprise following key steps [23]. Table 11.3 outlines the pivotal stages in conceptual design, accompanied by the application of digital twin (DT) in each phase.

DT has been applied to robot vacuum cleaners (RVCs) [23], enhancing the design process by enabling virtual modeling, simulation, and optimization of various aspects before physical prototypes are developed.

TABLE 11.3
Stages of Digital Twin Conceptual Design

Stage	DT Application
Problem Definition and Requirement Specification	DT assist in capturing and analyzing requirements. DT can gather data related to the problem context, user needs, and environmental conditions.
Concept Generation	Leverage DT for modification-based concept generation. Simulate and optimize different design alternatives virtually. Explore variations by adding, replacing, merging, or removing Design Parameters (DPs).
Contextual Information Integration	Utilize DT to include time, location, user activities, profiles, and environmental conditions. Compare virtual and physical contexts to enhance understanding and adaptability.
Learning and Simulation	Exploiting DT's learning ability to understand desirable and undesirable behaviors within specific contexts. Simulate and optimize performance in the virtual space beforehand implementing changes in the physical entity.
Evaluation and Selection	Applying axiomatic design theory and use DT to adjust design ranges, aiding in the understanding of system and design curves. Evaluate concepts based on the Independence Axiom and the Information Axiom.
Contradiction Resolution	Use DT to visualize and analyze data corresponding to different FRs. Identify contradictions through continuous correlation analysis, allowing for informed decision-making.
Constraint Management	Utilize DT to monitor real-time status, predict approaching limits, and provide warnings for identified design constraints. Distinguish between internal and external constraints.
Complexity Management	Make use of DT to monitor, gather, and archive data about functional performance, user feedback, and current product status. Leverage DT's self-learning ability to categorize and address different types of complexities.

11.3.1.2 Detailed Design

DT is useful during the detailed design stage, when a variety of variables and stakeholders are involved. Cheng (2020) introduced a DT-"enhanced industrial internet approach" for collecting diverse factors, such as market requirements and user reviews, to enhance product drafts [24]. Blockchain-based data management approaches [25] and optimization using DT for material selection [26] are also explored. User-oriented involvement is crucial, as demonstrated by models for the entire lifecycle of cruise ships [27]and reference models differentiating between conceptual and virtual representations [7].

11.3.1.3 Design Verification

Efforts to predict and prevent product failures are paramount, with DT employed for quality monitoring and root cause analysis [28]. Design verification using DT includes novel approaches like flexible material models (FlexMM) for predicting product shape changes [29] and virtual prototyping aids in reducing time and costs for design verification, as shown by various simulations [30], [31].

11.3.1.4 Redesign

DT serves not only in concept generation but also in product improvement and redesign. Approaches like tri-model-based development [32] and DT-driven frameworks for optimizing the redesign process of existing products [23] showcase the role of DT in enhancing product design throughout its lifecycle. The continuous collection, analysis, and application of data from physical to virtual spaces contribute to ongoing improvement processes.

11.3.1.5 Virtual Prototyping

Digital twins allow for the creation of virtual prototypes, enabling designers and engineers to simulate the performance of a product in a digital environment before the physical product is manufactured.

11.3.1.6 Design Optimization

DT facilitates the optimization of designs by providing insights into how different design elements interact, leading to improved efficiency and performance [33] [34].

11.3.2 Process Design

By creating virtual replicas of physical processes, DTs enable engineers to visualize, simulate, and analyze process behavior in unprecedented detail, leading to significant improvements in efficiency, quality, and cost-effectiveness. Digital twin simulation of a manufacturing process helps in pinpointing potential bottlenecks and areas for improvement before they manifest in real-world operations, saving time and resources [35]. Further DTs enable engineers to fine-tune process parameters to achieve optimal performance, reducing waste, energy consumption, and lead times [36].Utilizing a process twin enables the modeling of the consequences of adjusting various inputs, such as raw material feed rates, temperature, unit vibrations, etc.

11.3.3 Production Planning and Control

Digital twins offer exciting possibilities for transforming production planning and control (PPC) in modern manufacturing using cyber physical system [37], [38] . DTs can be applied in various domains of PPC as covered in following sub-sections.

11.3.3.1 Simulating Scenarios and Optimizing Production [39]

- **Virtual models**: create high-fidelity virtual models of your entire production system, including machines, processes, material flows, and even external factors like weather.
- **Scenario testing**: use the model to simulate different production plans and scenarios, testing changes in demand, resource allocation, scheduling, and maintenance strategies.
- **Predictive analytics**: integrate instantaneous data from sensors and machines into the model to continuously monitor and predict potential issues, allowing for proactive adjustments.

11.3.3.2 Improved Scheduling and Resource Allocation [40]

- **Real-time optimization**: dynamically adjust production schedules built on instantaneous data from the physical system, optimizing resource utilization and minimizing downtime.
- **Bottleneck identification**: identify bottlenecks and constraints in the production process through simulations, allowing for targeted interventions and capacity adjustments.
- **Improved resource allocation**: allocate resources (machines, manpower, materials) efficiently based on predicted demand and production requirements.

11.3.3.3 Enhanced Decision-making and Collaboration [41]

- **Visualization and communication**: utilize 3D visualization tools to communicate the current state and future performance of the production system to stakeholders.
- **Data-driven decision-making**: base your PPC decisions on instantaneous data and insights gleaned from the digital twin, rather than intuition or historical data alone.
- **Collaborative planning**: facilitate collaboration between production planners, engineers, and other stakeholders by providing a shared digital depiction of the system.

Possible advantages that digital twins could offer the sawmill industry have been investigated [42]. Production planning for aluminum extruded components has also made use of digital twin technology [43].

11.3.4 Asset Management

Digital twins (DT) find applications across various domains of asset management in manufacturing [44], and these applications are detailed in the following sub-sections.

11.3.4.1 Asset Configuration

DT models and simulates a production line, assessing its "Reliability, Availability, Maintainability performance" (RAM performance) to predict "Total Cost of Ownership" (TCO) [45]. This assists in choosing the most suitable design solution for the production line.

11.3.4.2 Asset Reconfiguration

DT models and simulates a complicated process production plant, much like the first scenario, assessing its systemic RAM performance to forecast total cost of ownership. This makes evaluating reconfiguration options to improve plant availability easier.

11.3.4.3 Asset Reconfiguration and Planning

In a web service-based production control system, DT functions as a semantic data model. It lays the groundwork for an "Open knowledge-driven Manufacturing Execution System" architecture, facilitating rapid configurations changes of the system.

11.3.4.4 Asset Commissioning

Using advanced simulation and data analytics, DT operates as a semantic data model to virtually commission the manufacturing system. This simulation is built on a semantic data model and software framework capable of analyzing data in real-time, supporting a speedy system commissioning.

11.3.4.5 Asset Condition Monitoring and Health Assessment

DT is utilized to diagnose assets, assisting in the evaluation of their state of health based on conditions that are tracked [46]. Through data analytics, the DT extracts features necessary for diagnosis, helping to mitigate unreliability situations [47].

11.3.4.6 Asset and Facility Safety

Ensuring safety in facilities engineering is of paramount importance, and digital twins offer valuable assistance in achieving compliance with regulations, standards, and best practices. By identifying and preventing potential failures, malfunctions, or breakdowns, digital twins contribute to damage and injury prevention [48]. In emergency situations or evacuations, digital twins prove invaluable by delivering alerts, notifications, and guidance to ensure a swift and effective response [49].

11.3.4.7 Supply Chain [50]

Digital twin (DT) technology is regarded as a catalyst for companies aiming to transition from a conventional, stable supply chain to an agile and adaptable model capable of rapid evolution. DT embedded within the supply chain boasts diverse applications, with primary focuses on bottleneck identification, inventory management, transportation mapping, and planning and forecasting [51].

11.3.4.7.1 Bottleneck Identification

Digital twins play a pivotal role in pinpointing and addressing issues within the supply chain by continuously collecting and analyzing real-time data. For example, in the context of shipments, sensors can be employed to track the transportation process, sending data to digital twins that can swiftly identify performance bottlenecks.

11.3.4.7.2 Inventory Management

Within the supply chain, digital twins offer instantaneous visibility into inventory levels, facilitating more precise and efficient inventory management. They can simulate diverse scenarios, providing analytics to optimize inventory levels and minimize waste, production costs, and storage expenses. Additionally, supply chain digital twins enable automated inventory tracking, issuing signals for replenishments [51].

11.3.4.7.3 Transportation Mapping

Digital twins empower organizations to measure the influence of changes in demand and supply on the supply chain's dynamics. This includes evaluating the requisite number of vehicles for transportation, assessing their load capacities, and determining whether their quantity should be increased or decreased based on fluctuating demand and supply conditions [52], [53].

11.3.4.7.4 Dynamic Supply Chain

A supply chain DT utilizes instantaneous data to create a comprehensive simulation model, allowing engineers to understand and test diverse aspects of the supply chain. This enables identification of abnormalities and the implementation of corrective actions. The technology enhances visibility, identifies patterns, and streamlines processes, providing opportunities for improvement and efficiency gains [54], [55].

11.3.4.7.5 Stress Testing

Serving as a digital representation of the supply chain, DT allows companies to simulate and stress-test their supply chains, identifying potential issues before they manifest [56].

11.3.4.7.6 Risk Management

By improving supply chain visibility, DT enables engineers to spot trends and run scenarios to determine how possible disruptions could affect things. Businesses may lessen the impact of disruptions and preserve business continuity by anticipating and reducing risks [57], [58].

11.3.4.7.7 Collaborative Supply Chain

Collaboration and coordination are crucial for sustainable supply chains, yielding benefits such as increased efficiency, reduced environmental impact, and enhanced innovation. DT serves as a powerful tool for stakeholders, providing a public platform for data sharing and communication [59]. This enhances transparency, reduces miscommunication, and facilitates faster decision-making, ultimately improving the efficiency of supply chain management [60].

Table 11.4 outlines the significant benefits of DT for supply chain optimization. It highlights how these virtual models act as valuable tools, helping businesses identify constraints, test changes, and improve efficiency in every stage, from design and planning to delivery and fulfillment [50].

11.3.5 Energy Management

For consumers to consume energy safely and affordably, a robust smart energy management framework requires an "Appliance Load Monitoring (ALM)" system. [61]. The primary goal of ALM is to achieve accurate energy sensing and furnish detailed information on energy usage [62]. Applications of energy digital twins (EDT) as reported in the literature encompass the entire lifecycle of a product or asset[63]. The delineation provided by Liu et al. [64] categorizes these applications into three phases: design, manufacturing, and service. These phases are covered in following sub-sections.

11.3.5.1 Design Phase

- **Optimization**: EDT aids designers in determining optimal design parameters for processes, minimizing energy usage. It supports new process/plant design, retrofitting existing processes for improved quality, production rate, and energy efficiency. Future applications involve integrating on-site and local area renewable energy.
- **Data generation**: EDT assists in generating required data, especially when insufficient information is available, rebalancing datasets, and estimating untested operating states. Machine learning can be employed to build accurate models, contributing to energy optimization.
- **Virtual evaluation, verification, and validation**: An economical tool for testing designed processes/plants, EDT identifies bottleneck units early in design and tests extreme operating conditions.

11.3.5.2 Operation Phase

- **Process monitoring**: EDT offers real-time monitoring superior to traditional technology. It offers a platform for evaluating past, present, and combine real-time data from both virtual and physical twins by integrating instantaneous data with a three-dimensional model.
- **Production control**: enhancing traditional production control, EDT connects virtual and physical parts, rejecting disturbances, and maintaining tight quality specifications.
- **Process prediction**: EDT predicts the impact of variations on physical processes, using soft sensing technology for predictions of critical variables difficult or expensive to measure directly.

TABLE 11.4
Role of DT in Supply Chain

Role of DT	Supply Chain Link	Digital Twin Action
Identify bottlenecks	Production, Logistics	Simulate different scenarios to spot choke points in material flow, production capacity, or transportation.
Design & develop changes	Strategy, Infrastructure	Test variations in supplier networks, warehouse locations, or transportation routes before implementation.
Understand behavior	Visibility, Planning	Analyze real-time data to predict demand fluctuations, material shortages, and potential disruptions.
Test design changes	Risk Management, Agility	Evaluate new systems or processes virtually to minimize real-world disruptions and cost overruns.
Optimize JIT/JIS production	Inventory Control, Efficiency	Precisely coordinate raw material arrival and production timing to avoid waste and storage costs.
Improve product lifecycle	Innovation, Collaboration	Track energy use, carbon footprint, and resource consumption across all stages for sustainable optimization.
Create contingency plans	Resilience, Flexibility	Simulate disruptive events like weather delays or supplier outages to prepare mitigation strategies and stock adjustments.
Forecast resource availability	Delivery Times, Visibility	Predict equipment maintenance needs, material deliveries, and labor availability to ensure on-time production and deliveries.
Boost procurement decisions	Volatility, Visibility	Analyze real-time market data and supply chain conditions to inform strategic sourcing and vendor contracts.
Stress-test before failure	Risk Mitigation, Proactiveness	Identify vulnerabilities in the supply chain before they cause real-world disruptions and financial losses.
Manage long lead times	Visibility, Planning	Monitor the effects of extended lead times for components on production schedules to mitigate delays and inventory shortages.
Monitor risks & contingencies	Risk Management, Resilience	Continuously monitor potential disruptions and test mitigation strategies to ensure business continuity.
Optimize transportation & inventory	Logistics, Efficiency	Plan optimal routes, vehicle utilization, and warehouse layout based on real-time data and forecasts.
Improve short-term forecasts	Demand Planning, Adaptability	Predict demand fluctuations over days and weeks to adjust production schedules and avoid overproduction or stockouts.

- **Process optimization and production planning**: going beyond design-centric optimization, energy DT substantially improves process optimization and production planning by utilizing extensive monitoring and predictive data.
- **Process training**: recognizing the potential, both industrial engineers and researchers advocate for EDT implementation in training.

11.3.5.3 Service Phase

- **Predictive maintenance**: similar to process prediction, EDT provides more accurate maintenance times by handling disturbances and uncertainties effectively.
- **Fault detection and diagnosis**: EDT locates faults rapidly by comparing data from virtual and physical operations. Finding the different sorts of faults or the resources involved helps with fault diagnosis.
- **Virtual testing**: as the virtual counterpart of the physical twin, EDT enables testing of operations in scenarios where the breakdown of the physical twin could result in substantial loss and harm.
- **Energy consumption analysis**: energy consumption monitoring and computation machines can be done by DT in production environment. [65]

11.3.6 Quality Control

In manufacturing, a digital twin employs data from sensors and cameras to instantly identify product quality issues and defects. This real-time analysis involves comparing data from the physical asset with its DT, enabling the prompt identification of deviations and the implementation of corrective measures.

DT-based quality control system is developed, as demonstrated by centroid data from the cabin assembly process of an aeronautical product [66]. Digital twin technology offers various applications to enhance quality control across different sectors. Some key uses include:

- **Product and process behavior simulation**: digital twin technology replicates product and manufacturing process behavior in a virtual environment, allowing early detection of potential faults and errors. Engineers can optimize the manufacturing process, reducing the likelihood of defects by modeling different scenarios [67].
- **Predictive maintenance**: leveraging digital twin technology for continuous monitoring of machinery and equipment allows for the anticipation of maintenance needs and the early identification of potential issues, averting them before they arise. This proactive approach enhances quality control by addressing concerns before they lead to product flaws.
- **Real-time quality monitoring**: DT technology facilitates instantaneous tracking of product quality, enabling early detection of faults and errors in the manufacturing process [68]. Engineers can analyze instantaneous to identify trends, contributing to an improved quality control process and reducing the risk of defects in the final product.

- **Process optimization**: utilizing DT technology, engineers can enhance the efficiency of the production process by modeling various scenarios. This exploration of different production methods helps lower the probability of flaws and errors, ultimately improving overall product quality.
- **Root cause analysis**: elements of a DT intended for data analytics-based root cause investigation and monitoring of product quality [28].

11.3.7 Factory Design

Unlike product design, which can rely on physical experiments, factory design necessitates a dominant reliance on digital simulation due to its intricate nature. In this context, the DT emerges as a valuable tool, aiding designers in gaining a deeper understanding of the system being designed and unveiling potential hidden design flaws. Through simulation, coupled with targeted solutions for identified defects, an iterative optimization process can be carried out to refine the overall design [69], [70].

11.3.7.1 Layout Planning

Digital twins play a crucial role in optimizing and streamlining both the process of planning new factory layouts and refining existing ones. Below is an overview of how digital twins are employed in this context [71], [72].

- **Virtual replica creation:** DT involve creating a virtual copy of the physical manufacturing facility. This replica includes detailed 3D models of equipment, machinery, and other assets.
- **Data integration:** digital twins integrate various data sources, such as 3D laser scanning, CAD models, and simulation data, to provide a detailed and accurate depiction of the facility.
- **Simulation and analysis:** the DT allows for the simulation of different layout scenarios. This simulation enables in-depth analysis of factors like material flow, equipment utilization, and overall efficiency.
- **Optimization algorithms**: optimization algorithms are applied within the digital twin to explore and refine different layout configurations. This aids in pinpointing the most efficient and effective allocation of resources.
- **Collaboration and stakeholder involvement**: DT facilitate collaboration by enabling stakeholders, including production managers and engineers, to interact with the virtual replica. This involvement ensures that multiple perspectives are considered during the planning phase.
- **Real-time monitoring:** once the final layout is determined, the DT can be used for instantenous monitoring of the manufacturing processes. This helps in identifying any discrepancies between the planned and actual operations.
- **Space utilization and ergonomics**: DT aids in visualizing and analyzing space utilization, allowing for adjustments that improve the ergonomic aspects of the factory layout.
- **Integration with manufacturing execution systems (MES)**: digital twins can be integrated with MES or other manufacturing software systems, allowing for

seamless communication and data exchange between the virtual and physical environments.

- **Implementation planning**: DT serves as a planning tool for the phased implementation of the finalized layout, assisting in minimizing disruptions to ongoing operations.
- **Risk reduction and informed decision-making**: the use of DT in factory layout planning reduces risks associated with changes by providing a platform for informed decision-making. Stakeholders can assess potential outcomes in a virtual environment before committing to physical alterations.

The application of DT in factory design has been extended to the production of "Body-in-White" (BiW) [73] and the factory dedicated to punching plastic bumpers [71].

11.3.8 Shop Floor

Manufacturing shop-floor stands as the pivotal unit in production. To expedite the digital transformation of the shop-floor, the concept of a "digital twin shop-floor (DTS)" has been introduced [12].

The "physical shop floor (PS)", "virtual shop floor (VS)", "shop floor service system (SSS)", "shop floor digital twin data (SFDTD)" [74], and the linkages between these elements are all included in DTS, which is based on the "five-dimensional digital twin framework" [9].

The primary objective of DTS is to enhance monitoring, control, optimization, and management of the shop-floor through the integration of data and models. DTS has found successful applications in shop-floor design [75], scheduling [76], management [77], and control [78].

- **DTS phases**: the digital twin shop (DTS) operates through three distinct phases [12], pre- production, production, and post-production.
    - In the **pre-production phase**, DTS facilitates the generation of a comprehensive plan by assimilating orders and collating information from various sources, including materials, machinery, and operational plans. This amalgamated information forms the basis for an intelligent production plan.
    - During the **production phase**, DTS oversees the execution of the smart plan, ensuring real-time synchronization with the physical manufacturing processes. It actively monitors operations and intervenes if discrepancies arise, providing timely recommendations for adjustments to maintain alignment with the predefined plan.
    - In the **post-production phase**, DTS continues its role by archiving the historical data pertaining to the production process. This data serves as a valuable resource for analysis, enabling the system to glean insights and refine its operational strategies for subsequent production cycles.

11.3.9 Virtual Commissioning

Virtual commissioning (VC) is a technique used in the manufacturing and automation industry to simulate the performance of physical systems using virtual models before entering into the expensive process of physical integration.

It involves testing and validating automation processes and control systems using digital twins, which are dynamic, virtual replicas of matching actual items. By giving engineers the opportunity to recognize and resolve integration issues in a virtual environment prior to implementing them in the physical system, virtual commissioning (VC) seeks to significantly minimize the delays and costs associated with system integration and commissioning [79], [80].

11.3.9.1 Commissioning Procedure

In the realm of machine commissioning, various strategies aim to overcome challenges associated with tangible verification delays. The concept of virtual commissioning involves utilizing innovative approaches such as "hardware-in-the-loop (HIL)", "reality-in-the-loop (RIL)", and "software-in-the-loop (SIL)" [81].

- **Hardware-in-the-loop (HIL)**: This approach replaces the mechanical system with a virtual model, emulating its physical behavior. The real controller, comprising a "computer numerical controller (CNC)" and programmable logic controller (PLC), governs the virtual mechanical system. HIL enables early verification and insights before the actual hardware is completed, reducing time delays.
- **Reality-in-the-loop (RIL)**: RIL involves using the real mechanical system while employing a simulated control system. This strategy allows for the assessment of the mechanical system's performance with a simulated control setup, aiding in early problem identification and resolution.
- **Software-in-the-loop (SIL)**: SIL takes a fully virtual approach, encompassing both the control system and the mechanical system, along with sensors, actuators, and the entire operational process. SIL, alongside HIL and RIL, facilitates comprehensive verification before the final commissioning, minimizing overall time delays and costs.

A DT-based virtual commissioning approach is suggested for "computerized numerical control machine tool" virtual prototype design built on DT [82].

Figure 11.4 illustrates a summarized overview of the diverse use cases of digital twin technology in the manufacturing sector.

11.4 DT ENABLING TECHNOLOGIES APPLICATIONS IN SMART MANUFACTURING

Digital twin technology relies on combining cutting edge technologies to bridge the gap, between the digital worlds [83], [84]. The internet of things (IoT) plays a role by using sensors and actuators to collect real-time data from assets. Communication technologies enable exchange of data between the realm and the digital twin.

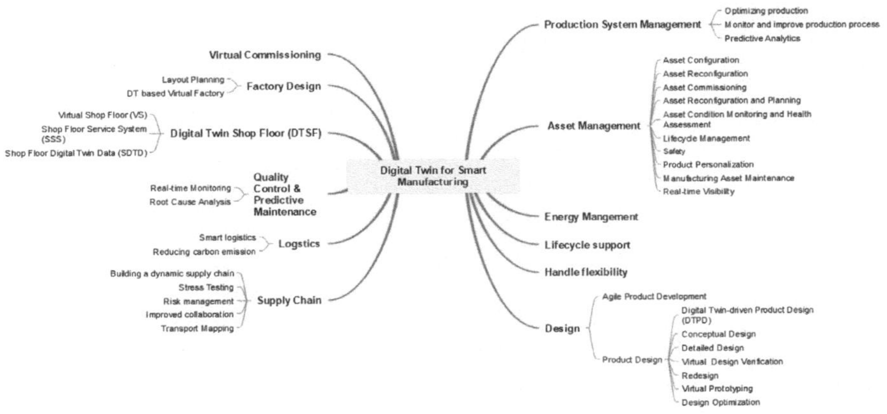

FIGURE 11.4 Use Cases of Digital Twin in Smart Manufacturing.

Both cloud computing and edge computing provide the power and storage capacity to process vast amounts of data ensuring scalability and real-time responsiveness. Artificial intelligence (AI) and machine learning (ML) algorithms analyze the data to extract insights predict outcomes and optimize performance. Augmented reality (AR) and virtual reality (VR) enhance visualization and interaction, with twins in training and maintenance scenarios [85].

Incorporating Blockchain ensures data security and integrity while simulation software enables creation and manipulation of twin models.

The internet of things (IoT) provides the foundational connectivity, with sensors and actuators collecting real-time data from physical assets [86].

Communication technologies facilitate seamless data exchange between the physical world and the digital twin. The application of these technologies across various domains of smart manufacturing is summarized in Table 11.5

11.5 DIGITAL TWIN CASES ACROSS INDUSTRIES

11.5.1 Asset Management

Airbus adopted the cosmo tech digital twin simulation solution, simulating 120,000 operations to identify optimal asset replacement strategies. Airbus plans to extend digital twin simulations to other plants, emphasizing continued commitment to enhanced asset management [87].

11.5.2 Energy Management

- IES, a key contributor, developed an integrated tool for sustainable manufacturing, forming the basis of their current digital twin for manufacturing. Collaborating with **Airbus**, IES identified and reduced energy consumption in large-scale manufacturing processes, optimizing equipment usage [88].

TABLE 11.5
Application of Enabling Technologies of DT in Smart Manufacturing

		Tools													
Area of Application	**Description**	**CAD/ CAE**	**IoT**	**AI/ ML**	**Analytics**	**IIoT**	**Vision**	**Blockchain**	**CAD Models**	**3D Laser Scanning**	**RFID**	**PLC**	**AR/VR/ XR**	**AI/ ML**	**Cloud Computing**
Product Design	Creating virtual prototypes, simulating performance, and optimizing designs.	✓												✓	✓
Process Optimization	Analyzing and enhancing manufacturing processes for efficiency and quality.		✓	✓	✓									✓	✓
Predictive Maintenance	Monitoring equipment health to predict and prevent potential breakdowns.		✓	✓	✓									✓	✓
Quality Control	Implementing real-time monitoring to ensure and enhance product quality.					✓	✓							✓	✓
Supply Chain Management	Enhancing visibility and control across the supply chain for improved efficiency.		✓	✓	✓	✓		✓			✓			✓	✓

Virtual Commissioning	Simulating and testing control systems before actual deployment in the plant.							✓	✓		✓	✓
Worker Training	Providing virtual training environments for employees to enhance skills.								✓		✓	✓
Energy Management	Optimizing energy usage and identifying areas for energy conservation.	✓	✓	✓							✓	✓
Asset Lifecycle Management	Managing the entire lifecycle of manufacturing assets, from inception to decommissioning..										✓	✓
Inventory Optimization	Improving inventory management by analyzing real-time data and demand forecasts.	✓		✓	✓			✓			✓	✓
Factory design / layout	Optimize factory design and layout.					✓	✓			✓	✓	✓

- **Coca-Cola İçecek (CCI)**, based in Istanbul and a Coca-Cola bottler, embraced digital twins to enhance asset optimization, minimize downtime, enhance manufacturing intelligence, and promote sustainability. This optimization led to increased efficiency, reduced errors in time measurement, and notable reductions of 20% in annual energy consumption and 9% in annual water usage [89].

11.5.3 Predictive Maintenance

- **Rolls-Royce** has revolutionized aerospace engine tracking and maintenance using digital twins, setting a new industry standard. Leveraging integrated engine sensors, the digital twin acts as an early warning system, enabling proactive and accurate maintenance scheduling. This has significantly reduced unplanned downtime, enhancing engine reliability and performance. The implementation of digital twin technology in aviation has been a transformative technological advancement for Rolls-Royce [90].
- **Kaeser Compressors**, a U.S.-based compressor manufacturer, implemented digital twins to shift from product-centric to value-centric offerings. Unlike the traditional approach of deploying hardware and transferring ownership to the customer, Kaeser now maintains ownership throughout the product lifecycle, charging based on actual usage rather than a fixed rate. This data-driven approach ensures equipment uptime, and precise billing for each billing cycle. Through digital twins, Kaeser has achieved a 30% reduction in product costs and successfully onboarded 50% of major vendors [91], [92].
- **GE's** adoption of digital twin extends anomaly prediction lead time from 20–30 days to 60 days, aiding in proactive corrective measures. This technology is also utilized to highly accurately forecast the lifespan of turbine blades in specific aircraft engines. GE's digital twins for industrial facilities simulate complex system interactions, considering various future scenarios and optimizing key performance parameters, leveraging extensive data sources for weather, performance, and operations [92].

11.5.4 Asset Management

- **Boeing** implemented a digital twin in aviation to enhance safety measures for the 787 Dreamliner's battery system, exemplifying proactive risk management in the aviation industry. The incorporation of DT in the design and development process allowed early identification and resolution of issues, reflecting Boeing's commitment to safety [93].
- **Lufthansa Technik** has conducted "Performance Restoration Shop Visit" for a "LEAP-1A" engine owned by Nova Airlines, enabling real-time analysis and design improvements. Lufthansa Technik plans to enhance services based on this groundbreaking experience and their extensive prior work with the LEAP-1A engine [94].

- **General Engineering(GE)** developed a digital twin for the GE 90 engine focusing on aircraft engine blade maintenance to predict blade degradation, enabling timely maintenance recommendations to prevent issues [95].
- In the endeavor to energize the world's largest aluminum smelter in Bahrain **GE's HA-turbine** faced scrutiny due to its limited operating hours. Leveraging the "Digital Power Plant (DPP)", GE's software suite, GE showcased a digital twin created from the operational data of a similar turbine in France. The DPP demonstrated how the turbines would respond to various scenarios, reassuring Aluminum Bahrain (Alba) about grid stability [96].
- **GE's gas turbine** power plant in Bouchain, France, incorporated a digital twin system for optimizing operational and maintenance procedures. This advanced system utilizes sophisticated algorithms to collect real-time data from turbine-installed sensors, providing insights for proactive maintenance and performance enhancements [97].
- The **United States Air Force** utilized a DT of the Lockheed Martin "GPS IIR" satellite for cybersecurity evaluation via penetration testing. Through the creation of a digital replica, potential cybersecurity issues were examined [98].
- **Stara**, a tractor manufacturer based in Brazil, utilizes digital twins to revolutionize agriculture. Digital twins enable Stara to provide agricultural services to farmers with real-time insights on optimal crop planting conditions, leading to a significant reduction in seed (21%) and fertilizer (19%) usage, attributed to Stara's guidance [92].
- Digital twin technology is used by American automaker **Tesla** in each and every vehicle it produces.

The seamless and uninterrupted exchange of essential data between the cars and the manufacturer elevates the overall standard of Tesla's output [99], [100].

11.5.5 Process Design

- In Brazil, at Unilever's Indaiatuba facility, renowned as the world's largest laundry detergent powder factory, innovative technologies including digital twinning and AI have been deployed. These advancements are intended to improve cost efficiency, agility, and significantly reduce the environmental footprint of operations [101]. This digital replica predicts optimal process parameters for new formulations, eliminating the need for physical trials and significantly accelerating the launch of innovations, such as the introduction of their initial anti-residue detergent.
- Takeda Pharmaceutical's development team in Japan has successfully built comprehensive digital models of their manufacturing processes, creating individual digital twins for each step and linking them together through a unified digital twin. This interconnected system allows for a more efficient and streamlined manufacturing process [102]. This comprehensive simulation enhances automation possibilities for complex chemical and biochemical processes, especially when real-time monitoring is impractical.

- A **chemical manufacturer** created a digital twin for factory processes to enhance enzyme production, catering to biofuel, agriculture, and consumer goods industries. Initially focusing on key parameters like average pH, airflows, and dissolved oxygen levels, the process twin facilitated engineers in refining set points and schedules for enzyme production optimization. Expanding the digital twin, the company incorporated additional data through agile two-week development sprints, including live feeds of production status from sensor data. Within a year, the digital twin facilitated a production increase equivalent to the addition of an entire new production line [103].

11.5.6 Product Development

- **Bridgestone** utilizes virtual tyre development, using digital twin, to design, test, and fine-tune digital tire prototypes before physical production. Approximately 200 tires are saved in each project, resulting in a 60% reduction in raw materials and CO_2 emissions during development. Virtual tyre development enables flexible testing of tire variants, reducing product development time and outdoor testing [104].
- With the use of a digital twin asset development model, **Boeing** has achieved a notable improvement of up to 40% in the initial quality of aircraft components and systems. This model works by creating a virtual replica of physical airplane parts, allowing for simulation of their lifecycle. It is expected that this model will lead to substantial production efficiency improvements for Boeing over the next ten years, as it aids in streamlining the production process, improving accuracy, and reducing waste. By incorporating this digital twin model, Boeing is taking strides in modernizing aircraft production and staying competitive in the industry [105].
- **Rolls-Royce** employs digital twin technology for fan blades in the construction of their "UltraFan" jet engines, aiming to achieve a remarkable 25% improvement in fuel consumption efficiency. Every blade in the testing phase is associated with a digital twin. Collected data during testing is integrated into these digital twins, allowing engineers to anticipate the performance of each blade during actual service [106].

11.5.7 Virtual Commissioning

- **Ferrero**, in collaboration with **Siemens**, leverages virtual commissioning for high-bay warehouse projects to enhance efficiency. The goal was to reduce commissioning time by 30%, and the project, including 8000 pallet spaces and four stacker cranes. Project achieved significant success with a 13-week timeline, compared to the conventional 19 weeks needed for a similar project in 2019. The adoption of virtual commissioning showcased an 88% reduction in achieving the target system availability within three weeks, as opposed to the usual six months [107].
- **Siemens software** utilizes a DT, of the robotics system and its environment, engineers can conduct offline programming and test without disrupting ongoing

production. This approach helps identify and rectify issues like programming bugs and mechanical interference, ensuring that the production area aligns with expectations. The process utilizes robotics simulation software, accommodating various robot brands and supporting multiple robots in a shared workspace.

- **Wipro PARI**, an automation systems design and manufacturing company, which achieved a remarkable 70% reduction in commissioning time through the implementation of VC. The company utilized VC to construct a digital twin of a production line, enabling comprehensive testing and validation before the physical assembly. The early detection and resolution of issues during this virtual phase resulted in substantial time and cost savings. Furthermore, VC facilitated enhanced collaboration between design and manufacturing teams [108].

11.5.8 Quality Control

- **Nestle** employs digital twinning to oversee its manufacturing processes and enhance quality control measures. The company has generated digital replicas of its production lines, simulating the entire production journey, encompassing raw materials to final products. These DT leverage instantaneous data from sensors and cameras to actively monitor the production process, identifying potential quality concerns promptly. Through this approach, Nestle can enact real-time corrective measures, thereby enhancing overall product quality [109].

11.5.9 Supply Chain

- **Nestle** has successfully implemented digital twin technology and AI-optimization techniques to continuously adapt its supply chain, ensuring cost-effectiveness and resilience during challenging times. A significant part of this strategy involves the use of digitally-enabled transport control towers, called T-Hubs, which have been implemented in over 20 locations. These T-Hubs oversee the daily management of over 16,000 trucks, accounting for 85% of the company's total revenue. This approach allows Nestle to have a more efficient and streamlined supply chain [109].
- **Kraft Heinz,** a well-known company in the food and beverage industry, utilizes real-time data streaming from various sources such as sensors, human interactions, and operational, financial, and commercial data. In an effort to enhance its supply chain, Kraft Heinz has partnered with Microsoft to create a supply chain control tower for real-time visibility into plant operations and automate the distribution of its various product categories. Furthermore, Kraft Heinz has plans to develop digital twins for its 34 manufacturing facilities located in North America, which will allow the company to simulate and optimize its operations [110].
- **Ocado Group** uses virtual versions of its operating supermarket fulfilment center's called digital twins to simulate the effects of future layout modifications in real-time. Using this approach, Ocado can make informed and cost-efficient decisions without disrupting the daily functions of the facility [111].

11.5.10 Factory Design

Cold-chain warehousing and logistics company **Lineage Logistics** utilizes digital twins to streamline the design process before incurring significant costs related to constructing real warehouses. Digital twins assist the organization generate individually suitable designs for each warehouse by simulating multiple what-if situations and taking into account aspects like facility location, client profile, and demand characteristics [112].

Tetra Pak, a packaging company with a valuation of $13 billion, is at the forefront of integrating digital twin technology into warehouse management. The company has implemented a digital replica of one of its warehouses in Southeast Asia, which receives operational data from the physical facility's internet of things (IoT) sensing infrastructure. This digital twin assists Tetra Pak in overseeing stock locations, managing inventory, optimizing workflows, and allocating warehouse equipment [113].

11.5.11 Logistics

UPS uses a variety of sensors on its hubs, vehicles, and packages. It also gathers information from drivers, staff members, and clients to continuously compile operational data for its vast logistics network. Digital twins play a crucial role by tracking packages in instantaneous throughout the entire process, from the creation of shipping labels to final delivery [114].

11.6 CHALLENGES

In navigating the landscape of digital twin (DT) technology, a host of challenges parallels its promising advantages, aligning with the hurdles encountered in AI and IoT domains. Shared challenges include data standardization, management complexities, and security concerns, as well as obstacles hindering the smooth implementation and modernization of legacy systems [115].

11.6.1 Data Integration Complexity

A significant complexity arises from integrating data from diverse sources, such as IoT sensors, CAD models, historical records, and real-time data, all stored in varied formats into a unified model. This integration process is intricate, and any inaccuracies in the data can jeopardize the efficacy of digital twin (DT) solutions. Ensuring the accuracy and completeness of this integrated data is a persistent challenge.

11.6.2 IT Infrastructure

The effectiveness of a DT relies on an infrastructure supporting IoT and data analytics. A well-designed and connected IT framework is essential for the digital twin to fulfill its intended objectives; without it, the digital twin may struggle to achieve its goals [83].

11.6.3 Data Quality

The data requirements for a digital twin pose a challenge, as they necessitate high-quality, uninterrupted data streams that are free from noise. In the event of poor and inconsistent data, there is a possibility of suboptimal performance from the digital twin, as it operates based on potentially flawed and incomplete information.

11.6.4 Standardization in Modelling

When it comes to creating digital twins, there are several challenges related to the lack of a standardized approach. These challenges can arise in various stages of digital twin development, including the initial design and simulation phases. Having a standardized methodology, whether it is physics-based or design-based, is crucial. This is because standardization ensures that the digital twin is easily understood by those in the relevant domain, and enables the smooth flow of information throughout the development and implementation process. This not only helps to improve the efficiency of the development process, but also ensures that the digital twin is more effective and accurate in its representation of the physical system it is intended to model.

11.6.5 Data Privacy and Security

In an industrial setting, ensuring privacy and security for digital twins is a major challenge, due to the large amount of data used and the potential risk to sensitive system information. To overcome this hurdle, it is essential to align key technologies that enable digital twins, such as data analytics and internet of things (IoT), with current security and privacy regulations. By prioritizing security and privacy in the management of digital twins' data, trust issues related to their implementation can be effectively addressed.

11.7 CONCLUSION

In recent times, digital twin technology has attracted substantial attention from industry and academia, with diverse definitions found in the literature, reflecting its application across different disciplines. The fundamental concept involves seamless data integration between a physical and virtual entity in both directions. Initially employed in astronautics and aerospace, NASA utilized digital twin technology for missions like Apollo 13 and Mars Rover Curiosity. A literature review indicates the ongoing expansion of digital twins' scope and influence, positioning it as a rapidly growing IT solution across various industries. The paper highlights the absence of precise definitions for digital twins, underscoring the consistent lack of differentiation in its definition since its initial coinage in 2012. This paper extensively covers the use cases of digital twin technology in the manufacturing domain, encompassing diverse aspects such as product design, process design, factory design, and quality control. It provides in-depth insights into how digital twins are applied across these domains to enhance efficiency, optimize processes, and ensure quality in manufacturing operations

The real-life applications of digital twins in top-level manufacturing companies are vividly illustrated across various domains. In the aerospace sector, companies such as Boeing, Rolls-Royce, GE, Airbus, and Lufthansa Technik have seamlessly integrated digital twins to enhance efficiency, maintenance, and performance monitoring of aircraft. In the FMCG sector, renowned companies like Coca-Cola and Unilever have leveraged digital twins for optimizing production processes, ensuring supply chain efficiency, and maintaining product quality. Similarly, in the automobile manufacturing domain, companies like Tesla and Saturn have implemented digital twins for detailed simulations, predictive maintenance, and streamlined production processes. The intricate details of these applications showcase the versatility and impact of digital twins across diverse industries and manufacturing domains.

The paper also comprehensively addresses the challenges associated with the implementation of digital twins in the current environment. It sheds light on the obstacles and considerations that organizations face when integrating digital twins into their operations, providing a holistic view of the opportunities and challenges in the current landscape.

REFERENCES

1 Y. Borole, P. Borkar, R. Raut, V. P. Balpande, and P. Chatterjee, "Digital Twins: Internet of Things, Machine Learning, and Smart Manufacturing," in *Digital Twins*, De Gruyter, 2023. doi: 10.1515/9783110778861.

2 S. Neethirajan and B. Kemp, "Digital twins in livestock Farming," *Animals*, vol. 11, no. 4, Art. no. 4, Apr. 2021, doi: 10.3390/ani11041008.

3 I. Onaji, D. Tiwari, P. Soulatiantork, B. Song, and A. Tiwari, "Digital twin in manufacturing: Conceptual framework and case studies," *International Journal of Computer Integrated Manufacturing*, vol. 35, no. 8, pp. 831–858, Aug. 2022, doi: 10.1080/0951192X.2022.2027014.

4 B. D. Allen, "Digital Twins and Living Models at NASA," Accessed: Jan. 07, 2024. [Online]. Available: https://ntrs.nasa.gov/citations/20210023699

5 M. Grieves and J. Vickers, "Digital twin: Mitigating unpredictable, undesirable emergent behavior in complex systems," in *Transdisciplinary Perspectives on Complex Systems: New Findings and Approaches*, Springer, 2017, pp. 85–113. doi: 10.1007/978-3-319-38756-7_4.

6 R. Stark, C. Fresemann, and K. Lindow, "Development and operation of digital twins for technical systems and services," *CIRP Annals*, vol. 68, no. 1, pp. 129–132, Jan. 2019, doi: 10.1016/j.cirp.2019.04.024.

7 B. Schleich, N. Anwer, L. Mathieu, and S. Wartzack, "Shaping the digital twin for design and production engineering," *CIRP Annals*, vol. 66, no. 1, pp. 141–144, Jan. 2017, doi: 10.1016/j.cirp.2017.04.040.

8 D.-Y. Jeong *et al.*, "Digital twin: Technology evolution stages and implementation layers with technology elements," *IEEE Access*, vol. 10, pp. 52609–52620, 2022, doi: 10.1109/ACCESS.2022.3174220.

9 F. Tao, M. Zhang, Y. Liu, and A. Y. C. Nee, "Digital twin driven prognostics and health management for complex equipment," *CIRP Annals*, vol. 67, no. 1, pp. 169–172, 2018, doi: 10.1016/j.cirp.2018.04.055.

10 F. Tao, J. Cheng, Q. Qi, M. Zhang, H. Zhang, and F. Sui, "Digital twin-driven product design, manufacturing and service with big data," *International Journal of Advanced Manufacturing Technology*, vol. 94, no. 9, pp. 3563–3576, Feb. 2018, doi: 10.1007/s00170-017-0233-1.

11 F. Tao, Q. Qi, L. Wang, and A. Y. C. Nee, "Digital twins and cyber–physical systems toward smart manufacturing and Industry 4.0: Correlation and comparison," *Engineering*, vol. 5, no. 4, pp. 653–661, Aug. 2019, doi: 10.1016/j.eng.2019.01.014.

12 F. Tao and M. Zhang, "Digital twin shop-floor: A new shop-floor paradigm towards smart manufacturing," *IEEE Access*, vol. 5, pp. 20418–20427, 2017, doi: 10.1109/ACCESS.2017.2756069.

13 Q. Qi, F. Tao, Y. Zuo, and D. Zhao, "Digital twin service towards smart manufacturing," *Procedia CIRP*, vol. 72, pp. 237–242, Jan. 2018, doi: 10.1016/j.procir.2018.03.103.

14 "Definition of a Digital Twin," Digital Twin Consortium. Accessed: Jan. 07, 2024. [Online]. Available: www.digitaltwinconsortium.org/initiatives/the-definition-of-a-digital-twin/

15 E. Glaessgen and D. Stargel, "The digital twin paradigm for future NASA and U.S. air force vehicles," in *53rd AIAA/ASME/ASCE/AHS/ASC Structures, Structural Dynamics and Materials Conference
20th AIAA/ASME/AHS Adaptive Structures Conference
14th AIAA*, Honolulu, Hawaii: American Institute of Aeronautics and Astronautics, Apr. 2012. doi: 10.2514/6.2012-1818.

16 "Definition of Digital Twin - Gartner Information Technology Glossary," Gartner. Accessed: Jan. 07, 2024. [Online]. Available: www.gartner.com/en/information-technology/glossary/digital-twin

17 A. Stanford-Clark, E. Frank-Schultz, "What are digital twins?," IBM Developer. Accessed: Jan. 07, 2024. [Online]. Available: https://developer.ibm.com/articles/what-are-digital-twins/

18 "Digital Twin | Siemens," Siemens Digital Industries Software. Accessed: Jan. 07, 2024. [Online]. Available: www.plm.automation.siemens.com/global/en/our-story/glossary/digital-twin/24465

19 "Azure Digital Twins." Accessed: Jan. 07, 2024. [Online]. Available: https://microsoft.github.io

20 "What is a Digital Twin?," IBM. Accessed: Jan. 07, 2024. [Online]. Available: www.ibm.com/topics/what-is-a-digital-twin

21 A. Andrade, "The 3 Levels of the Digital Twin Technology," Vidya. Accessed: Jan. 07, 2024. [Online]. Available: https://vidyatec.com/blog/the-3-levels-of-the-digital-twin-technology-2/

22 K. T. Ulrich and S. D. Eppinger, *Product design and development*, 6th edition. New York, NY: McGraw-Hill Education, 2016.

23 Y. Wang, A. Liu, F. Tao, and A. Y. C. Nee, "Digital twin driven conceptual design," in *Digital Twin Driven Smart Design*, Elsevier, 2020, pp. 33–66. doi: 10.1016/B978-0-12-818918-4.00002-6.

24 J. Cheng, H. Zhang, F. Tao, and C.-F. Juang, "DT-II: Digital twin enhanced Industrial Internet reference framework towards smart manufacturing," *Robotics and Computer-Integrated Manufacturing*, vol. 62, p. 101881, Apr. 2020, doi: 10.1016/j.rcim.2019.101881.

25 S. Huang, G. Wang, Y. Yan, and X. Fang, "Blockchain-based data management for digital twin of product," *Journal of Manufacturing Systems*, vol. 54, pp. 361–371, Jan. 2020, doi: 10.1016/j.jmsy.2020.01.009.

26 F. Xiang, Z. Zhang, Y. Zuo, and F. Tao, "Digital twin driven green material optimal-selection towards sustainable manufacturing," *Procedia CIRP*, vol. 81, pp. 1290–1294, Jan. 2019, doi: 10.1016/j.procir.2019.04.015.

27 V. Arrichiello and P. Gualeni, "Systems engineering and digital twin: a vision for the future of cruise ships design, production and operations," *International Journal on Interactive Design and Manufacturing*, vol. 14, no. 1, pp. 115–122, Mar. 2020, doi: 10.1007/s12008-019-00621-3.

28 A. Detzner and M. Eigner, "A digital twin for root cause analysis and product quality monitoring," in *DS 92: Proceedings of the DESIGN 2018 15th International Design Conference*, 2018, pp. 1547–1558. doi: 10.21278/idc.2018.0418.

29 M. Groen, G. Zijlstra, D. San-Martin, J. Post, and J. Th. M. De Hosson, "Product shape change by internal stresses," *Materials & Design*, vol. 157, pp. 492–500, Nov. 2018, doi: 10.1016/j.matdes.2018.08.013.

30 A. Patrikeev, A. Tarasov, A. Borovkov, M. Aleshin, and O. Klyavin, "NVH analysis of offroad vehicle frame. Evaluation of mutual influence of body-frame system components," *Materials Physics and Mechanics*, vol. 34, no. 1, pp. 70–75, 2017, doi: 10.18720/MPM.3412017_8.

31 A. Vuruskan and S. P. Ashdown, "Modeling of half-scale human bodies in active body positions for apparel design and testing," *International Journal of Clothing Science and Technology*, vol. 29, no. 6, pp. 807–821, Jan. 2017, doi: 10.1108/IJCST-12-2016-0141.

32 P. Zheng and A. S. Sivabalan, "A generic tri-model-based approach for product-level digital twin development in a smart manufacturing environment," *Robotics and Computer-Integrated Manufacturing*, vol. 64, p. 101958, Aug. 2020, doi: 10.1016/j.rcim.2020.101958.

33 K. Y. H. Lim, P. Zheng, C.-H. Chen, and L. Huang, "A digital twin-enhanced system for engineering product family design and optimization," *Journal of Manufacturing Systems*, vol. 57, pp. 82–93, Oct. 2020, doi: 10.1016/j.jmsy.2020.08.011.

34 F. Bellalouna, "Case study for design optimization using the digital twin approach," *Procedia CIRP*, vol. 100, pp. 595–600, Jan. 2021, doi: 10.1016/j.procir.2021.05.129.

35 J. Vachalek, L. Bartalsky, O. Rovny, D. Sismisova, M. Morhac, and M. Loksik, "The digital twin of an industrial production line within the industry 4.0 concept," in *2017 21st International Conference on Process Control (PC)*, Strbske Pleso, Slovakia: IEEE, Jun. 2017, pp. 258–262. doi: 10.1109/PC.2017.7976223.

36 Microsoft, "The Process Digital Twin: A Step Towards Operational Excellence," 2017. [Online]. Available: https://info.microsoft.com/rs/157-GQE-382/images/Digital%20Twin%20Vision.pdf

37 F. Biesinger, D. Meike, B. Kraß, and M. Weyrich, "A digital twin for production planning based on cyber-physical systems: A case study for a cyber-physical system-based creation of a digital twin," *Procedia CIRP*, vol. 79, pp. 355–360, Jan. 2019, doi: 10.1016/j.procir.2019.02.087.

38 V. Bhatia, V. Jaglan, S. Kumawat, and K. S. Kaswan, "Real-Life Applications of Soft Computing in Cyber-Physical System: A Compressive Review," in *Soft Computing: Theories and Applications*, T. K. Sharma, C. W. Ahn, O. P. Verma, and B. K. Panigrahi, Eds., Singapore: Springer, 2022, pp. 501–514. doi: 10.1007/978-981-16-1740-9_41.

39 R. Mallach, "A Digital Twin Framework for Production Planning Optimization: Applications for Make-To-Order Manufacturers," University of Massachusetts Amherst. doi: 10.7275/34368754.
40 Í. R. S. Agostino, E. Broda, E. M. Frazzon, and M. Freitag, "Using a Digital Twin for Production Planning and Control in Industry 4.0," in *Scheduling in Industry 4.0 and Cloud Manufacturing*, B. Sokolov, D. Ivanov, and A. Dolgui, Eds., in International Series in Operations Research & Management Science., Cham: Springer International Publishing, 2020, pp. 39–60. doi: 10.1007/978-3-030-43177-8_3.
41 P. Novak, J. Vyskočil, and B. Wally, "The digital twin as a core component for Industry 4.0 smart production planning," *IFAC-Papers OnLine*, vol. 53, pp. 10803–10809, Jan. 2020, doi: 10.1016/j.ifacol.2020.12.2865.
42 S. Chabanet, H. Bril El-Haouzi, M. Morin, J. Gaudreault, and P. Thomas, "Toward digital twins for sawmill production planning and control: Benefits, opportunities, and challenges," *International Journal of Production Research*, vol. 61, no. 7, pp. 2190–2213, Apr. 2023, doi: 10.1080/00207543.2022.2068086.
43 S. Zahno *et al.*, "Dynamic project planning with digital twin," *Frontiers in Manufacturing Technology*, vol. 3, pp. 1–8, 2023, Accessed: Jan. 10, 2024. [Online]. Available: www.frontiersin.org/articles/10.3389/fmtec.2023.1009633
44 M. Macchi, I. Roda, E. Negri, and L. Fumagalli, "Exploring the role of digital twin for asset lifecycle management," *IFAC-Papers OnLine*, vol. 51, no. 11, pp. 790–795, Jan. 2018, doi: 10.1016/j.ifacol.2018.08.415.
45 V. A. Dolgov, P. A. Nikishechkin, V. E. Arkhangelskii, P. I. Umnov, and A. A. Podkidyshev, "Models for managing production systems of machine-building enterprises based on the development and using of their digital twins," *EPJ Web Conference*, vol. 248, p. 04015, 2021, doi: 10.1051/epjconf/202124804015.
46 L. C. W. Kong, S. Harper, D. Mitchell, J. Blanche, T. Lim, and D. Flynn, "Interactive Digital Twins Framework for Asset Management Through Internet," in *2020 IEEE Global Conference on Artificial Intelligence and Internet of Things (GCAIoT)*, Dec. 2020, pp. 1–7. doi: 10.1109/GCAIoT51063.2020.9345890.
47 "Digital Twin - Making Your Asset Smarter with the Digital Twin," DNV. Accessed: Jan. 10, 2024. [Online]. Available: www.dnv.com/article/making-your-asset-smarter-with-the-digital-twin-63328
48 S. Khajavi, M. Tetik, Z. Liu, P. Korhonen, and J. Holmström, "Digital twin for safety and security: Perspectives on building lifecycle," *IEEE Access*, vol. 11, pp. 52339–52356, Jun. 2023, doi: 10.1109/ACCESS.2023.3278267.
49 J. S. Lackey Mike, "Digital Twins Offer a Safer Future," Machine Design. Accessed: Jan. 10, 2024. [Online]. Available: www.machinedesign.com/automation-iiot/article/21838343/digital-twins-offer-a-safer-future
50 Mohsen Attaran and Sharmin Attaran, "The impact of digital twins on the evolution of intelligent manufacturing and Industry 4.0," *Advances in Computational Intelligence*, vol. 3, no. 3, p. 11, Jun. 2023, doi: 10.1007/s43674-023-00058-y.
51 Anastasiya Haritonova, "Digital Twins in Supply Chain: Benefits & Applications," PixelPlex. Accessed: Jan. 10, 2024. [Online]. Available: https://pixelplex.io/blog/digital-twin-supply-chain/
52 M. Kosacka-Olejnik, M. Kostrzewski, M. Marczewska, B. Mrówczyńska, and P. Pawlewski, "How digital twin concept supports internal transport systems?—Literature review," *Energies*, vol. 14, no. 16, Art. no. 16, Jan. 2021, doi: 10.3390/en14164919.
53 V. Bhatia, V. Jaglan, S. Kumawat, V. Siwach, and H. Sehrawat, "Intelligent Transportation System Applications: A Traffic Management Perspective," in

Intelligent Sustainable Systems, J. S. Raj, R. Palanisamy, I. Perikos, and Y. Shi, Eds., Singapore: Springer, 2022, pp. 419–433. doi: 10.1007/978-981-16-2422-3_33.
54 T. D. Moshood, G. Nawanir, S. Sorooshian, and O. Okfalisa, "Digital twins driven supply chain visibility within logistics: A new paradigm for future logistics," *Applied System Innovation*, vol. 4, no. 2, p. 29, Jun. 2021, doi: 10.3390/asi4020029.
55 Sathyanarayanan B, "Digital Twin-Driven Adaptive Supply Chain." Accessed: Jan. 10, 2024. [Online]. Available: www.tcs.com/insights/blogs/supply-chain-transformation-digital-twin
56 D. Ivanov and A. Dolgui, "Stress testing supply chains and creating viable ecosystems," *Operation Management Research*, vol. 15, no. 1, pp. 475–486, Jun. 2022, doi: 10.1007/s12063-021-00194-z.
57 D. Ivanov and A. Dolgui, "New disruption risk management perspectives in supply chains: digital twins, the ripple effect, and resileanness," *IFAC-Papers OnLine*, vol. 52, no. 13, pp. 337–342, Jan. 2019, doi: 10.1016/j.ifacol.2019.11.138.
58 R. P. B. Edlund, "Usage of Digital Twins in Supply Chain Risk Management," 2022. Accessed: Jan. 10, 2024. [Online]. Available: https://aaltodoc.aalto.fi/handle/123456789/115137
59 X. Lu, W. Wang, W. Li, Y. Jing, and X. Li, "A Generic Digital Twin Framework for Collaborative Supply Chain Development," in *2022 5th International Conference on Computing and Big Data (ICCBD)*, Dec. 2022, pp. 177–181. doi: 10.1109/ICCBD56965.2022.10080555.
60 Z. Chen and L. Huang, "Digital twins for information-sharing in remanufacturing supply chain: A review," *Energy*, vol. 220, p. 119712, Apr. 2021, doi: 10.1016/j.energy.2020.119712.
61 M. M. Hasan, D. Chowdhury, and M. Z. R. Khan, "Non-intrusive load monitoring using current shapelets," *Applied Sciences*, vol. 9, no. 24, p. 5363, Jan. 2019, doi: 10.3390/app9245363.
62 D.-E. A. Mansour *et al.*, "Applications of IoT and digital twin in electrical power systems: A comprehensive survey," *IET Generation, Transmission & Distribution*, vol. 17, no. 20, pp. 4457–4479, 2023, doi: 10.1049/gtd2.12940.
63 W. Yu, P. Patros, B. Young, E. Klinac, and T. G. Walmsley, "Energy digital twin technology for industrial energy management: Classification, challenges and future," *Renewable and Sustainable Energy Reviews*, vol. 161, p. 112407, Jun. 2022, doi: 10.1016/j.rser.2022.112407.
64 M. Liu, S. Fang, H. Dong, and C. Xu, "Review of digital twin about concepts, technologies, and industrial applications," *Journal of Manufacturing Systems*, vol. 58, pp. 346–361, Jan. 2021, doi: 10.1016/j.jmsy.2020.06.017.
65 C. Cimino, E. Negri, and L. Fumagalli, "Review of digital twin applications in manufacturing," *Computers in Industry*, vol. 113, p. 103130, Dec. 2019, doi: 10.1016/j.compind.2019.103130.
66 C. Zhuang, Z. Liu, J. Liu, H. Ma, S. Zhai, and Y. Wu, "Digital twin-based quality management method for the assembly process of aerospace products with the Grey-Markov model and Apriori Algorithm," *Chinese Journal of Mechanical Engineering*, vol. 35, no. 1, p. 105, Aug. 2022, doi: 10.1186/s10033-022-00763-8.
67 R. Luo, B. Sheng, Y. Lu, Y. Huang, G. Fu, and X. Yin, "Digital twin model quality optimization and control methods based on workflow management," *Applied Sciences*, vol. 13, no. 5, Art. no. 5, Jan. 2023, doi: 10.3390/app13052884.
68 X. Zhu and Y. Ji, "A digital twin–driven method for online quality control in process industry," *International Journal of Advance Manufacture Technology*, vol. 119, no. 5, pp. 3045–3064, Mar. 2022, doi: 10.1007/s00170-021-08369-5.

69 N. Zhao, J. Guo, and H. Zhao, "Chapter 8 - Digital Twin Driven Factory Design," in *Digital Twin Driven Smart Design*, F. Tao, A. Liu, T. Hu, and A. Y. C. Nee, Eds., Academic Press, 2020, pp. 205–235. doi: 10.1016/B978-0-12-818918-4.00008-7.
70 J. Guo, N. Zhao, L. Sun, and S. Zhang, "Modular based flexible digital twin for factory design," *Journal of Ambient Intelligence and Humanized Computing*, vol. 10, no. 3, pp. 1189–1200, Mar. 2019, doi: 10.1007/s12652-018-0953-6.
71 D. Nåfors, B. Johansson, P. Gullander, and S. Erixon, "Simulation in Hybrid Digital Twins for Factory Layout Planning," in *2020 Winter Simulation Conference (WSC)*, Dec. 2020, pp. 1619–1630. doi: 10.1109/WSC48552.2020.9384075.
72 "Digital Twin for Factory Layout," Prevu3D. Accessed: Jan. 11, 2024. [Online]. Available: www.prevu3d.com/digital-twin-use-cases/factory-layout/
73 H. Li and H. Wang, "Chapter 3 - Digital Twin-Driven Production Line Custom Design Service," in *Digital Twin Driven Service*, F. Tao, Q. Qi, and A. Y. C. Nee, Eds., Academic Press, 2022, pp. 59–88. doi: 10.1016/B978-0-323-91300-3.00001-2.
74 F. Tao, H. Zhang, A. Liu, and A. Y. C. Nee, "Digital twin in industry: State-of-the-art," *IEEE Transactions on Industrial Informatics*, vol. 15, no. 4, pp. 2405–2415, Apr. 2019, doi: 10.1109/TII.2018.2873186.
75 J. Leng *et al.*, "Digital twin-driven rapid reconfiguration of the automated manufacturing system via an open architecture model," *Robotics and Computer-Integrated Manufacturing*, vol. 63, p. 101895, Jun. 2020, doi: 10.1016/j.rcim.2019.101895.
76 M. Zhang, F. Tao, and A. Y. C. Nee, "Digital Twin Enhanced Dynamic Job-Shop Scheduling," *Journal of Manufacturing Systems*, vol. 58, pp. 146–156, Jan. 2021, doi: 10.1016/j.jmsy.2020.04.008.
77 C. Zhuang, J. Liu, and H. Xiong, "Digital twin-based smart production management and control framework for the complex product assembly shop-floor," *Int J Adv Manuf Technol*, vol. 96, no. 1, pp. 1149–1163, Apr. 2018, doi: 10.1007/s00170-018-1617-6.
78 K. Zhang *et al.*, "Digital twin-based opti-state control method for a synchronized production operation system," *Robotics and Computer-Integrated Manufacturing*, vol. 63, p. 101892, Jun. 2020, doi: 10.1016/j.rcim.2019.101892.
79 Maprlsoft, "Virtual Commissioning with a Model-Driven Digital Twin," Waterloo Canada. [Online]. Available: https://altair.com/docs/default-source/resource-library/vpmodeldriventwin_maplesoft.pdf?sfvrsn=6657f7c4_3
80 G. Barbieri *et al.*, "A virtual commissioning based methodology to integrate digital twins into manufacturing systems," *Production Engineering-Research and Development*, vol. 15, no. 3, pp. 397–412, Jun. 2021, doi: 10.1007/s11740-021-01037-3.
81 Autodesk, "The Virtual Commissioning in the Factory | Autodesk University." Accessed: Jan. 11, 2024. [Online]. Available: www.autodesk.com/autodesk-university/class/Virtual-Commissioning-Factory-2019
82 W. Shen, T. Hu, Y. Yin, J. He, F. Tao, and A. Y. C. Nee, "Digital Twin Based Virtual Commissioning for Computerized Numerical Control Machine Tools," in *Digital Twin Driven Smart Design*, F. Tao, A. Liu, T. Hu, and A. Y. C. Nee, Eds., Academic Press, 2020, pp. 289–307. doi: 10.1016/B978-0-12-818918-4.00011-7.
83 A. Fuller, Z. Fan, C. Day, and C. Barlow, "Digital twin: Enabling technologies, challenges and open research," *IEEE Access*, vol. 8, pp. 108952–108971, 2020, doi: 10.1109/ACCESS.2020.2998358.
84 W. Hu, T. Zhang, X. Deng, Z. Liu, and J. Tan, "Digital twin: A state-of-the-art review of its enabling technologies, applications and challenges," *Journal of Intelligent Manufacturing and Special Equipment*, vol. 2, no. 1, pp. 1–34, Jan. 2021, doi: 10.1108/JIMSE-12-2020-010.

85 V. Bhatia and B. Bhatia, "Machine Learning-Based Solutions for Internet of Things-Based Applications," in *Automated Secure Computing for Next-Generation Systems*, John Wiley & Sons, Ltd, 2024, pp. 295–318. doi: 10.1002/9781394213948.ch15.
86 V. Bhatia, S. Kumawat, and V. Jaglan, "Overview of the role of the internet of things and cyber-physical systems in various applications," in *Handbook of Research of Internet of Things and Cyber-Physical Systems*, Apple Academic Press, 2022.
87 P. Wong, "Asset Management at Airbus: Driving Change with Digital Twin Simulation," Cosmo Tech. Accessed: Jan. 12, 2024. [Online]. Available: https://cosmotech.com/resources/article/fyf-asset-management-at-airbus-driving-change-with-digital-twin-simulation/
88 "Digital Twin R&D Pilot Projects for Airbus and Toyota Pave Way for Sustainable Manufacturing." Accessed: Jan. 12, 2024. [Online]. Available: www.iesve.com/discoveries/view/34707/toyota-airbus
89 "Improving Operational Performance Using AWS IoT SiteWise | Coca-Cola İçecek Case Study | AWS," Amazon Web Services, Inc. Accessed: Jan. 13, 2024. [Online]. Available: https://aws.amazon.com/solutions/case-studies/coca-cola-iot-sitewise/
90 "Digital Twin:Rolls-Royce." Accessed: Jan. 12, 2024. [Online]. Available: www.rolls-royce.com/innovation/digital/digital-twin.aspx
91 Kaeser Compressors, "Complete Connectivity for Maximum Compressed air Supply - KAESER COMPRESSORS New Zealand," KAESER. Accessed: Jan. 13, 2024. [Online]. Available: https://nz.kaeser.com/company/press/press-releases/n-complete-connectivity.aspx
92 P. Augustine, "The Industry Use Cases for the Digital Twin Idea," *Advances in Computers*, vol. 117, 2020, pp. 79–105. doi: 10.1016/bs.adcom.2019.10.008.
93 S. Srivastava, "The Role of Digital Twin in Aerospace To Enhance Safety and Efficiency," Appinventiv. Accessed: Jan. 12, 2024. [Online]. Available: https://appinventiv.com/blog/digital-twin-in-aerospace/
94 "World's First Performance Restoration Shop Visit of a LEAP-1A Carried Out." Accessed: Jan. 12, 2024. [Online]. Available: www.lufthansa-technik.com/en/world-s-first-performance-restoration-shop-visit-of-a-leap-1a-carried-out-dba4b8a903a8bf43
95 "Digital Twin Creation | GE Research." Accessed: Jan. 12, 2024. [Online]. Available: www.ge.com/research/offering/digital-twin-creation
96 GE, "When Hardware Met Software: The Digital Twin Of This Huge Gas Turbine Will Drive The World's Largest Aluminum Plant | GE News." Accessed: Jan. 12, 2024. [Online]. Available: www.ge.com/news/reports/guinness-work-worlds-largest-aluminum-smelter-meets-worlds-efficient-gas-turbine
97 C. E. C. Directors, "GE's Digital Power Plant Helps French Gas Plant Achieve Record-Setting Efficiency Rate," Power Engineering. Accessed: Jan. 12, 2024. [Online]. Available: www.power-eng.com/gas/ge-s-digital-power-plant-helps-french-gas-plant-achieve-record-setting-efficiency-rate/
98 G. Qiao, Y. Zhuang, T. Ye, and Y. Qiao, "A Digital-Twin-Based Detection and Protection Framework for SDC-Induced Sinkhole and Grayhole Nodes in Satellite Networks," *Aerospace*, vol. 10, no. 9, Art. no. 9, Sep. 2023, doi: 10.3390/aerospace10090788.
99 "Digital Twins in Automotive Industry." Accessed: Jan. 13, 2024. [Online]. Available: https://encyclopedia.pub/entry/25054
100 D. Piromalis and A. Kantaros, "Digital Twins in the Automotive Industry: The Road toward Physical-Digital Convergence," *Applied System Innovation*, vol. 5, no. 4, Art. no. 4, Aug. 2022, doi: 10.3390/asi5040065.

101 Unilever, "Unilever Sites Join Network of World's Most Digitally Advanced Factories," Unilever. Accessed: Jan. 12, 2024. [Online]. Available: www.unilever.com/news/news-search/2023/unilever-sites-join-network-of-worlds-most-digitally-advanced-factories/
102 S. Caganof, "Digital Twins | Deloitte," *Tech Trends*, 2020. Accessed: Jan. 12, 2024. [Online]. Available: www.deloitte.com/conf/modern/settings/wcm/templates/modern--di-research-template/initial.html
103 "Digital Twins: From One Twin to the Enterprise Metaverse | McKinsey." Accessed: Jan. 13, 2024. [Online]. Available: www.mckinsey.com/capabilities/mckinsey-digital/our-insights/digital-twins-from-one-twin-to-the-enterprise-metaverse
104 BRIDGESTONE, "How Bridgestone's Virtual Tyre Modelling is Revolutionising Tyre Development." Accessed: Jan. 12, 2024. [Online]. Available: www.bridgestone.co.in/stories-1/mobility/how-bridgestones-virtual-tyre-modelling-is-revolutionising-tyre-development
105 Woodrow Bellamy, "Boeing CEO Talks 'Digital Twin' Era of Aviation," Avionics International. Accessed: Jan. 13, 2024. [Online]. Available: www.aviationtoday.com/2018/09/14/boeing-ceo-talks-digital-twin-era-aviation/
106 B. Sampson, "Rolls-Royce Tests Low Pressure System for Ultrafan," Aerospace Testing International. Accessed: Jan. 13, 2024. [Online]. Available: www.aerospacetestinginternational.com/news/engine-testing/rolls-royce-tests-low-pressure-system-for-ultrafan.html
107 D. Lenk, "Ferrero: 'Using a comprehensive digital twin to reduce high-bay warehouse commissioning time by 30 percent' ".
108 SIEMENS, "Using Virtual Commissioning to Reduce Commissioning Time by 70 Percent," Siemens Resource Center. Accessed: Jan. 13, 2024. [Online]. Available: https://resources.sw.siemens.com/en-US/case-study-wipro-pari
109 "Accelerating Data-Driven Digitalization | Nestlé Annual Report." Accessed: Jan. 13, 2024. [Online]. Available: www.nestle.com/investors/annual-report/digitalization
110 "Kraft Heinz's Latest Tech Investments Tackle Control Tower and Digital Twins," Consumer Goods Technology. Accessed: Jan. 13, 2024. [Online]. Available: https://consumergoods.com/kraft-heinzs-latest-tech-investments-tackle-control-tower-and-digital-twins
111 "What is a digital twin at Ocado Group?," Ocado Group. Accessed: Jan. 13, 2024. [Online]. Available: www.ocadogroup.com/about-us/what-we-do/digital-twins/
112 "Lineage | Automation is Using Software to Transform the Cold Chain." Accessed: Jan. 13, 2024. [Online]. Available: www.onelineage.com/news-stories/automation-using-software-transform-cold-chain
113 "Tetra Pak collaborates with Hexagon in Smart Plant Engineering." Accessed: Jan. 13, 2024. [Online]. Available: www.tetrapak.com/en-in/about-tetra-pak/news-and-events/newsarchive/digital-plant-engineering-platform-with-hexagon
114 UPS, "Data & Digital Twin Drive Real-Time Predictive Analytics," Redis. Accessed: Jan. 13, 2024. [Online]. Available: https://redis.com/the-data-economy-podcast/episode-6/
115 M. Attaran and B. G. Celik, "Digital twin: Benefits, use cases, challenges, and opportunities," *Decision Analytics Journal*, vol. 6, p. 100165, Mar. 2023, doi: 10.1016/j.dajour.2023.100165.

12 Data Analytics and Visualization in Smart Manufacturing Using AI-based Digital Twins

M. Sivakumar, M. Maranco, N. Krishnaraj, and U. Srinivasulu Reddy

12.1 INTRODUCTION

Smart manufacturing incorporates cutting-edge technology, including the internet of Things (IoT), artificial intelligence (AI), and data analytics, to optimize industrial processes and improve efficiency. A fundamental element of intelligent manufacturing is the notion of digital twins (Tyagi & Richa, 2023), which are virtual duplicates of tangible resources. Data analytics approaches applied in smart manufacturing through AI-based digital twins have a wide range of applications and significant effects. AI-powered digital replicas assess live sensor data and past maintenance records to forecast equipment malfunctions in advance. Manufacturers can proactively schedule maintenance, minimize downtime, and prevent expensive unplanned shutdowns by detecting early signals of degradation or irregularities in machine behavior. Utilizing data analytics techniques on digital twins allows for the continuous monitoring of production processes and the assessment of product quality in real-time. Manufacturers can utilize sensor data and production parameters to analyze and identify deviations from quality standards, as well as potential faults, at an early stage in the manufacturing process. This helps to minimize waste and maintain consistent product quality. AI-powered digital replicas replicate and enhance manufacturing procedures to enhance effectiveness, productivity, and resource allocation (Xia et al., 2023). Manufacturers may achieve optimal production schedules, decrease energy usage, and streamline operations for maximum throughput by analyzing data from digital twins and discovering opportunities for optimization. Applying data analytics approaches to digital twins enhances the efficiency of supply chain operations, encompassing inventory management, demand forecasting, and logistics planning. Manufacturers can enhance overall supply chain efficiency by utilizing digital twins and external data sources to analyze and manage inventory levels, predict demand variations, and optimize transportation routes, resulting in cost reduction.

DOI: 10.1201/9781003480860-12

AI-powered digital replicas enable virtual prototyping and product design by replicating the performance of a product under different circumstances (Mihai et al., 2022). Manufacturers can enhance product design and optimization by analyzing data from digital twins and conducting virtual trials. This iterative process enables them to meet performance criteria, shorten time-to-market, and expedite innovation cycles. Applying data analytics approaches to digital twins enhances the efficiency of resource allocation and capacity planning in manufacturing plants. Manufacturers can enhance production schedules, distribute resources effectively, and optimize production capacity by scrutinizing production data and demand estimates. This enables them to satisfy consumer demand while saving expenses. AI-powered digital replicas assess data on energy use and production factors to maximize energy utilization and minimize ecological footprint. Manufacturers can decrease carbon emissions, enhance sustainability, and meet regulatory requirements by identifying energy-intensive procedures and applying energy-saving strategies. Applying data analytics approaches to digital twins enables the real-time monitoring and provision of decision assistance for manufacturing operations. Operators and managers may enhance production results by utilizing interactive dashboards to visualize key performance indicators (KPIs) and performance metrics. This allows them to make informed decisions, discover patterns, and take timely measures. In summary, the utilization of data analytics approaches in smart manufacturing using AI-based digital twins allows producers to attain enhanced efficiency, productivity, and adaptability in the current competitive market. Manufacturers may exploit the potential of data analytics and AI-driven insights to discover fresh prospects for innovation, expansion, and sustainability.

Visualization approaches are essential in smart manufacturing using AI-based digital twins since they offer clear and practical insights into intricate manufacturing processes (Wang et al., 2024). Visualization dashboards present up-to-date information from AI-powered digital replicas, enabling operators to visually and easily track key performance indicators (KPIs), equipment conditions, and production metrics. Real-time feedback on production performance is provided through interactive charts, graphs, and gauges. This allows operators to swiftly spot issues, trends, and anomalies and promptly take necessary steps. Visualization tools facilitate the detection of anomalies and underlying causes in industrial processes by visually representing patterns, trends, and correlations within the data. Heatmaps, scatter plots, and time series charts are used to emphasize deviations from typical operating conditions. This allows engineers to examine the root causes and apply remedial actions to avoid future occurrences. Visualization tools display predictive maintenance data produced by AI-powered digital replicas, including equipment health scores, forecasts of remaining useful life, and recommendations for maintenance. Predictive maintenance dashboards present information on the likelihood of equipment failure, timetables for maintenance, and evaluations of risk. This allows maintenance teams to prioritize jobs, allocate resources effectively, and optimize maintenance processes.

Visualization techniques are employed to display simulation findings and optimization recommendations generated by AI-based digital twins, with the aim of enhancing industrial processes. 3D models, animations, and virtual reality (VR) representations replicate production operations, material flows, and resource consumption, enabling engineers to detect bottlenecks, inefficiencies, and possibilities for enhancement. Visualization tools facilitate "what-if" scenario research, enabling engineers to investigate different process setups, layouts, and parameters in order to enhance performance and save expenses. Visualization dashboards display quality control data produced by AI-powered digital replicas, including defect rates, quality measures, and inspection outcomes. Pareto charts, histograms, and defect maps are graphical tools that display defects categorized by kind, location, and frequency. These tools allow quality control teams to detect patterns, determine underlying causes, and pinpoint areas that require enhancement. Utilizing real-time defect tracking and visualization aids in diminishing faults, enhancing product quality, and augmenting client satisfaction. Visualization techniques are used to represent supply chain data and logistics information that is created by AI-powered digital replicas, known as digital twins. This includes visualizing inventory levels, demand predictions, and transit routes. Supply chain dashboards exhibit supply chain performance indicators, such as rates of order fulfillment, turnover of inventory, and durations of lead times, allowing supply chain managers to detect inefficiencies, hazards, and possibilities for enhancement. Geographic maps, network diagrams, and flowcharts provide visual representations of supply chain networks, allowing managers to optimize transportation routes, warehouse placements, and inventory quantities in order to save expenses and enhance operational effectiveness. In summary, the utilization of visualization approaches in smart manufacturing through the implementation of AI-based digital twins allows manufacturers to acquire more profound understanding, make well-informed choices, and enhance performance throughout the entirety of the production process. Visualization technologies enable operators, engineers, and managers to promote continuous improvement, increase efficiency, and achieve operational excellence by presenting visual representations of intricate data and insights.

12.2 SMART MANUFACTURING AND DIGITAL TWINS

Smart manufacturing is the integration of cutting-edge technologies, data-driven analysis, and intelligent automation to enhance manufacturing processes and achieve operational excellence (Leng et al., 2021). The fundamental objective of smart manufacturing is to optimize efficiency, productivity, and flexibility throughout the whole production process.

12.2.1 Key Concepts of Smart Manufacturing

Smart manufacturing is an innovative method of production that utilizes cutting-edge technologies and data-driven analysis to enhance manufacturing processes, enhance operational efficiency, and foster innovation along the whole value chain, as depicted in Figure 12.1.

- **Interconnectivity**: smart manufacturing utilizes the internet of things (IoT) to establish connections between physical objects, sensors, and machinery in the production setting. This network of interconnected systems allows for the continuous monitoring, gathering of information, and communication among different parts of the production ecosystem.

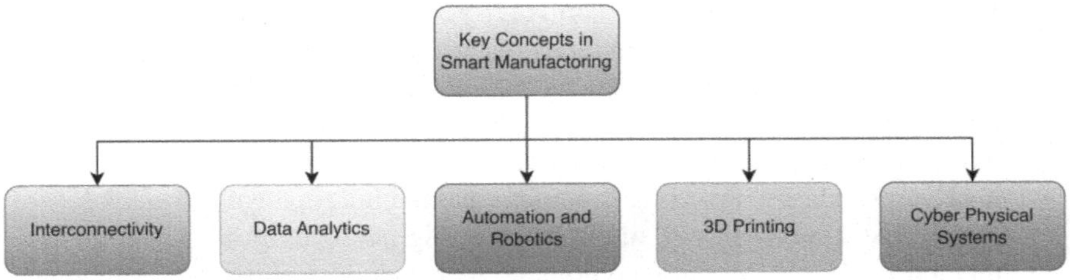

FIGURE 12.1 Key Concepts of Smart Manufacturing.

- **Data analytics**: data analytics plays a crucial role in smart manufacturing by analyzing large volumes of data collected from sensors, production equipment, and other sources. Manufacturers can employ advanced analytics techniques such as machine learning and predictive analytics to acquire actionable insights into production processes, identify patterns, and improve performance.
- **Automation and robotics**: automation and robotics technologies are employed to automate repetitive tasks, streamline manufacturing processes, and enhance operational efficiency in smart industrial environments. Advanced robotics systems, such as collaborative robots (cobots) and autonomous guided vehicles (AGVs), work together with human operators to do tasks in a safe and efficient way (Tyagi et al., 2021).
- **Additive manufacturing (3D printing)**: 3D printing and other additive manufacturing technologies enable the rapid production, customization, and fabrication of complex parts and components. Additive manufacturing, through the elimination of traditional manufacturing methods, offers more flexibility, cost-effectiveness, and design independence in the field of smart manufacturing.
- **Cyber-physical systems**: smart manufacturing integrates cyber-physical systems (CPS) that establish a connection between physical processes and digital technologies. CPS combine sensors, actuators, and control systems with computer intelligence to actively monitor, assess, and optimize industrial operations in real-time.

12.2.2 Role of Digital Twins in Optimizing Manufacturing Operations

Digital twins are virtual replicas of physical assets, processes, or systems that accurately mimic their real-world counterparts within a digital setting (Friederich et al., 2022). Within the realm of smart manufacturing, digital twins are crucial for optimizing production operations due to the following reasons:

- **Simulation and modeling**: digital twins simulate and model manufacturing processes, enabling engineers to test different scenarios, optimize parameters, and predict outcomes before implementation in the physical world. By simulating production workflows, material flows, and resource utilization, digital twins help identify inefficiencies, bottlenecks, and opportunities for improvement.

- **Real-time monitoring and control**: digital twins enable the continuous monitoring and management of industrial activities through the collection of data from sensors, production equipment, and internet of things (IoT) devices. Digital twins facilitate predictive maintenance, quality control, and performance optimization by evaluating real-time data streams. This capability empowers manufacturers to promptly adapt to changing conditions and minimize potential hazards.
- **Predictive analytics and maintenance**: digital twins utilize data analytics methodologies to forecast equipment malfunctions, anticipate maintenance requirements, and optimize maintenance timetables. Digital twins utilize historical data, sensor readings, and equipment health indicators to detect early indications of deterioration or abnormalities in machine performance. This allows for preventive maintenance and reduces the amount of time that the unit is out of service.
- **Performance optimization**: digital twins enhance industrial performance by analyzing production data, identifying areas for optimization, and providing recommendations for improvement. By simulating production scenarios and monitoring performance indicators, digital twins help improve production schedules, resource allocation, and workflow efficiency to increase throughput and decrease costs.

Smart manufacturing concepts utilize cutting-edge technologies and data-driven analysis to optimize manufacturing operations along the whole value chain. Digital twins are essential in smart manufacturing since they create virtual models of physical assets and processes. This allows for the modeling, prediction, and optimization of manufacturing activities, leading to enhanced efficiency, productivity, and competitiveness.

12.2.3 Benefits of Digital Twins

Digital twins offer a range of benefits in monitoring, analysis, prediction, and optimization of manufacturing processes:

- **Real-time monitoring**: digital twins facilitate the continuous monitoring of manufacturing processes, allowing producers to monitor key performance indicators (KPIs), equipment status, and output metrics in real-time. Manufacturers can promptly intervene and take corrective steps by visually analyzing data obtained from sensors, IoT devices, and production equipment. This enables them to detect errors, trends, and anomalies in real-time.
- **Predictive maintenance**: digital twins utilize historical data, sensor readings, and machine learning algorithms to forecast equipment malfunctions in advance. Digital twins can proactively schedule maintenance, limit downtime, and prevent costly unplanned shutdowns by evaluating patterns and trends in equipment performance to detect early symptoms of degradation or anomalies.
- **Process optimization**: digital twins enable the simulation and optimization of production processes, enabling firms to detect and address inefficiencies, bottlenecks, and areas for improvement (Min et al., 2019). Digital twins utilize production data analysis and scenario simulation to enhance production

schedules, resource allocation, and workflow efficiency, hence maximizing throughput and minimizing costs.

- **Quality control**: digital twins provide instantaneous monitoring and analysis of product quality, enabling manufacturers to promptly identify flaws, deviations, and non-conformities at the initial stages of the manufacturing process. Digital twins utilize sensor data, production parameters, and quality measures to guarantee product uniformity, adherence to quality standards, and customer contentment.
- **Supply chain optimization**: digital twins provide a visual representation and enhance the efficiency of supply chain operations, encompassing tasks such as managing inventory, predicting demand, and planning logistics. Digital twins utilize supply chain data analysis and scenario simulations to optimize inventory levels, predict demand variations, and enhance transportation routes for cost reduction and efficiency enhancement.
- **Energy management**: digital twins utilize energy consumption data and production characteristics to monitor and optimize energy usage, hence minimizing environmental effect. Digital twins contribute to cost reduction, carbon emission minimization, and sustainability improvement by recognizing energy-intensive operations and applying energy-saving strategies.
- **Decision support**: digital twins offer decision support tools and insights to operators, engineers, and managers, facilitating well-informed decision-making and strategic planning. Digital twins enable stakeholders in industrial operations to make data-driven decisions and achieve continuous improvement by visualizing key performance indicators, performance measurements, and optimization recommendations.

12.3 DATA COLLECTION AND INTEGRATION

Data collection and integration are essential procedures in the data management lifecycle. They encompass the acquisition and merging of data from many sources to establish a cohesive dataset for the purposes of analysis, reporting, and decision-making.

12.3.1 Sensor Data Acquisition

The procedure of gathering sensor data from manufacturing equipment and gadgets encompasses multiple stages to guarantee precise and dependable data acquisition. Below is a summary of the standard procedure illustrated in Figure 12.2.

- **Identify sensor requirements**: to begin, it is essential to determine the specific types of sensors required for monitoring the pertinent characteristics or variables inside the manufacturing process. These may encompass temperature sensors, pressure sensors, flow meters, vibration sensors, proximity sensors, and other similar devices. The choice of sensors is contingent upon the particular demands of the manufacturing process and the variables that need to be monitored.

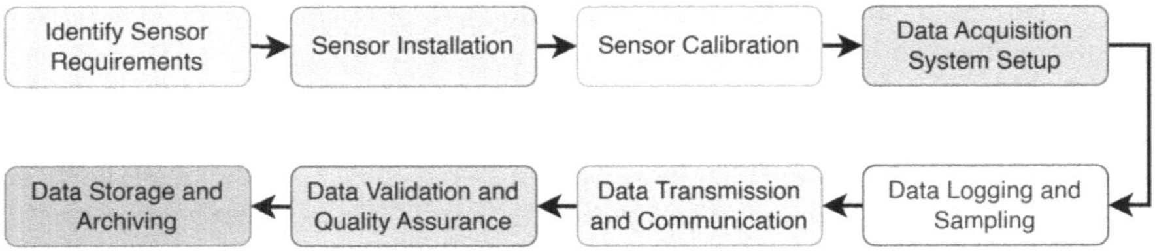

FIGURE 12.2 Overall Process of Sensor Data Acquisition.

- **Sensor installation**: after choosing the sensors, they must be properly put on the production equipment or devices in suitable positions to accurately capture the needed data. Accurate installation is crucial to guarantee the sensors' effective and dependable measurement of the variables of interest.
- **Sensor calibration**: following the installation, it is imperative to calibrate the sensors in order to guarantee precision and uniformity in the results. Calibration entails the process of aligning the sensor readings with established reference values or standards. This aids in the elimination of any systematic inaccuracies or gradual changes in the sensor values as time progresses.
- **Data acquisition system setup**: subsequently, a data gathering system is established to gather and retain the sensor data. This system generally comprises of data acquisition hardware, such as data loggers or programmable logic controllers, along with software designed for data logging and display. The data acquisition system can be either linked with the control system of the manufacturing equipment or operated separately.
- **Data logging and sampling**: the data collection system gathers sensor data at consistent intervals, referred to as the sampling frequency or rate. The sampling frequency may vary depending on the production process needs and the monitored variables. Data logging entails the act of saving the gathered sensor data in a database or data repository for subsequent analysis.
- **Data transmission and communication**: occasionally, it may be necessary to transfer sensor data instantaneously to a central monitoring system or control center for prompt analysis and decision-making. To do this, it is necessary to build dependable communication channels, such as wired or wireless networks, in order to communicate the data collected by the sensors to the data gathering system.
- **Data validation and quality assurance**: after the collection of sensor data, it goes through validation and quality assurance procedures to verify its accuracy, comprehensiveness, and dependability. This task entails the identification and rectification of any inaccuracies or irregularities in the data, such as outliers, missing numbers, or measurement problems.
- **Data storage and archiving**: ultimately, the sensor data is preserved and cataloged for subsequent consultation, examination, and documentation. Adhering to appropriate data storage techniques is crucial for preserving data integrity, safeguarding against security breaches, and complying with regulatory mandates.

12.3.2 Types of Sensors

- **Temperature sensors**: these sensors measure temperature variations in the manufacturing environment and are crucial for processes where temperature control is essential, such as heat treatment or cooling processes.
- **Pressure sensors**: pressure sensors monitor variations in pressure levels within manufacturing equipment, which is vital for ensuring safety and efficiency in processes like hydraulic systems or pneumatic machinery.
- **Flow sensors**: flow sensors measure the rate of fluid flow through pipes or channels, commonly used in applications such as monitoring liquid or gas flow rates in manufacturing processes.
- **Vibration sensors**: vibration sensors detect mechanical vibrations in equipment, helping to identify potential faults or abnormalities in rotating machinery, motors, or bearings.
- **Proximity sensors**: proximity sensors detect the presence or absence of objects within a certain range, often used in automated manufacturing systems for object detection or position sensing.
- **Level sensors**: level sensors measure the level of liquids, solids, or powders in tanks or containers, essential for inventory management, batching, or material handling processes.
- **Humidity sensors**: humidity sensors measure the moisture content in the air or within materials, critical for processes where humidity control is necessary, such as in cleanrooms or humidity-sensitive manufacturing environments.

12.3.3 Data Transmission Protocols

- **Modbus**: a widely used protocol for serial communication between industrial devices, Modbus facilitates data exchange between sensors, controllers, and supervisory systems in manufacturing environments.
- **OPC UA (unified architecture)**: OPC UA is a standardized protocol for secure and reliable data exchange in industrial automation and control systems, supporting interoperability between diverse devices and platforms.
- **Ethernet/IP**: Ethernet/IP is an industrial Ethernet protocol used for real-time communication and control in manufacturing environments, enabling high-speed data transmission and seamless integration with Ethernet-based networks.
- **Profibus**: Profibus is a fieldbus communication protocol commonly used in process automation and control systems, providing reliable data exchange between sensors, actuators, and control devices.
- **WirelessHART**: WirelessHART is a wireless communication protocol based on the HART (highway addressable remote transducer) protocol, designed for industrial applications where wired communication is impractical or costly.

12.3.4 Challenges Associated with Sensor Data Acquisition

- **Data quality**: ensuring the accuracy, reliability, and consistency of sensor data can be challenging due to factors such as sensor drift, calibration errors, environmental conditions, or electromagnetic interference.

- **Data integration**: integrating data from heterogeneous sensors and devices into a unified data acquisition system can be complex, requiring compatibility between different communication protocols, data formats, and hardware interfaces.
- **Data security**: protecting sensor data from unauthorized access, manipulation, or cyber-attacks is crucial for maintaining data integrity, confidentiality, and availability in manufacturing environments where sensitive information is transmitted and stored.
- **Scalability**: scaling up sensor networks to accommodate large-scale manufacturing operations may pose challenges in terms of network infrastructure, bandwidth requirements, and data processing capabilities, particularly in distributed or remote locations.
- **Interference and noise**: interference from electromagnetic fields, electrical noise, or other sources can distort sensor readings and affect data accuracy, requiring noise filtering or shielding measures to mitigate their impact on data acquisition.
- **Power supply**: providing reliable power sources for sensors, especially in remote or harsh environments, can be challenging, requiring solutions such as battery-powered sensors, solar panels, or energy harvesting techniques to ensure continuous operation.
- **Cost**: the cost of sensors, data acquisition systems, and communication infrastructure can be significant, particularly for deploying large-scale sensor networks or upgrading existing systems to support advanced data acquisition capabilities.

12.3.5 Data Integration and Preprocessing

The process of integrating data from numerous sources within the manufacturing ecosystem entails consolidating data from diverse sources, including sensors, equipment, databases, and enterprise systems, into a cohesive data infrastructure. This infrastructure is then used for the purposes of analysis, visualization, and decision-making. There are multiple techniques available for incorporating data from various sources into the industrial ecosystem:

- **Data standardization**: standardizing data formats, schemas, and protocols helps ensure compatibility and consistency across diverse data sources. Adopting industry-standard formats such as XML, JSON, or CSV facilitates data exchange and interoperability between different systems and applications (Tyagi et al., 2023).
- **Data aggregation**: aggregating data from multiple sources involves consolidating data into a centralized repository or data warehouse for storage and analysis. This approach simplifies data management and enables comprehensive analysis of the entire manufacturing ecosystem.
- **Data transformation**: transforming data involves converting data from different formats or structures into a common format or schema to facilitate integration. This may involve data normalization, cleansing, or enrichment to ensure data consistency and quality across diverse sources.

- **APIs and web services**: application programming interfaces (APIs) and web services provide standardized interfaces for accessing and exchanging data between different systems and applications. Using APIs allows seamless integration between disparate systems, enabling data sharing and collaboration across the manufacturing ecosystem.
- **Middleware platforms**: middleware platforms act as intermediaries between different systems and applications, facilitating data integration, communication, and interoperability. These platforms provide tools and services for data transformation, routing, and orchestration, enabling seamless integration of diverse data sources.
- **Data integration tools**: data integration tools and platforms offer capabilities for extracting, transforming, and loading (ETL) data from diverse sources into a centralized data repository. These tools provide graphical interfaces, workflow automation, and data mapping functionalities to streamline the integration process.
- **Event-driven architecture**: event-driven architecture (EDA) enables real-time data integration and processing by reacting to events or triggers generated by diverse data sources. This approach allows for dynamic, event-driven integration of data streams from sensors, equipment, and other sources, enabling timely insights and responses to changing conditions.
- **Cloud-based integration**: cloud-based integration platforms provide scalable and flexible solutions for integrating data from diverse sources across distributed environments. These platforms offer cloud-native services, such as data lakes, data warehouses, and integration-as-a-service (iPaaS), to facilitate seamless integration and analysis of manufacturing data.
- **Semantic interoperability**: semantic interoperability involves defining common data models, ontologies, and vocabularies to ensure shared understanding and interpretation of data across different systems and domains. By establishing semantic interoperability standards, manufacturers can enable seamless integration and interoperability of diverse data sources within the manufacturing ecosystem.
- **Federated data architectures**: federated data architectures enable distributed data integration by maintaining data sovereignty and autonomy across different systems and organizations. This approach allows for decentralized data management while enabling seamless access and integration of data from diverse sources within the manufacturing ecosystem.

12.3.6 Techniques Commonly Used for Each of These Preprocessing Tasks

- **Data cleaning**
  - **Handling missing values**: *imputation* involves substituting missing values with a statistical measure, such as the mean, median, or mode. *Deletion* eliminates rows or columns containing missing values if they are not significant or if there is an excessive number of missing values. *Interpolation* involves estimating missing values by utilizing neighboring data points through techniques such as linear interpolation or spline interpolation.

- **Outlier detection and removal**: *statistical methods* detect and eliminate outliers by utilizing statistical measurements such as Z-score, standard deviation, or percentiles. *Visualization* utilizes data distributions and scatterplots to visually detect outliers and eliminate them depending on subject expertise.
- **Noise removal**: *smoothing techniques* applies smoothing filters such as moving averages or median filters to remove noise from time series or signal data. *Clustering* identifies and removes noisy data points by clustering similar data points together and removing outliers.

- **Data normalization**
  - **Min-max scaling**: normalize the data by subtracting the minimum value and dividing by the range to achieve a fixed range, such as [0, 1].
  - **Z-score standardization**: standardize the data by subtracting the mean and dividing by the standard deviation to transform it into a standard normal distribution.
  - **Robust scaling**: scale the data using robust statistics like median and interquartile range to mitigate the influence of outliers.
  - **Normalization to unit vector**: scale each feature to have a unit norm, which is useful for algorithms that rely on the magnitude of feature vectors, such as clustering algorithms or distance-based methods.
- **Feature extraction**
  - **Principal component analysis (PCA)**: diminish the dimensionality of the dataset by converting it into a fresh collection of orthogonal variables (principal components) that capture the utmost volatility in the data.
  - **Linear discriminant analysis (LDA)**: find linear combinations of features that best separate different classes or groups in the data while maximizing class separability.
  - **Feature selection**: use statistical tests (e.g. chi-square test, ANOVA) or algorithms (e.g. recursive feature elimination) to select the most relevant features based on their importance or predictive power.
  - **Text processing techniques**: *tokenization* refers to the process of dividing text data into individual tokens, which might be either words or phrases. *Stemming and lemmatization* employ techniques to transform words into their base form in order to standardize text data. The *bag-of-words* (BoW) technique involves transforming text data into numerical vectors by considering the frequency of words in documents. The *TF-IDF* (term frequency-inverse document frequency) method assigns weights to words in documents based on their frequency across texts, hence determining their relevance.
  - **Feature engineering**: generate novel functionalities by employing mathematical operations, merging existing attributes, or extracting domain-specific data.

12.4 AI-BASED ANALYTICS

AI-based analytics, also known as artificial intelligence (AI) analytics, refers to the utilization of artificial intelligence and machine learning techniques to examine large volumes of data, extract significant insights, and offer data-driven predictions or decisions. This technique employs advanced algorithms and computer models to uncover patterns, trends, and relationships within complex information, enabling organizations to get deep insights and make more informed decisions.

12.4.1 Predictive Maintenance

AI algorithms analyze data from digital twins to predict equipment failures and schedule maintenance proactively through a combination of data processing, pattern recognition, and predictive modeling techniques (Li et al., 2014). Below is a concise summary of the procedure:

- **Data collection**: the first step involves collecting data from sensors embedded in physical equipment and devices, as well as virtual sensors within the digital twin environment. This data may include operational parameters, environmental conditions, performance metrics, maintenance logs, and historical records.
- **Data integration and preprocessing**: the collected data is integrated and preprocessed to ensure consistency, completeness, and quality. This may involve tasks such as data cleaning, normalization, outlier detection, and feature engineering to prepare the data for analysis.
- **Feature extraction and selection**: the data is analyzed to identify significant attributes and patterns that are associated with the condition and functioning of the equipment. Feature selection approaches can be utilized to discover the most useful and predictive features for modeling.
- **Model training**: AI algorithms, such as machine learning and deep learning models, are trained using historical data from the digital twin to learn patterns and relationships between input features and equipment failures. Supervised learning algorithms, such as classification or regression, are commonly used for predictive maintenance tasks.
- **Failure prediction**: after the models have been trained, they are implemented to assess real-time data that is being sent from the digital twin environment. The algorithms utilize this data to generate forecasts on future equipment failures or decline in performance, relying on acquired patterns and correlations.
- **Probabilistic risk assessment**: predictive maintenance algorithms evaluate the likelihood of equipment failures happening within a specific period, including criteria such equipment condition, operating conditions, environmental factors, and past failure patterns.
- **Maintenance scheduling**: based on the predicted risk of failure, maintenance activities are scheduled proactively to address potential issues before they escalate into costly downtime or breakdowns. Maintenance scheduling algorithms optimize the timing and prioritization of maintenance tasks to minimize disruption to operations and maximize equipment availability.
- **Feedback loop and model updating**: predictive maintenance systems incorporate a feedback loop to continuously monitor equipment performance and collect new data from the digital twin environment. This data is used to update and refine the predictive models over time, improving their accuracy and effectiveness.

Manufacturers can utilize AI algorithms to examine data from digital twins, enabling them to obtain valuable insights on the condition and performance of equipment. This allows for more precise predictions of probable failures and the ability to proactively schedule maintenance tasks, resulting in improved asset reliability and productivity optimization. Implementing this proactive maintenance strategy aids in decreasing operational downtime, minimizing expenses, and prolonging the lifespan of vital equipment and infrastructure.

12.4.2 Machine Learning and Deep Learning Techniques

Predictive maintenance in smart manufacturing employs a range of machine learning (ML) and deep learning (DL) methods to anticipate equipment malfunctions and enhance maintenance timetables (Kotsiopoulos et al., 2021). Below are many frequently employed methodologies:

- **Machine learning techniques**
  - **Regression analysis**: regression models, such as linear regression or logistic regression, are used to predict continuous or categorical outcomes based on input features. They are suitable for tasks like estimating remaining useful life (RUL) or predicting the likelihood of equipment failure within a specified time frame.
  - **Classification algorithms**: classification algorithms, such as support vector machines (SVM), decision trees, random forests, and k-nearest neighbors (KNN), are used to categorize the health states of equipment (e.g. normal, degraded, failed) using sensor data and operating characteristics.
  - **Time series forecasting**: time series forecasting methods, such as autoregressive integrated moving average (ARIMA), exponential smoothing, and seasonal decomposition, are employed to anticipate future values of sensor readings or performance measures across time. This allows for the timely identification of abnormalities or deterioration patterns.
  - **Anomaly detection**: anomaly detection algorithms, such as isolation forest, one-class SVM, or autoencoders, identify unusual patterns or outliers in sensor data that may indicate equipment faults, deviations from normal operation, or impending failures.
  - **Deep learning techniques**
  - **Recurrent neural networks (RNNs)**: recurrent neural networks (RNNs), namely long short-term memory (LSTM) and gated recurrent unit (GRU), are highly efficient at analyzing sequential data. They excel in tasks involving time series sensor readings or maintenance logs, since they can effectively capture temporal dependencies. This enables them to make accurate predictions regarding future equipment states or failures.
  - **Convolutional neural networks (CNNs)**: convolutional neural networks (CNNs) are frequently employed for predictive maintenance jobs that involve analyzing images, specifically for identifying faults or abnormalities during visual inspections of equipment components. Additionally, they can be utilized to examine spatial data obtained from sensors, such as vibrations or audio waves.

- **Auto-encoders**: auto-encoders are models for unsupervised learning that aim to reconstruct input data while minimizing information loss. These methods can be employed to acquire knowledge about features, decrease the number of dimensions, and discover abnormalities by reconstructing data from sensors and detecting departures from typical patterns.
- **Generative adversarial networks (GANs)**: GANs comprise a pair of neural networks, namely a generator and a discriminator, which engage in a competitive process to produce authentic data samples. GANs are a useful tool for generating artificial sensor data that can be used to train predictive maintenance models in situations when there is a lack of or insufficient real-world data.

- **Hybrid models**
  - **Ensemble methods**: ensemble learning techniques, such as bagging, boosting, and stacking, combine multiple base models to improve predictive performance and robustness. They are used to mitigate overfitting, reduce prediction variance, and enhance the accuracy of predictive maintenance models.
  - **Hybrid approaches**: hybrid models combine physics-based modeling with data-driven ML techniques to leverage domain knowledge and capture complex interactions between equipment components. These models integrate sensor data with physical models of equipment behavior to improve predictive accuracy and interpretability.

By leveraging these ML and DL techniques, organizations can develop advanced predictive maintenance solutions that enable proactive identification of equipment failures, optimization of maintenance schedules, and maximization of asset reliability and performance in smart manufacturing environments.

12.4.3 Process Optimization

AI-based analytics are applied to digital twins in smart manufacturing to optimize processes, improve resource utilization, and enhance efficiency through data-driven insights and predictive capabilities (Ding et al., 2020). AI-based analytics are used in conjunction with digital twins to enhance their functionality and capabilities as follows:

- **Real-time monitoring and control**: AI systems process live data streams from physical equipment and devices, along with virtual sensors in the digital twin environment. These algorithms identify patterns, abnormalities, and deviations from anticipated behavior, enabling prompt intervention and remedial measures to enhance process efficiency and avoid interruptions.
- **Predictive maintenance**: AI-based predictive maintenance models analyze historical and real-time data from digital twins to forecast equipment failures and schedule maintenance proactively. By predicting potential issues before they occur, organizations can minimize unplanned downtime, reduce maintenance costs, and extend the lifespan of critical assets.

- **Process optimization**: AI algorithms analyze data from digital twins to identify inefficiencies, bottlenecks, and optimization opportunities within manufacturing processes. These insights enable organizations to streamline workflows, optimize production schedules, and improve resource allocation to maximize productivity and minimize waste.
- **Quality control and defect detection**: AI-based analytics analyze sensor data and image data from digital twins to detect defects, anomalies, and quality issues in real-time. By identifying quality deviations early in the production process, organizations can take corrective actions to reduce scrap, rework, and product defects, ensuring higher product quality and customer satisfaction.
- **Supply chain optimization**: AI algorithms analyze data from digital twins to optimize supply chain operations, including inventory management, demand forecasting, and logistics planning. By leveraging AI-based analytics, organizations can optimize inventory levels, reduce lead times, and improve supply chain resilience, ensuring timely delivery of materials and components to support manufacturing operations.
- **Energy management and sustainability**: AI-based analytics analyze energy consumption data from digital twins to identify opportunities for energy efficiency improvements and cost savings. By optimizing energy usage, organizations can reduce operational costs, minimize environmental impact, and enhance sustainability across manufacturing facilities.
- **Product lifecycle management**: AI algorithms analyze data from digital twins throughout the product lifecycle, from design and prototyping to production and maintenance. These insights enable organizations to optimize product design, improve manufacturing processes, and enhance product reliability and performance over time.

AI-based analytics applied to digital twins enable organizations to gain valuable insights into manufacturing processes, optimize resource utilization, and enhance operational efficiency, leading to improved productivity, quality, and competitiveness in the smart manufacturing landscape.

12.4.4 Optimization Algorithms

In smart manufacturing environments, optimization algorithms are used to improve various aspects of operations, such as production scheduling, resource allocation, inventory management, and logistics planning (Hu et al., 2019). Below are several frequently utilized optimization methods and their application in smart manufacturing, as seen in Figure 12.3.

- **Linear programming (LP)**: linear programming (LP) is a mathematical method employed to optimize a linear objective function while adhering to linear equality and inequality constraints. LP, or linear programming, is employed in smart manufacturing to optimize production planning, scheduling, and resource allocation. This involves establishing the most efficient production schedules,

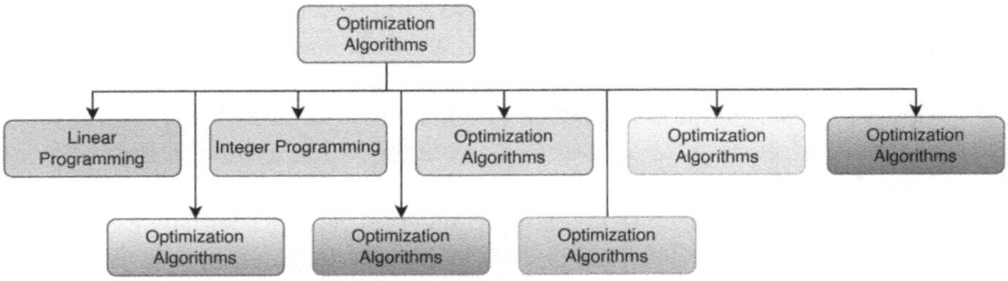

FIGURE 12.3 Optimization Algorithms.

workforce assignments, and inventory levels in order to save costs or maximize throughput.

- **Integer programming (IP)**: integer programming (IP) expands upon linear programming by allowing choice variables to be restricted to integer values. This capability enables the representation of discrete decision variables and the solution of combinatorial optimization problems. IP, or integer programming, is utilized in smart manufacturing for several activities like production scheduling with discrete work assignments, facility location and layout optimization, and equipment selection and configuration.
- **Mixed-integer linear programming (MILP)**: MILP combines elements of linear and integer programming, allowing for the optimization of models with both continuous and discrete decision variables. In smart manufacturing, MILP is applied to complex optimization problems, such as production planning with discrete production quantities, resource constraints, and setup costs.
- **Genetic algorithms (GA)**: genetic algorithm (GA) is an optimization technique that draws inspiration from the principles of natural selection and evolution. Smart manufacturing use genetic algorithms (GA) for several purposes, including production scheduling, layout optimization, and parameter tuning in machine learning models.
- **Particle swarm optimization (PSO)**: PSO is an optimization technique that is based on the collective behavior of birds flocking or fish schooling. PSO, or particle swarm optimization, is utilized in smart manufacturing to address many challenges, including optimizing facility layout, scheduling production, and optimizing the supply chain.
- **Simulated annealing (SA)**: simulated annealing (SA) is a stochastic optimization method that draws inspiration from the annealing process in metallurgy. In metallurgy, materials are heated and gradually cooled to reduce imperfections. Smart manufacturing use SA for many tasks like production scheduling, facility layout optimization, and parameter tuning in machine learning models.
- **Tabu search (TS)**: TS is a metaheuristic optimization technique that systematically searches the search space by iteratively examining solutions and keeping track of previously visited solutions to prevent redundant visits. Within the realm of intelligent manufacturing, TS is utilized to address issues such as the arrangement of production schedules, allocation of resources, and optimization of the supply chain.

- **Ant colony optimization (ACO)**: the ACO algorithm is a form of optimization that draws inspiration from the foraging behavior of ants. In this algorithm, ants communicate and locate the most efficient paths to food sources by depositing pheromone trails. ACO, or ant colony optimization, is employed in smart manufacturing for several tasks including truck routing, scheduling, and work shop optimization.
- **Reinforcement learning (RL)**: reinforcement learning (RL) is a type of machine learning in which an intelligent agent learns how to interact with its environment in order to maximize the total rewards it receives over a period of time. Reinforcement learning (RL) is utilized in smart manufacturing for many tasks like dynamic scheduling, resource allocation, and control of autonomous systems.

These optimization algorithms are used in combination with mathematical models, simulation techniques, and domain-specific constraints to address complex optimization problems and improve efficiency, productivity, and competitiveness in smart manufacturing environments.

12.5 VISUALIZATION TECHNIQUES

Visualization techniques refer to methods and tools used to represent data visually, allowing users to explore, understand, and communicate insights from complex datasets. Visualization techniques aim to transform raw data into graphical representations that are intuitive, informative, and actionable.

12.5.1 Dashboard Design

Creating effective dashboards to visualize manufacturing data involves adhering to design principles and best practices that ensure clarity, usability, and actionable insights (Jwo et al., 2021). Below are essential design considerations and optimal strategies for developing manufacturing dashboards:

- **Understand user needs**: identify the target audience for the dashboard, including operators, engineers, managers, and executives, and understand their specific information needs and goals. Tailor the dashboard design to address the requirements of different user roles, providing relevant metrics, KPIs, and visualizations that support their decision-making processes.
- **Focus on key metrics**: prioritize the display of key performance indicators (KPIs) and metrics that are critical to monitoring manufacturing operations and achieving business objectives. Avoid cluttering the dashboard with unnecessary information and focus on presenting the most relevant and actionable insights to users.
- **Use clear and intuitive visualizations**: choose appropriate visualization types (e.g. line charts, bar charts, pie charts) that effectively communicate the underlying data patterns and relationships. Ensure that visualizations are clear, concise, and easy to interpret, using labels, legends, and annotations to provide context and guidance to users.

- **Provide context and drill-down capability**: include contextual information, such as historical trends, benchmarks, and targets, to help users interpret the data and understand its significance. Enable drill-down functionality that allows users to explore detailed information and underlying data sources behind summary metrics and visualizations.
- **Ensure responsiveness and interactivity**: design dashboards that are responsive and adaptable to different screen sizes and devices, including desktops, laptops, tablets, and mobile phones. Incorporate interactive elements, such as filters, sliders, and selectors, that allow users to customize views, explore data from different perspectives, and perform ad-hoc analysis.
- **Maintain consistency and simplicity**: maintain a consistent visual style, layout, and color scheme across the dashboard to enhance usability and user experience. Avoid unnecessary complexity and visual clutter by simplifying the design, removing redundant elements, and focusing on the most important information.
- **Provide real-time updates**: incorporate real-time data updates and streaming capabilities to ensure that the dashboard reflects the latest information and changes in manufacturing operations. Implement automated data refresh mechanisms and alerts to notify users of significant events or deviations from expected performance.
- **Include contextual information**: provide contextual information, such as equipment status, production schedules, and downtime reasons, alongside performance metrics to help users understand the factors influencing manufacturing performance. Use annotations, callouts, and tooltips to provide additional context and insights within the dashboard interface.
- **Iterate and gather feedback**: continuously iterate on the dashboard design based on user feedback, usability testing, and performance evaluations to address usability issues and improve effectiveness. Solicit input from stakeholders and end-users throughout the design process to ensure that the dashboard meets their needs and expectations.

In manufacturing, key performance indicators (KPIs), data visualization techniques, and interactive dashboard features play crucial roles in monitoring and optimizing operations.

12.5.2 Key Performance Indicators (KPIs)

Key performance indicators (KPIs) are measurable parameters utilized to assess the achievement or effectiveness of an organization, department, process, project, or individual in relation to certain objectives, goals, or targets. Key performance indicators (KPIs) are crucial for tracking advancements, pinpointing locations in need of enhancement, and making well-informed choices to propel organizational triumph.

- **Overall equipment effectiveness (OEE)**: measures the performance, availability, and quality of equipment, providing insights into equipment utilization and efficiency.

- **Downtime**: tracks the amount of time equipment is offline or not in operation, helping to identify causes of downtime and prioritize maintenance efforts.
- **Production yield**: calculates the ratio of good-quality output to total output, indicating the efficiency and effectiveness of the manufacturing process.
- **Cycle time**: measures the time it takes to complete one production cycle or unit, helping to identify bottlenecks and optimize production throughput.
- **Scrap and rework rate**: tracks the percentage of defective or non-conforming products, highlighting areas for quality improvement and waste reduction.
- **Inventory levels**: monitors the amount of raw materials, work-in-progress, and finished goods in inventory, ensuring optimal inventory management and minimizing stockouts or overstock situations.
- **Energy consumption**: measures the energy usage of manufacturing operations, helping to identify opportunities for energy efficiency improvements and cost savings.
- **Maintenance costs**: tracks the cost of preventive and corrective maintenance activities, helping to assess the effectiveness of maintenance strategies and control maintenance expenses.

12.5.3 Data Visualization Techniques

Data visualization techniques are methods used to represent data visually, making complex information more accessible, understandable, and actionable.

- **Line charts**: display trends and variations in KPIs over time, such as production output, equipment downtime, or energy consumption.
- **Bar charts**: compare performance metrics across different categories, such as comparing OEE for different production lines or comparing scrap rates for different product types.
- **Pie charts**: show the distribution of a total value into its component parts, such as the proportion of production downtime attributed to different downtime reasons.
- **Heatmaps**: visualize complex data patterns and correlations, such as analyzing equipment performance across different shifts or identifying hotspots of energy consumption.
- **Gauges**: provide a visual representation of a single KPI or metric, such as displaying OEE as a percentage or equipment uptime as a fraction.
- **Stacked area charts**: illustrate the cumulative contribution of different factors to a total value, such as showing the breakdown of production downtime by downtime reason over time.
- **Scatter plots**: explore relationships between two variables, such as analyzing the correlation between equipment age and maintenance costs.
- **Histograms**: display the distribution of a continuous variable, such as analyzing the distribution of cycle times or defect sizes in production.

12.5.4 Interactive Dashboard Features

Interactive dashboards are powerful tools for data visualization and analysis, allowing users to explore and interact with data dynamically (Holjevac et al., 2020).

- **Filtering**: allow users to filter data based on specific criteria, such as date ranges, production lines, or product categories.
- **Drill-down**: enable users to drill down from summary KPIs to detailed data and underlying metrics for further analysis.
- **Hover-over tooltips**: provide additional context and details about data points or visualizations when users hover over them with their cursor.
- **Data brushing**: allow users to select and interact with data points or regions of interest in visualizations, facilitating exploratory analysis and comparison.
- **Linked visualizations**: create connections between different visualizations so that interactions in one visualization update or filter data in related visualizations.
- **Dynamic updates**: automatically refresh data and visualizations at predefined intervals to ensure users have access to the latest information.
- **Alerts and notifications**: trigger alerts or notifications when KPIs or metrics exceed predefined thresholds or when significant events occur.
- **Export and share**: Enable users to export data or visualizations in various formats (e.g. CSV, PDF) and share dashboards with colleagues or stakeholders.

12.5.5 Augmented Reality (AR) and Virtual Reality (VR)

Augmented reality (AR) and virtual reality (VR) technologies provide immersive visualization capabilities that improve the comprehension and engagement with digital twins in intelligent manufacturing settings (Büttner et al., 2017). Below are the different applications of augmented reality (AR) and virtual reality (VR) technologies for seeing digital twins in immersive environments.

- **Design and prototyping**: engineers and designers can use AR and VR to visualize digital twins of products and manufacturing facilities during the design and prototyping stages. This allows them to assess the spatial layout, ergonomics, and functionality of equipment and production processes in a virtual environment before physical implementation.
- **Training and simulation**: AR and VR simulations provide realistic training environments for operators, technicians, and maintenance personnel to familiarize themselves with equipment operation, maintenance procedures, and emergency scenarios. Users can interact with virtual replicas of equipment and practice troubleshooting procedures in a safe and controlled environment.
- **Maintenance and repair**: AR technologies overlay digital twin data, such as equipment status, sensor readings, and maintenance instructions, onto the physical environment in real-time. Maintenance technicians wearing AR glasses or headsets can visualize equipment health indicators, diagnostic information, and

step-by-step repair instructions overlaid onto the physical equipment, facilitating faster and more accurate maintenance tasks.

- **Remote assistance**: AR enables remote experts to provide real-time guidance and support to on-site personnel by overlaying annotations, instructions, and annotations onto the live video feed from AR-enabled devices. This allows experts to remotely diagnose problems, provide troubleshooting advice, and guide maintenance procedures, reducing the need for travel and minimizing downtime.
- **Data visualization and analysis**: VR environments provide immersive data visualization capabilities for exploring complex datasets and digital twin representations. Users can navigate through 3D models of manufacturing facilities, equipment layouts, and process flows, interact with data visualizations, and gain insights into operational performance and optimization opportunities.
- **Collaborative decision-making**: AR and VR technologies facilitate remote collaboration and decision-making by enabling many stakeholders to perceive and interact with digital replicas in shared virtual worlds. This promotes cross-functional collaboration, the exchange of knowledge, and the establishment of agreement among teams engaged in the design, planning, and implementation of manufacturing processes.
- **Customer engagement and sales**: VR experiences allow customers to virtually explore and interact with digital twins of products, manufacturing facilities, and production processes. This immersive visualization enhances customer engagement, facilitates product demonstrations, and supports sales and marketing efforts by providing a compelling and interactive way to showcase capabilities and features.

In smart manufacturing environments, the use of AR and VR technologies greatly improves the visualization of digital twins. This allows for immersive experiences in various areas like as design, training, maintenance, collaboration, and customer interaction. Through the utilization of these technologies, organizations can enhance efficiency, productivity, and decision-making throughout the whole product lifecycle. augmented reality (AR) and virtual reality (VR) are extensively used in training, maintenance, and design review processes across several industries such as manufacturing, construction, healthcare, and automotive (Kaplan et al., 2021; Brown et al., 2023; Nee et al., 2012). Here are many precise applications:

- **Training**
  - **Equipment operation**: AR and VR simulations can provide realistic training environments for operators to learn how to operate complex machinery, equipment, and systems. Users can practice tasks such as equipment setup, calibration, and operation in a safe and controlled virtual environment.
  - **Safety training**: AR and VR can simulate hazardous or high-risk scenarios, such as fire emergencies, chemical spills, or electrical hazards, to train

personnel on safety protocols, evacuation procedures, and emergency response techniques.
  - **Soft skills training**: VR simulations can be used to train soft skills, such as communication, teamwork, and leadership, by simulating interactive scenarios and social interactions in virtual environments.
  - **Onboarding and familiarization**: new employees can use AR and VR to familiarize themselves with the workplace environment, equipment layout, and standard operating procedures before starting their job roles, improving efficiency and reducing the learning curve.
- **Maintenance**
  - **Remote assistance**: augmented reality (AR) allows maintenance technicians to obtain immediate guidance and assistance from remote specialists by superimposing instructions, notes, and digital twin data onto the physical surroundings. This enhances the effectiveness of problem-solving, decreases the amount of time that systems are not functioning, and eliminates the necessity for in-person visits.
  - **Interactive manuals**: AR applications can provide interactive maintenance manuals and guides that overlay step-by-step instructions, diagrams, and videos onto the physical equipment, assisting technicians in performing routine maintenance tasks and repairs.
  - **Training simulations**: virtual reality simulations enable technicians to practice diagnostic procedures, equipment inspections, and repair jobs in a virtual environment, replicating maintenance scenarios and processes. This allows them to gain experience and proficiency before carrying out these duties on real equipment.
  - **Predictive maintenance visualization**: AR and VR can visualize data from digital twins and predictive maintenance models, such as equipment health indicators, sensor readings, and fault predictions, enabling maintenance technicians to monitor equipment status and plan maintenance activities proactively.
- **Design review processes**
  - **Collaborative design reviews**: VR environments enable multidisciplinary teams to collaborate and review 3D models, CAD drawings, and design prototypes in immersive virtual spaces. Participants can interact with the models, annotate design changes, and make real-time decisions in a collaborative setting.
  - **Spatial visualization**: AR and VR provide spatial visualization capabilities that allow designers and engineers to assess the scale, proportion, and spatial relationships of design elements, components, and assemblies in a virtual environment.
  - **Ergonomic analysis**: VR simulations can simulate human interactions with products and environments to evaluate ergonomics, human factors, and user experience. Designers can assess factors such as reachability, visibility, and comfort to optimize product design and usability.

- **Conceptual design exploration**: VR environments enable designers to explore multiple design concepts, configurations, and alternatives in a virtual space, facilitating rapid iteration and decision-making in the design process.

12.6 CASE STUDIES AND APPLICATIONS

A collection of relevant case studies and practical applications are presented below that demonstrate the use of data analytics and visualization in smart manufacturing, specifically employing AI-powered digital twins.

12.6.1 Case Study-1: Siemens Digital Twin for Automotive Manufacturing

Siemens employed a digital twin solution in car manufacture, generating virtual duplicates of production lines to model and enhance manufacturing processes (Annanth et al., 2021). The digital twin utilizes real-time data from sensors, IoT devices, and industrial equipment to actively monitor performance, identify anomalies, and forecast equipment malfunctions. Data analytics methodologies were utilized to examine past and current data, detect areas for improvement, and enhance operational effectiveness and quality. The visualization tools included interactive dashboards and 3D simulations for visualizing industrial activities, analyzing performance data, and optimizing production schedules..

12.6.2 Case Study-2: GE Aviation's Predix Platform for Aircraft Engines

GE Aviation created a digital twin platform named Predix specifically for aviation engines. This platform utilizes AI-driven analytics to oversee engine performance, forecast maintenance requirements, and enhance fuel efficiency (Bouzidi et al., 2020). The digital twin utilizes sensor data, flight data, maintenance records, and environmental conditions to generate virtual replicas of aircraft engines and evaluate their condition and operational state. The data analytics algorithms examined the engine data to detect any abnormalities, identify patterns of performance decline, and forecast upcoming breakdowns. This allowed for preventative maintenance measures to be implemented. The utilization of visualization techniques offers user-friendly dashboards and interactive visualizations for the purpose of monitoring engine health, analyzing performance parameters, and optimizing maintenance plans to enhance reliability and achieve cost savings.

12.6.3 Case Study-3: ABB Ability™ Digital Twin for Industrial Plants

ABB created the ABB Ability™ digital twin platform for industrial facilities, incorporating AI-driven analytics and visualization capabilities to enhance plant operations and maximize asset performance (Platenius-Mohr et al., 2019). The digital twin integrates real-time data from sensors, control systems, and production processes to generate virtual reproductions of industrial assets and systems. The operational data was examined by machine learning algorithms to detect inefficiencies, optimize process parameters, and forecast equipment breakdowns. This allowed for proactive maintenance and performance optimization. The visualization tools offered encompassed immersive 3D models, interactive dashboards, and augmented reality overlays. These techniques were employed to visually represent plant operations, track essential performance indicators, and aid plant operators and maintenance teams in making informed decisions.

12.6.4 Case Study-4: Bosch Rexroth's IoT Gateway and Analytics Solution

Bosch Rexroth developed an IoT gateway and analytics solution for smart manufacturing, which integrated AI-based analytics and visualization tools to optimize production processes and enhance equipment reliability (Madasamy et al., 2021). The IoT gateway collected data from sensors, actuators, and control systems installed on manufacturing equipment to monitor performance, detect anomalies, and predict maintenance needs. Data analytics algorithms analyzed sensor data to identify patterns, trends, and correlations related to equipment health, energy consumption, and production efficiency. Visualization tools provided real-time dashboards, trend charts, and heatmaps to visualize manufacturing data, analyze performance metrics, and optimize production workflows for improved productivity and cost savings.

12.6.5 Insights Based on the Use Cases and Industry Trends

The implementation of digital twin solutions entails the integration of diverse technologies, including IoT sensors, data analytics platforms, AI algorithms, and visualization tools, into pre-existing manufacturing processes. Adapting digital twin solutions is frequently essential to fulfill certain industry demands, manufacturing settings, and corporate goals. Successful implementation requires vital collaboration across cross-functional teams, encompassing engineers, data scientists, IT professionals, and domain specialists. Digital twins provide the continuous monitoring, analysis, and improvement of production processes, resulting in heightened productivity, less downtime, and improved resource allocation. Utilizing AI-driven analytics on digital replicas enables firms to anticipate equipment malfunctions, reduce unexpected periods of inactivity, and prolong the lifespan of assets through proactive maintenance. Visualization solutions offer user-friendly dashboards, interactive visualizations, and augmented reality overlays that empower stakeholders to make informed decisions based on data, optimize production schedules, and promptly adapt to evolving market demands. Organizations can attain substantial cost savings and operational efficiencies by optimizing processes, minimizing maintenance costs, and enhancing asset performance.

Ensuring data accuracy, consistency, and interoperability across disparate systems and data sources is a common challenge in digital twin projects. Organizations need to invest in data governance, data integration, and data quality management practices. As digital twin implementations scale to encompass larger production facilities or more complex systems, managing the scalability and complexity of the underlying technologies becomes increasingly important. Modular and flexible architectures are essential to accommodate evolving requirements. Adopting digital twin technologies often requires cultural and organizational changes to embrace data-driven decision-making, collaboration across departments, and continuous improvement initiatives. Change management strategies and employee training programs are critical for successful adoption. Protecting sensitive data, intellectual property, and proprietary information from cybersecurity threats and unauthorized access is a paramount concern in digital twin projects. Implementing robust security measures, encryption protocols, and access controls is essential to mitigate risks. Digital twin projects should be viewed as iterative processes that evolve over time through continuous improvement and feedback loops. Organizations should leverage lessons learned from previous implementations to refine their strategies, optimize performance, and achieve greater business value.

12.7 CHALLENGES AND FUTURE DIRECTIONS

Challenges and future directions in data analytics and visualization for smart manufacturing using AI-based digital twins revolve around addressing technical, organizational, and societal challenges while leveraging emerging technologies and trends (Rathore et al., 2021).

12.7.1 Challenges

Ensuring the quality, consistency, and interoperability of data from heterogeneous sources, including sensors, IoT devices, and legacy systems, remains a significant challenge. Scaling digital twin implementations to encompass larger production facilities, more complex systems, and interconnected supply chains presents scalability and complexity challenges. Ensuring the interpretability, transparency, and trustworthiness of AI models used in digital twin applications is essential for gaining user acceptance, regulatory compliance, and stakeholder trust. Protecting sensitive data, intellectual property, and proprietary information from cybersecurity threats, data breaches, and unauthorized access is a paramount concern in smart manufacturing environments. Enhancing human-machine interaction, collaboration, and decision-making in digital twin environments requires intuitive visualization tools, user-friendly interfaces, and collaborative platforms. Addressing ethical, societal, and regulatory implications of AI-based analytics and digital twin technologies, including concerns about job displacement, algorithmic bias, and data privacy. Embracing continuous innovation, adaptation, and agility to keep pace with rapid technological advancements, market dynamics, and evolving customer needs

12.7.2 Future Direction

The implementation of digital twin solutions necessitates the integration of diverse technologies, including IoT sensors, data analytics platforms, AI algorithms, and visualization tools, into pre-existing manufacturing processes. Personalization of digital twin solutions is frequently essential to fulfill certain industry demands, manufacturing settings, and company goals. Successful implementation requires collaboration among cross-functional teams, comprising engineers, data scientists, IT professionals, and domain specialists. Digital twins provide the continuous monitoring, analysis, and optimization of production processes, resulting in heightened productivity, less downtime, and improved resource use. Utilizing AI-powered analytics on digital twins facilitates preventative maintenance, enabling firms to anticipate equipment malfunctions, reduce unexpected periods of inactivity, and prolong the lifespan of assets. Visualization technologies offer user-friendly interfaces, dynamic visual representations, and augmented reality overlays that empower stakeholders to make informed decisions based on data, improve production schedules, and promptly adapt to shifting market needs. Organizations can realize substantial cost savings and operational efficiencies by streamlining processes, decreasing maintenance costs, and enhancing asset performance.

12.7.3 Emerging Trends

Emerging trends such as edge computing, federated learning, and advanced visualization techniques have the potential to revolutionize the future of manufacturing by enabling real-time data processing, distributed intelligence, and immersive analytics capabilities.

- **Edge computing**: edge computing brings computing resources closer to the data source, enabling real-time processing, analysis, and decision-making at the edge of the network. In manufacturing, edge computing allows for localized data processing and analytics within industrial equipment, sensors, and devices, reducing latency, bandwidth usage, and dependence on centralized infrastructure. Edge computing enables predictive maintenance, anomaly detection, and optimization algorithms to run directly on manufacturing equipment, leading to faster response times, improved reliability, and reduced downtime. Edge computing also supports edge AI applications, such as machine learning inference and pattern recognition, enabling intelligent automation, quality control, and adaptive manufacturing processes.
- **Federated learning**: federated learning is a decentralized machine learning approach where model training is performed collaboratively across multiple edge devices or data sources without centralized data aggregation. In manufacturing, federated learning enables collaborative model training using data from distributed sensors, machines, and production facilities while preserving data privacy and security. Manufacturers can leverage federated learning to train predictive maintenance models, quality control algorithms, and production optimization models using data from diverse sources without exposing sensitive information or proprietary knowledge. Federated learning promotes data sovereignty, compliance with data regulations, and trust among stakeholders, fostering collaboration and knowledge sharing while preserving data ownership and privacy.

- **Advanced visualization techniques**: advanced visualization techniques, such as augmented reality (AR), virtual reality (VR), and mixed reality (MR), provide immersive and interactive ways to visualize, analyze, and interact with manufacturing data and digital twins. In manufacturing, AR and VR technologies enable technicians, engineers, and operators to visualize equipment layouts, assembly instructions, and process flows in 3D space, improving comprehension, decision-making, and training effectiveness. Advanced visualization techniques enhance situational awareness, spatial understanding, and human-machine interaction in manufacturing environments, facilitating intuitive data exploration, troubleshooting, and collaborative decision-making. AR and VR applications also support remote assistance, training simulations, and design reviews, enabling distributed teams to collaborate, learn, and innovate in virtual environments, regardless of geographical location.

The convergence of edge computing, federated learning, and advanced visualization techniques holds tremendous potential to transform manufacturing operations, enabling intelligent, agile, and resilient production systems that can adapt to changing market demands, optimize resource utilization, and drive innovation in the digital age. As these technologies continue to mature and proliferate, manufacturers will increasingly leverage them to unlock new opportunities for efficiency, competitiveness, and sustainability in the global marketplace.

12.8 CONCLUSION

Data analytics and visualization in smart manufacturing using AI-based digital twins hold immense potential for revolutionizing manufacturing operations. Addressing challenges such as data integration, scalability, complexity, real-time analytics, and security will be crucial for realizing this potential. Current research results demonstrate promising applications in predictive maintenance, process optimization, quality control, and supply chain management, while future research directions include advancements in AI algorithms, edge computing, interoperability standards, cybersecurity, and human-machine collaboration.

REFERENCES

Annanth, V. K., Abinash, M., & Rao, L. B. (2021, July). Intelligent manufacturing in the context of industry 4.0: A case study of siemens industry. *Journal of Physics: Conference Series*, 1969(1), 012019.

Bouzidi, Z., Terrissa, L. S., Zerhouni, N., & Ayad, S. (2020). An efficient cloud prognostic approach for aircraft engines fleet trending. *International Journal of Computers and Applications*, *42*(5), 514–529.

Brown, C., Hicks, J., Rinaudo, C. H., & Burch, R. (2023). The use of augmented reality and virtual reality in ergonomic applications for education, aviation, and maintenance. *Ergonomics in Design*, *31*(4), 23–31.

Büttner, S., Mucha, H., Funk, M., Kosch, T., Aehnelt, M., Robert, S., & Röcker, C. (2017, June). The design space of augmented and virtual reality applications for assistive environments in manufacturing: a visual approach. In *Proceedings of the 10th International Conference on PErvasive Technologies Related to Assistive Environments* (pp. 433–440). New York, NY. https://doi.org/10.1145/3056540.3076193

Ding, H., Gao, R. X., Isaksson, A. J., Landers, R. G., Parisini, T., & Yuan, Y. (2020). State of AI-based monitoring in smart manufacturing and introduction to focused section. *IEEE/ASME Transactions on Mechatronics*, *25*(5), 2143–2154.

Friederich, J., Francis, D. P., Lazarova-Molnar, S., & Mohamed, N. (2022). A framework for data-driven digital twins of smart manufacturing systems. *Computers in Industry*, *136*, 103586.

Holjevac, M., & Jakopec, T. (2020, September). Web application dashboards as a tool for data visualization and enrichment. In *2020 43rd International Convention on Information, Communication and Electronic Technology (MIPRO)* (pp. 1740–1745). IEEE.

Hu, Y., Zhu, F., Zhang, L., Lui, Y., & Wang, Z. (2019). Scheduling of manufacturers based on chaos optimization algorithm in cloud manufacturing. *Robotics and Computer-Integrated Manufacturing*, *58*, 13–20.

Jwo, J. S., Lin, C. S., & Lee, C. H. (2021). An interactive dashboard using a virtual assistant for visualizing smart manufacturing. *Mobile Information Systems*, *2021*, 1–9.

Kaplan, A. D., Cruit, J., Endsley, M., Beers, S. M., Sawyer, B. D., & Hancock, P. A. (2021). The effects of virtual reality, augmented reality, and mixed reality as training enhancement methods: A meta-analysis. *Human factors*, *63*(4), 706–726.

Kotsiopoulos, T., Sarigiannidis, P., Ioannidis, D., & Tzovaras, D. (2021). Machine learning and deep learning in smart manufacturing: The smart grid paradigm. *Computer Science Review*, *40*, 100341.

Leng, J., Wang, D., Shen, W., Li, X., Liu, Q., & Chen, X. (2021). Digital twins-based smart manufacturing system design in Industry 4.0: A review. *Journal of manufacturing systems*, *60*, 119–137.

Li, H., Parikh, D., He, Q., Qian, B., Li, Z., Fang, D., & Hampapur, A. (2014). Improving rail network velocity: A machine learning approach to predictive maintenance. *Transportation Research Part C: Emerging Technologies*, *45*, 17–26.

Madasamy, N., Aishvarya, B., Kumar, K., & Vinith, K. (2020). Android application for accessing bosch rexroth PLC. *International Journal of Research in Engineering, Science and Management*, *3*(4), 612–613.

Mihai, S., Yaqoob, M., Hung, D. V., Davis, W., Towakel, P., Raza, M., ... & Nguyen, H. X. (2022). Digital twins: A survey on enabling technologies, challenges, trends and future prospects. *IEEE Communications Surveys & Tutorials*, *24*(4), 2255–2291.

Min, Q., Lu, Y., Liu, Z., Su, C., & Wang, B. (2019). Machine learning based digital twin framework for production optimization in petrochemical industry. *International Journal of Information Management*, *49*, 502–519.

Nee, A. Y., Ong, S. K., Chryssolouris, G., & Mourtzis, D. (2012). Augmented reality applications in design and manufacturing. *CIRP annals*, *61*(2), 657–679.

Platenius-Mohr, M., Malakuti, S., Grüner, S., & Goldschmidt, T. (2019, October). Interoperable digital twins in iiot systems by transformation of information models: A case study with asset administration shell. In *Proceedings of the 9th International Conference on the Internet of Things* (pp. 1–8). Association for Computing Machinery.

Rathore, M. M., Shah, S. A., Shukla, D., Bentafat, E., & Bakiras, S. (2021). The role of ai, machine learning, and big data in digital twinning: A systematic literature review, challenges, and opportunities. *IEEE Access*, *9*, 32030–32052.

Tyagi, A.K., Fernandez, T.F., Mishra, S., & Kumari, S. (2021) Intelligent Automation Systems at the Core of Industry 4.0. In *Intelligent Systems Design and Applications*, Abraham A., Piuri V., Gandhi N., Siarry P., Kaklauskas A., Madureira A. (eds). ISDA 2020. Advances in Intelligent Systems and Computing, vol 1351. Springer. https://doi.org/10.1007/978-3-030-71187-0_1

Tyagi, A.K., Lakshmi Priya, R, Mishra, A.K., & Balamurugan, G. (2023). Industry 5.0: Potentials, Issues, Opportunities, and Challenges for Society 5.0. In *Privacy Preservation of Genomic and Medical*, 409–32. Wiley. https://doi.org/10.1002/9781394213726.ch17

Tyagi, A.K. & Richa (2023). Digital Twin Technology: Opportunities and Challenges for Smart Era's Applications. *In Proceedings of the 2023 Fifteenth International Conference on Contemporary Computing (IC3-2023).* Association for Computing Machinery (pp. 328–336). https://doi.org/10.1145/3607947.3608015

Wang, B., Zhou, H., Li, X., Yang, G., Zheng, P., Song, C., & Wang, L. (2024). Human Digital Twin in the context of Industry 5.0. *Robotics and Computer-Integrated Manufacturing*, *85*, 102626.

Xia, D., Shi, J., Wan, K., Wan, J., Martínez-García, M., & Guan, X. (2023). Digital Twin and Artificial Intelligence for Intelligent Planning and Energy-Efficient Deployment of 6G Networks in Smart Factories. *IEEE Wireless Communications*, *30*(3), 171–179.

13 Business Analytics, Business Intelligence, and Paradigm Shift in Organizational Structure

Adeyemi Omolade S., Timilehin O. Olubiyi, Venkatesha Nayak, and Bhupendra Kumar

13.1 INTRODUCTION

The contemporary business environment is rapidly globalising and evolving, making it challenging for businesses to adapt to their customers' constantly shifting needs and preferences. This is due to the astounding advancements in technology and the continuous demand from consumers for better products and services, which has led to an increase in the number of businesses worldwide attempting to meet these expectations (Olasoji & Akpa, 2023; Olubiyi, 2022a). These companies' ability to use business intelligence to swiftly and effectively adapt to the complex dynamics of the global market will determine their level of success or even survival. Businesses may offer their clients the kind of products and services they want by using business analytics and business intelligence to guide their decision-making. Obtaining crucial company knowledge is also beneficial. Moreover, it converts this vital information into a practical framework that the company can utilise to create decisions that will enhance operations and earnings. The operations and financial performance of tech companies across the globe are significantly impacted by business analytics and intelligence (Aydiner et al., 2019). There has been much research and discussion on the use of analytics and big data in business.

Data analysis can provide a competitive advantage due to the strategic significance of data (e.g. Dunk, 2004; Maelah, Auzair, Amir, & Ahmad, 2017). The development and use of competitive strategies may be impacted by information from data analysis (Kushwaha, 2011).

Even while these tools enhance performance, organisational structure, and business processes, tech companies and other organisations still struggle to turn business analytics and business intelligence into benefits for the company (Fink et al., 2017). According to research conducted by Ukhalkar, Phursule, Gadekar, &

DOI: 10.1201/9781003480860-13

Sable (2020), businesses also have to deal with problems related to big data's nature, quality, quantity, scale, and storage. The use of business intelligence and analytics by African businesses facilitates the development of data storage systems and analytical tools. They are recognised to use more clever corporate management strategies. Chen et al. (2012) assert that the use of business intelligence is growing in organisational procedures and inefficient decision-making.

Furthermore, as Adeborna and Siau (2014) noted, business analytics highlights the analytical aspect of business intelligence. As a result, combining business analytics and business intelligence results in comparatively little difference (Chen et al., 2012). Despite the crucial role these tools play in organisational decision-making, tech organisations still encounter difficulties when integrating business analytics and business intelligence into organisational decision-making processes or organisational structures. These difficulties include the price of employing a qualified expert to supervise the data collection procedure and the procedures for implementing business analytics and business intelligence to maintain a successful, profitable, and efficient organisational structure. Business analytics and business intelligence are vital in Nigeria, particularly for companies and individuals in the sector hoping to expand internationally. Businesses that use business information can enhance and grow their organisational structure with the unique support of business analytics (Arnott, Lizama, & Song, 2017).

Web analytics, sales data, business intelligence, and data centres are just a few of the many data sources that are included in these systems. Since various data sources frequently produce separate reports, they must be integrated in order to place them all in one environment (Côrte-Real, Ruivo, & Oliveira, 2020). In order to collect data, managers interact extensively with pertinent information sources such as books, newspapers, and trade journals. Even though this is an important factor, most Nigerian IT companies still place less emphasis on business analytics and business intelligence when deciding on their organisational structures, despite these efforts (Seddon, Constantinidis, Tamm, & Dod, 2016). The system is under a lot of stress from the process of collecting data that will subsequently be utilised to address issues with plans, strategies, and organisational structure. Furthermore, the work is typically completed by skilled experts who are aware of the protocol and can gather data covertly (Ghasemaghaei, Hassanein, & Turel, 2017).

Furthermore, these experts are used to gathering as much information as they can because anything can be a source of knowledge. Most research on corporate intelligence and analytics has been conducted in developed countries (Ukhalkar et al., 2020; Bordeleau, Mosconi, & de Santa-Eulalia, 2020; Arnott et al., 2017; Moro, Cortez, & Rita, 2015; Shollo & Galliers, 2015). Few organisations have implemented business analytics and intelligence projects, especially startups, small and medium-sized enterprises, and online corporations. Because of the expenses, knowledge, and skilled labor required to carry out business analytics and business intelligence duties, most of these businesses are still reluctant to do so (Bordeleau, Mosconi, & de Santa-Eulalia, 2020; Moro, Cortez, & Rita, 2015; Shollo & Galliers, 2015). Despite the fact that business analytics and intelligence have been adopted by the majority of businesses globally, local organisations in developing nations (such as Nigeria) have not done much research in this area. This study aims to close the existing research gap by investigating how business intelligence and analytics can lead to a paradigm change in organisational structure. The purpose of this study is to investigate how paradigm shifts, business analytics, and business intelligence have affected the organisational structures of a few Nigerian IT organisations. With the help of case studies of specific tech companies in Nigeria, this study seeks to assess the relationship between business analytics and the paradigm shift in organisational structure, as well as the relationship between business intelligence and the latter.

Hypotheses

The following hypotheses were developed:

H_{01}: business analytics has no significant effect on the paradigm shift in the organisational structure of selected tech firms in Nigeria.

H_{02}: business intelligence has no significant effect on the paradigm shift in the organisational structure of selected tech firms in Nigeria.

13.2 LITERATURE

13.2.1 Business Analytics

Business analytics, also known as analytics, is described as "the application of data, quantitative methods, statistical analysis, mathematical or computer-based models, and information technology to help managers make better, fact-based decisions by improving their understanding of their business operations" (Evans, 2020). Business analytics is the process of improving business operations through the implementation of a thorough analytics plan. Business analytics is the process of figuring out the best course of action for the right individuals at the right moment.

Additionally, it comprises gathering data and transforming it such that both people and technology can support decision-making. Nowadays, practically every facet of strategic research, forecasting, and decision-making at major organisations is impacted by big data and business analytics (Griffin & Wright, 2015). Business executives are using analytics and accounting technologies more and more to aid in their decision-making. Financial analytics, forecasting, planning, budgeting, and even income generation are supported by these instruments. Businesses employ data and information to improve supply chain management, attract new consumers, and upsell current ones.

13.2.2 Business Intelligence

Business intelligence, according to Ukhalkar, Phursule, Gadekar, and Sable (2020), is the process of obtaining valuable information from a dataset to aid in decision-making and help a company maintain and expand its competitive advantage. Business intelligence is also described as a structured framework, protocols, strategies, and methodology that helps owners, managers, and decision-makers turn large amounts of data into information that will help them make well-informed decisions.

Managers, owners, and stakeholders can use business intelligence to turn information into valuable knowledge. In a similar vein, business intelligence is defined as a tool for carrying out efficient business analysis that raises an organisation's performance. Business intelligence is the act of acquiring, preserving, and structuring knowledge to make it easier to analyse complicated company data and enhance business performance.

13.2.3 Business Analytics, Business Intelligence, and Paradigm Shift in the Organisational Structure

Business analytics, often known as business intelligence, is the process of extracting meaning from data and using it to inform choices. According to a special edition of MIS Quarterly on transformative problems in big data and analytics in networked business, this has become increasingly important for organisations these days. Furthermore, assessing business intelligence and analytics efficacy is critical to an organization's comprehension of management operations and return on investment. Business and IT should work together on business analytics and intelligence projects to maximise enterprise value. Business executives are spending more money on analytics to boost efficiency. After evaluating the literature, it can be concluded that there is a lot of research on the relationship between improved organisational performance and B.I. investments, assets, and B.I. Not enough research has been done on the conditions or procedures required to link them.

Consequently, a study cannot fully explain the relationship between business intelligence, analytics, and corporate value. The two approaches to business analytics and business intelligence value creation strategic value and operational value depend on business analytics and business intelligence capabilities, according to a study by Fink et al. (2017). The use of business analytics and business intelligence technologies improves how well organisations operate. Big data can enhance decision-making and increasingly present alternative perspectives. Success with business intelligence and analytics is a complicated issue. More specifically, it depends on how well a company can use a variety of resources including data at the same time in a commercial setting. While business analytics and business intelligence capabilities are critical for improving an organization's performance, it is not always clear what and how to do this.

Theoretical review

System theory and organisational agility

According to the theory of systems, a system's special qualities are determined by the interactions between its subsystems. According to theories of systems, the universe is made up of different components, all properties have a value assigned to them at any given time, and systems possess attributes. The idea of systems has been extended by suggesting that some system attributes might be characteristics of the components, but with new values, and that other system attributes would be unique since no single component would have them. The last set of traits is referred to as emergent system attributes. Emergent characteristics are evaluated by how they interact with the constituent parts of a composition.

13.3 METHODOLOGY

13.3.1 Research Design

The research design employed in this study was a cross-sectional survey. This is so that they can provide more accurate and genuine feedback on the study's designs since it makes it easier to evaluate the ideas, views, and feelings of various populations. The cross-sectional survey study design is a more practical and cost-effective option than alternative research methodologies.

The research undertaken by Olubiyi (2019), Olasoji, and Akpa,(2023); Olubiyi, Lawal, and Adeoye (2022), Olubiyi (2022a), Olubiyi (2022b), Olubiyi, Jubril, Sojinu, and Ngari (2022), and Uwem, Oyedele, and Olubiyi (2021) also provided the foundation for the execution of this design. Additionally, the survey approach will enable the researcher to examine and clarify respondent comments to ascertain the relationship between the variables, making it suitable for this analysis.

13.3.2 Area of Study

This study was carried out at a few tech companies in Nigeria in addition to the technology sector as a whole. These tech companies were chosen for their beneficial contributions to the expansion and advancement of the national economy. Concurrently, these IT companies have been running their online operations.

13.3.3 Population of the Study

The target audience for the study is 1200 employees of specific IT companies in Nigeria. Based on their market share, customers, and worth in Nigeria, the chosen companies were determined. The chosen tech companies were sized using the purposive sample technique to create a limited population. The sampling formula is used to create the following statistical example:

$$n = \frac{N}{1 + N\left(e^2\right)}$$

where,

N = known population

e = error level or % percent confidence interval or alpha level. For 0.95 confidence interval, e = 0.05.

$$n = \frac{600}{1 + 600(0.05)^2}$$

$$n = 240$$

The distribution of the copies of questionnaire to respondents was arrived at as shown below:

TABLE 13.1
Distribution of Selected Tech Firms

S/N	Tech Firms	Number of Employees	Number of Administered Questionnaire
1	Jumia	145	145×240/600 = 58
2	Paystack	88	88×240/600 = 35
3	Interswitch	105	105×240/600 = 42
4	Flutterwave	120	120×240/600 = 48
5	Opay	74	74×240/600 = 30
6	Dataflex	68	68v240/600 = 27
		600	**240**

Source: Researchers Survey, 2024.

where,

$$nh = \frac{NHXn}{N}$$

nh = sample size for each tech firm picked
n = sample size
N = total population

Table 13.1 shows the distribution of selected tech firms.

13.3.4 Approach to Data Analysis

Using the mean and standard deviation to illustrate the differences in responses and opinions, the descriptive analysis of the data in this study was conducted using SPSS. Descriptive statistics allow the researcher to characterise the properties of the variables under investigation by covering the dependent and independent variable items. Furthermore, the influence and depth of regression were ascertained by regression analysis.

13.3.5 Method of Data Analysis

Table 13.2 shows the hypotheses and tools for the data analysis.

13.3.6 Model Specification

$$y = \beta 0 + \beta 1 x 1 + \beta 2 x 2 + \ldots \beta k x k + \mu$$

where,
y is the dependent variable and x is the independent variable.

TABLE 13.2
Hypotheses and Tools for the Data Analysis

	Hypothesis	Tools of Analysis
$H_0 1$	Business analytics has no significant effect on the paradigm shift in the organizational structure of selected tech firms in Nigeria	Multiple Linear Regression
$H_0 2$	Business intelligence has no significant effect on the paradigm shift in the organizational structure of selected tech firms in Nigeria	Multiple Linear Regression

Source: Researchers computation, 2024.

TABLE 13.3
Models Apriori Expectations

	Models	Apriori Expectations IF:
$H_0 1$	OS=(BA,BI)------------(i)	P≤0.05; H_{01} will be rejected
$H_0 2$	OS=(BA,BI)------------(ii)	P≤0.05;H_{01} will be rejected

Source: Researchers' computation, 2024.

In the models:
y = paradigm shift in the organizational structure (O.S.)
x = business analytics, business intelligence (BABI)
B.A. = business analytics
B.I. = business intelligence

13.3.7 Apriori Expectations

Based on selected tech businesses in Nigeria, the study examined the effects of business analytics, business intelligence, and a paradigm shift in organisational structure. However, it was discovered that business analytics and business intelligence have a favourable influence on the paradigm change in the organisational structure of the selected tech-based enterprises in Nigeria.

Table 13.3 shows the models apriori expectations.

13.3.8 Research Instrument

To gather data from respondents, a structured five-point Likert scale questionnaire was employed in this study. The responses ranged from Strongly Agree (5), Agree (4), Undecided (3), Disagree (2), and Strongly Disagree (1).

TABLE 13.4
Analysis of Questionnaire

	Frequency	Percentage
Retrieved	240	80
Not retrieved	60	20
Total	**300**	**100**

Source: Researchers' survey, 2024.

13.4 DATA ANALYSIS

Table 13.4 shows that 60 (20%) of the respondents did not complete and return the questionnaire, whereas 240 (80%) did. As a result, the proportion of completed questionnaires is sufficient to allow the study project's findings to be broadly applied.

Table 13.5 shows the demographic analysis of the respondents, the result shows that there are 150 (62.5%) men and 90 (37.5%) women. 144 (60.0%) of the respondents are single, and 96 (40.0%) of the respondents are married. The age distribution of the responders shows that 48 (20.0%) are younger than 25. 53 (22.1%) of the respondents are in the 25–29 age range; 46 (19.2%) are in the 30–35 age range; 61 (25.4%) are in the 36–40 age range; and 32 (13.3%) are above 40. Based on the educational backgrounds of the respondents, 21.3% had obtained their Senior School Certificate (SSCE), 37.9% had obtained a Bachelor's degree or Higher National Diploma, 28.3% had obtained a Master's degree, and 12.5% had obtained a PhD. With respect to the length of time spent at work, 18 (7.5%) of the participants had been employed for less than a year, 68 (28.3%) for 1–3 years, 73 (30.4%) for 4–6 years, 43 (18.0%) for 7–9 years, and 38 (15.8%) for more than ten years.

13.4.1 Regression Analysis

The study's findings showed that neither business analytics nor business intelligence ($\beta = 0.503$, $t = 8.937$, $p<0.05$) had a significant effect on the paradigm shift in the organisational structure of a group of Nigerian IT enterprises. The adjusted R square of 0.459 (Table 13.6) indicates that business intelligence and analytics together only explained 45.9% of the variation in the paradigm shift in organisational structure. This implies that 45.9% of the changes in a paradigm shift in organisational structure may be attributable to business analytics, business intelligence, and the paradigm shift in the organisational structure of certain tech businesses in Nigeria.

The $F = 102.553$ p-value (0.000) $p<0.05$ (Table 13.7) indicates that the model was statistically significant overall based on the results. The regression model indicates that if business analytics and business intelligence were held equal at zero, the paradigm shift in organisational structure would be 0.298 (Table 13.8). This implies that the paradigm shift would have been 0.298 in the absence of these notions. These

TABLE 13.5
Demographic Data of Respondents

Variables	Frequency	Percentage
	Gender	
Male	150	62.5
Female	90	37.5
Total	**240**	**100**
Marital Status		
Single	144	60.0
Married	96	40.0
Total	**240**	**100**
	Age	
Less than 25years	48	20.0
25-29	53	22.1
30-35	46	19.2
36-40	61	25.4
Above 40years	32	13.3
Total	**240**	**100**
	Educational Qualification	
SSCE	51	21.3
BSc/HND	91	37.9
MSc	68	28.3
PhD	30	12.5
Total	**240**	**100**
	Length of Employment	
Less than a year	18	7.5
1-3years	68	28.3
4-6years	73	30.4
7-9years	43	18.0
above10years	38	15.8
Total	**240**	**100**

Source: Researchers' survey, 2024.

TABLE 13.6
Model Summary

Model	R	R Square	Adjusted R Square	Std. Error of the Estimate
1	0.681[a]	0.464	0.459	0.59415

Note:
[a] Predictors (Constant), Business Analytics, Business Intelligence.

Source: Researchers' survey, 2024

TABLE 13.7
ANOVA[a]

Model	Sum of Squares	Df	Mean Square	F	Sig.
Regression	72.405	2	36.203	102.553	0.000[b]
1 Residual	83.664	237	0.353		
Total	156.069	239			

Source: Researchers' survey, 2024.

a. Dependent Variable: Paradigm shift in Organizational Structure.
b. Predictors (Constant), Business Analytics, Business Intelligence.

TABLE 13.8
Coefficients[a]

	Unstandardized Coefficients		Standardized Coefficients		
Model	B	Std.Error	Beta	T	Sig.
(Constant)	0.298	0.222		1.341	0.181
Business Analytics	0.521	0.058	0.503	8.937	0.000
Business Intelligence	0.318	0.068	0.264	4.690	0.000

Source: Researchers' survey, 2024.

a. Dependent Variable: Paradigm shift in Organizational Structure.

results lead to the rejection of the null hypotheses H_{01} and H_{02}, which argue that business analytics and business intelligence, respectively, have no discernible influence on the paradigm change in the organisational structure of specific IT companies in Nigeria.

13.4.2 Discussion

There is a substantial influence, according to the results of multiple linear regression analysis, between the organisational structure of particular tech enterprises in Nigeria and the effects of business analytics, business intelligence, and paradigm shift. This study's findings demonstrated a paradigm shift in business intelligence and analytics as well as in the organisational design of particular Nigerian IT companies.

The findings of this study are consistent with other research (Bordeleau, Mosconi, & de Santa-Eulalia, 2020; Moro, Cortez, & Rita, 2015; Shollo, & Galliers, 2015) on the impact of business analytics, business intelligence, and paradigm shifts on the organisational structures of particular tech companies in Nigeria. Conceptually recorded research indicates that a paradigm shift, business analytics, and business intelligence have affected the organisational structure of certain IT companies in Nigeria (Chen et al., 2012).

The impact of business intelligence, analytics, and paradigm shifts on particular tech companies in Nigeria's organisational structure has been the subject of numerous studies (Božič, & Dimovski, 2019). Even so, it is well recognised that paradigm shifts in the organisational structures of several IT firms in Nigeria are linked to business analytics and intelligence.

Fink, Yogev, and Even (2017) claim that the study did not demonstrate the extent to which business analytics, business intelligence, or a paradigm shift altered the organisational structure of particular IT companies in Nigeria. The authors argue that, although they may not agree with certain other authors on the relationship between profitability and HRM techniques, profitability can be a reliable indicator of performance level. In light of this, the results of this study, business analytics, business intelligence, and paradigm change in the organisational structure of several tech companies in Nigeria are supported by conceptual, empirical, and theoretical presentations in earlier literature.

13.4.3 Implication

This study makes multiple contributions to the body of knowledge. First, the influence of business analytics, business intelligence, and a paradigm shift in organisational structure is evaluated from the perspective of specific technology businesses (studies of selected tech companies in Nigeria). The literature reviews have contributed to the research study's understanding of the impact of business analytics, business intelligence, and paradigm shifts on the organisational structures of particular tech companies in Nigeria, and the study's conclusions align with the theoretical framework of the investigation. The investigation's foundation is the evaluation of relevant literature.

However, this research contributes to the corpus of knowledge and literature in the fields of business intelligence, business analytics, and paradigm change in the organisational structure of multiple Nigerian IT businesses. Furthermore, this work offers a strong foundation for further research in this field. The results might be applied in the actual world by managers, stakeholders, decision-makers, and academics. Additionally, this research advances the understanding of tech-based businesses with respect to the role that analytics and business intelligence play in guaranteeing the paradigm shift in the organisational structure of particular tech companies in Nigeria.

13.5 CONCLUSION AND RECOMMENDATION

In this research, the organisational structures of several IT companies in Nigeria were studied in relation to paradigm shifts, business analytics, and business intelligence. Business analytics and intelligence were the two independent factors in the study, and a shift in the paradigm of organisational structure was the dependent variable. Descriptive and multiple regression analyses were used to meet the study objectives by testing the hypotheses and evaluating the statistical connection between each independent variable and the dependent variables.

The results were presented in a way that addressed the research questions. After accounting for all pertinent variables, the study's conclusions demonstrated that business intelligence and analytics had a major role in the paradigm shift in the organisational structure of the chosen tech-based businesses in Nigeria. Consequently, the paradigm shift of the chosen Nigerian IT businesses is greatly influenced by business analytics and intelligence.

The study's findings lead to the recommendation that tech-based businesses in Nigeria should make use of the proxies' business analytics and business intelligence to maintain an organisational structure paradigm shift and succeed in the business sector. Tech companies in particular ought to consider novel concepts and guidelines that might assist in resolving organisational structure problems. To promote a paradigm shift in the structure of organisations and boost profitability and productivity in the business sector, they must also set organisational rules. The study suggests that to maintain a paradigm change in the organisational structure of these organisations, managers, lawmakers, and executives of particular tech companies need to find ways to remain involved while taking business intelligence and analytics into consideration.

REFERENCES

Adeborna, E., & Siau, K. (2014). An Approach to Sentiment Analysis – The Case of Airline Quality Rating. In *Proceedings of the Pacific Asia Conference on Information Systems (PACIS 2014)*. Academic Press.

Arnott, D., Lizama, F., & Song, Y. (2017). Patterns of business intelligence systems use in organizations, *Decision Support Systems,* 97, 58–68.

Aydiner, A. S., Tatoglu, E., & Bayraktar E, (2019). Business analytics and firm performance: the mediating role of business process performance. *Journal of Business Research 96*, 228–237.

Bordeleau, F. E., Mosconi, E., & de Santa-Eulalia, L. A. (2020). Business intelligence and analytics value creation in industry 4.0: a multiple case study in manufacturing medium enterprises, *Production Planning and Control, 31* (2/3), 173–185,

Božič, K., & Dimovski, V. (2019). Business intelligence and analytics for value creation: the role of absorptive capacity, *International Journal of Information Management 46*, 93–103.

Chen, H., Chiang, R. H. L., & Storey, V. C. (2012). Business intelligence and analytics: From big data to big impact. *Management Information Systems Quarterly, 36*(4), 1165–1188.

Côrte-Real, N., Ruivo, P., & Oliveira, T. (2020). Leveraging internet of things and big data analytics initiatives in European and American firms: is data quality a way to extract business value?, *Information and Management, 57* (1), 103141.

Dunk, A.S. (2004). Product life cycle cost analysis: The impact of customer profiling, competitive advantage, and quality of I.S. information. *Management Accounting Research, 15*, 401–414. doi:10.1016/j.mar.2004.04.00

Evans, J. R. (2020). Business Analytics: Methods, Models and Decisions, 3rd Edition, pg. 4, Pearson Education, Inc.

Fink. L., Yogev, N., & Even, A. (2017). Business intelligence and organizational learning: an empirical investigation of value creation processes, *Information & Management 54*(1), 38–56.

Ghasemaghaei, M., Hassanein, K., & Turel, O. (2017). Increasing firm agility through the use of data analytics: The role of fit. *Decision Support Systems, 101*, 95–105.

Griffin, P.A. & Wright, A.M. (2015). Commentaries on Big Data's Importance for Accounting and Auditing, Accounting Horizons, 29(2), 377–380.

Kushwaha, G.S. (2011). Competitive advantage through information and communication technology enabled supply chain management practices. *International Journal of Enterprise Computing and Business Systems, 1*, 1–13.

Maelah, R., Auzair, S., Amir, A., & Ahmad, A. (2017). Implementation process and lessons learned in the determination of educational cost using modified activity-based costing (ABC). *Social and Management Research Journal, 14*, 1–32. doi:10.24191/smrj.v14i1.5277

Moro, S.S., Cortez, P., & Rita, P. (2015). Business intelligence in banking: a literature analysis from 2002 to 2013 using text mining and latent Dirichlet allocation, *Expert Systems with Applications,* 42 (3), 1314–1324.

Olasoji, O. T.& Akpa,V. O.(2023). Does innovative capability really matter for the profitability of consumer goods companies in developing economies? *Sun Text Review of Economics & Business,4*(3) 191–204

Olubiyi, T.O., (2022a). Measuring technological capability and business performance post-COVID Era: Evidence from Small and Medium-Sized Enterprises (SMEs) in Nigeria, *Management & Marketing Journal*, vol xx (2),234- 248

Olubiyi, T. O. (2022b). An investigation of sustainable innovative strategy and customer satisfaction in Small and Medium-Sized Enterprises (SMEs) in Nigeria. *Covenant Journal of Business and Social Sciences, 13*(2).Retrieved from https://journals.covenantuniversity.edu.ng/index.php/cjbss/article/view/3274

Olubiyi, T. O. (2019). Knowledge management practices and family business profitability: Evidence from Lagos State, Nigeria. *Global Journal of Management and Business Research*, 19(A11), 21–31. Retrieved from https://journalofbusiness.org/index.php/GJMBR/article/view/2872

Olubiyi, O.T., Lawal, A, T., & Adeoye, O.O. (2022). Succession planning and family business continuity: Perspectives from Lagos State, Nigeria. Organization and Human Capital Development, 1(1), 40–52. https://doi.org/10.31098/orcadev.v1i1.865

Olubiyi, T.O., Jubril, B., Sojinu, O.S., & Ngari, R. (2022). Strengthening gender equality in small business and achieving sustainable development goals (SDGs), Comparative analysis of Kenya and Nigeria, *Sawala Jurnal Administrasi Negara, 10*(2) 168–186.

Seddon, P. B., Constantinidis, D., Tamm, T., & Dod, H. (2016). How does business analytics contribute to business value? *Information Systems Journal, 27* (3), 237–269.

Shollo, A., & Galliers, R. D. (2015). Towards an understanding of the role of business intelligence systems in organizational knowing, *Information Systems Journal, 26* (4), 339–367.

Ukhalkar, P. K., Phursule, R. N., Gadekar, D. P., & Sable, N. P. (2020). Business Intelligence and Analytics: Challenges and Opportunities, *International Journal of Advanced Science and Technology 29*, (12s), 2669–2676

Uwem, E. I., Oyedele, O. O., & Olubiyi, O. T. (2021). Workplace Green Behavior for Sustainable Competitive Advantage. In *Human Resource Management Practices for Promoting Sustainability*. IGI Global (248–263).

14 Applications of Human Computer Interaction, Explainable Artificial Intelligence and Conversational Artificial Intelligence in Real-life Sectors

Amit Kumar Tyagi

14.1 INTRODUCTION

Human-computer interaction (HCI) is a multidisciplinary field that focuses on the design, evaluation, and implementation of interactive computing systems for human use [1]. It involves studying how people interact with technology and designing interfaces that are intuitive, efficient, and satisfying for users. HCI encompasses a wide range of topics and approaches, including user interface design, usability evaluation, interaction techniques, information visualization, user experience (UX) design, and cognitive psychology. The goal of HCI is to create technology that is usable, effective, and enjoyable for people to use.

Key concepts in HCI:

- **User-centered design**: HCI emphasizes the importance of involving users throughout the design process. User-centered design approaches involve understanding user needs, preferences, and behaviors to create interfaces that meet their requirements.
- **Usability**: usability refers to the ease of use and effectiveness of a system. HCI researchers and designers employ various methods, such as usability testing and heuristic evaluation, to assess and improve the usability of interfaces.
- **Interaction techniques**: HCI discusss different interaction techniques that allow users to interact with technology, including graphical user interfaces (GUIs), touchscreens, voice recognition, gesture-based input, virtual reality (VR), and augmented reality (AR).

DOI: 10.1201/9781003480860-14

- **Information visualization**: HCI researchers investigate techniques for visually representing complex data and information to enhance comprehension and decision-making. This involves designing effective visualizations and interaction techniques for exploring and manipulating data.
- **User experience (UX)**: UX focuses on the overall experience and satisfaction that users have when interacting with a system. It encompasses various factors, such as aesthetics, ease of use, efficiency, learnability, and emotional aspects.
- **Accessibility**: HCI considers the needs of diverse user groups, including those with disabilities. It aims to design interfaces that are accessible and inclusive, ensuring that everyone can effectively use and benefit from technology.

Methods and techniques in HCI:

- **User research**: HCI researchers employ various methods, such as interviews, surveys, and observations, to understand user needs, behaviors, and preferences.
- **Prototyping**: creating prototypes allows designers to test and refine interface designs before full-scale development. Prototypes can range from low-fidelity paper sketches to high-fidelity interactive mockups.
- **Usability testing**: usability testing involves observing users as they interact with a system to identify usability issues and gather feedback. This method helps designers understand how well the interface supports user tasks and identify areas for improvement.
- **Heuristic evaluation**: experts evaluate an interface based on a set of usability heuristics or principles. This method helps identify potential usability problems without involving actual users.
- **User interface design**: HCI incorporates principles of visual design, information architecture, and interaction design to create interfaces that are visually appealing, organized, and easy to navigate.

Applications of HCI

HCI principles and techniques are applied in various domains [2, 3], including:

- **Software development**: HCI is essential in designing user-friendly software applications, websites, and mobile apps.
- **Human-robot interaction**: HCI plays an important role in designing interfaces and interactions for robots and autonomous systems to improve their usability and safety.
- **Health care**: HCI is used to design user-friendly interfaces for medical devices, electronic health records, and healthcare systems.
- **Education**: HCI is applied in the design of educational software, interactive learning environments, and educational games.
- **Entertainment**: HCI contributes to the design of video games, virtual reality experiences, and interactive media.
- **Smart devices and IoT**: HCI principles are used to design interfaces for smart devices, home automation systems, and internet of things (IoT) applications.

In summary, human-computer interaction focuses on understanding human needs, behaviors, and preferences to design interfaces and systems that are user-friendly, efficient, and enjoyable to use. By applying HCI principles, researchers and designers strive to create technology that enhances the overall user experience and

14.1.1 Explainable Artificial Intelligence

Explainable artificial intelligence (XAI) refers to the development of AI systems that can provide understandable and transparent explanations for their decisions or actions [4]. The goal of XAI is to bridge the gap between the "black box" nature of many AI algorithms and the need for humans to understand and trust the decisions made by these systems. In traditional AI approaches, such as deep learning and neural networks, the decision-making process can be complex and opaque. These models learn patterns and make predictions based on large amounts of data, but understanding how they arrive at a specific decision can be challenging. This lack of transparency raises concerns, especially in important domains like healthcare, finance, and legal systems, where the ability to explain AI decisions is important for accountability, fairness, and trust. XAI aims to address these concerns by developing AI models and techniques that are not only accurate but also interpretable. It involves designing algorithms and methods that can provide insights into how an AI system arrived at a particular decision, including the underlying factors, features, or rules used in the decision-making process. This transparency allows humans to understand and validate the system's reasoning and make informed judgments about its outputs.

Methods and Techniques in XAI:

- **Rule-based explanations**: rule-based approaches provide explanations based on a set of predefined rules or logical statements. These rules can be explicitly defined by experts or learned from data, making the decision-making process more transparent and understandable.
- **Feature importance**: this approach aims to identify the features or input variables that most strongly influence an AI system's decision. Techniques such as feature attribution, sensitivity analysis, and feature relevance ranking help identify the key factors driving the system's output.
- **Local explanations**: local explanation methods focus on providing explanations for individual predictions or decisions made by an AI system. By examining specific instances or cases, these methods aim to shed light on how the model arrived at its output for a particular input.
- **Model visualization**: model visualization techniques present AI models in a more intuitive and understandable way. This includes visualizing the structure of neural networks, decision trees, or other models, allowing users to comprehend the underlying decision-making process.
- **Natural language explanations**: generating explanations in natural language helps users understand AI decisions in a more human-readable form. This involves translating the model's reasoning into understandable sentences or narratives that convey the decision process.

- **Interactive explanations**: interactive approaches involve enabling users to actively discuss and interact with AI systems to gain insights and explanations. Users can query the system, request additional information, or manipulate input variables to observe the impact on the model's decisions.

Applications of XAI

XAI has applications across various domains [5, 6] where interpretability, transparency, and accountability are important:

- **Healthcare**: XAI can help medical professionals understand and trust AI systems used in diagnosis, treatment recommendation, and patient monitoring.
- **Finance**: XAI can provide explanations for credit scoring, fraud detection, and investment decisions, ensuring fairness and regulatory compliance.
- **Autonomous vehicles**: XAI enables drivers and regulators to understand and validate the decisions made by self-driving cars, enhancing safety and public acceptance.
- **Legal systems**: XAI can support legal professionals in understanding the reasoning behind AI-assisted decisions, such as legal document analysis or predicting case outcomes.
- **Customer service**: XAI can provide explanations for AI chatbots or virtual assistants, helping users understand why certain responses or recommendations are provided.
- **Public sector**: XAI can assist in explaining government AI systems used in areas like public safety, social services, and policy-making.

In summary, Explainable Artificial Intelligence focuses on developing AI systems that can provide transparent and understandable explanations for their decisions. By enabling humans to comprehend and trust AI outputs, XAI aims to enhance accountability, fairness, and ethical decision-making in various domains.

14.1.2 Conversational AI

Conversational AI refers to the use of artificial intelligence techniques [7, 8] to enable computers or machines to engage in natural and human-like conversations with users. It involves the development of chatbots, virtual assistants, and other conversational interfaces that can understand user input, generate appropriate responses, and simulate human-like conversation.

Key components of conversational AI:

- **Natural language understanding (NLU)**: NLU is the ability of a system to understand and interpret user input in natural language. It involves techniques such as language parsing, entity recognition, and intent classification to extract meaning from user queries or statements.
- **Natural language generation (NLG)**: NLG involves generating human-like responses or output in natural language. It includes techniques for text

generation, language modeling, and dialogue generation to create coherent and contextually relevant responses.

- **Dialogue management**: dialogue management focuses on the flow and context of a conversation. It involves maintaining the state of the conversation, tracking user intents and preferences, and determining the appropriate system response based on the dialogue history.
- **Machine learning and deep learning**: machine learning and deep learning techniques play a important role in Conversational AI. They enable systems to learn from large amounts of conversational data and improve their performance over time. Reinforcement learning is often used to train conversational agents through interactions with users.
- **Context awareness**: context awareness is essential in conversational AI to understand and respond appropriately to user queries. It involves considering the user's previous statements, the current conversation history, and the wider context to provide relevant and accurate responses.
- **Multimodal interaction**: conversational AI is expanding beyond text-based interactions to include voice, visuals, and other modalities. Multimodal interfaces allow users to interact with AI systems through speech, images, gestures, or a combination of these modalities, providing more natural and engaging conversations.

Applications of conversational AI

Conversational AI has a wide range of applications [9] across industries and domains, including:

- **Customer support**: chatbots and virtual assistants are used to provide automated customer support, answer frequently asked questions, and assist customers with their queries and issues.
- **E-commerce**: conversational AI is employed to enhance the shopping experience by providing personalized product recommendations, assisting with product search, and processing orders.
- **Healthcare**: virtual assistants can assist in medical appointment scheduling, provide basic medical information, and provide symptom triage to patients.
- **Banking and finance**: conversational AI is used for account inquiries, transaction history, money transfers, and personalized financial advice.
- **Travel and hospitality**: virtual assistants can assist with travel bookings, provide information about destinations, suggest travel itineraries, and provide customer support for travel-related queries.
- **Education**: conversational AI is employed in intelligent tutoring systems, language learning platforms, and educational chatbots to provide personalized learning experiences and assistance.
- **Smart homes and IoT**: conversational AI interfaces are used to control smart home devices, answer queries, and provide information or assistance within the connected home ecosystem.

- **Entertainment**: chatbots and virtual assistants can provide personalized recommendations for movies, music, and books, provide trivia or game experiences, and engage in interactive storytelling.

Conversational AI is continually evolving, driven by advancements in natural language processing, machine learning, and user experience design. As these technologies advance, conversational interfaces are becoming more sophisticated, natural, and capable of understanding and responding to users in increasingly human-like ways.

14.1.3 Emotional Intelligence

Emotional intelligence (EI) refers to the ability to recognize, understand, manage, and express emotions in oneself and others. It involves being aware of and responsive to emotions, both in oneself and in others, and using this awareness to navigate social interactions, build relationships, and make decisions.
Components of emotional intelligence:

- **Self-awareness**: self-awareness is the ability to recognize and understand one's own emotions, strengths, weaknesses, values, and motivations. It involves being in tune with one's emotional state and having a clear understanding of how emotions influence thoughts and behavior.
- **Self-regulation**: self-regulation is the ability to manage and control one's emotions, impulses, and reactions. It involves being able to adapt to changing circumstances, maintain composure in stressful situations, and effectively handle challenges and setbacks.
- **Social awareness**: social awareness involves being attuned to and understanding the emotions, needs, and perspectives of others. It includes empathy, the ability to understand and share the feelings of others, as well as the capacity to accurately perceive and interpret nonverbal cues and social dynamics.
- **Relationship management**: Relationship management refers to the ability to establish and maintain healthy and meaningful relationships with others. It involves effective communication, conflict resolution, teamwork, collaboration, and the ability to inspire and influence others positively.

Benefits of emotional intelligence:

- **Improved interpersonal relationships**: emotional intelligence enhances communication, empathy, and understanding in relationships, leading to more effective and fulfilling interactions with others.
- **Effective leadership**: leaders with high emotional intelligence can inspire and motivate their teams, manage conflicts, and build strong relationships, resulting in improved team performance and employee satisfaction.

- **Stress management**: emotional intelligence enables individuals to effectively manage stress, handle pressure, and maintain emotional well-being in challenging situations.
- **Enhanced decision-making**: emotional intelligence helps individuals consider and integrate emotional factors into their decision-making process, leading to more informed and balanced decisions.
- **Improved conflict resolution**: individuals with high emotional intelligence are better equipped to manage conflicts, find common ground, and negotiate win-win solutions, fostering healthier and more productive relationships.
- **Greater emotional resilience**: emotional intelligence enhances one's ability to bounce back from setbacks, adapt to change, and maintain a positive outlook, contributing to emotional resilience and well-being.
- **Developing emotional intelligence**: emotional intelligence can be developed and improved through self-reflection, practice, and learning. Strategies for developing emotional intelligence include:
  - Increasing self-awareness through self-reflection, journaling, and seeking feedback from others.
  - Practicing mindfulness and self-regulation techniques to manage emotions effectively.
  - Cultivating empathy by actively listening, considering others' perspectives, and seeking to understand their emotions and experiences.
  - Enhancing social awareness through observing and interpreting nonverbal cues and practicing active observation in social interactions.
  - Building relationship management skills by improving communication, conflict resolution, and collaborative problem-solving.

In summary, emotional intelligence plays a important role in personal and professional success. It enhances self-awareness, self-regulation, social awareness, and relationship management skills, leading to improved interpersonal relationships, effective leadership, and emotional well-being.

14.1.4 Robotics

Robotics is a multidisciplinary field that combines computer science, engineering, and other disciplines to design, develop, and operate robots [10]. Robots are autonomous or semi-autonomous machines that can perform tasks with varying levels of complexity and sophistication. Robotics aims to create intelligent machines that can interact with the physical world, manipulate objects, and perform tasks that were traditionally carried out by humans.
Components of robotics:

- **Robot hardware**: the physical components of a robot, including the mechanical structure, actuators (such as motors and servos), sensors, and power systems. The hardware enables the robot to move, interact with the environment, and perceive its surroundings.

- **Robot software**: the software systems that control the behavior and operation of the robot. This includes the algorithms for perception, planning, and control, as well as the user interfaces and communication systems.
- **Perception**: perception refers to the ability of a robot to sense and understand its environment. Robots use various sensors, such as cameras, LIDAR (light detection and ranging), ultrasonic sensors, and tactile sensors, to gather data about the surrounding world. Perception algorithms analyze and interpret this data to create a representation of the environment.
- **Planning and control**: planning and control involve the algorithms and techniques that allow a robot to make decisions and execute actions. Planning algorithms generate sequences of actions to achieve a specific goal, while control algorithms determine the precise movements and actions of the robot's actuators.
- **Human-robot interaction**: human-robot interaction focuses on designing interfaces and interaction methods that enable effective communication and collaboration between humans and robots. This includes natural language interfaces, gesture recognition, and augmented reality systems.

Applications of robotics

Robotics has diverse applications in various industries and domains, including:

- **Manufacturing**: industrial robots are extensively used in manufacturing processes for tasks such as assembly, welding, painting, and material handling, increasing efficiency and precision.
- **Healthcare**: robots are employed in surgery, rehabilitation, patient assistance, and diagnostics, enabling minimally invasive procedures, precise movements, and enhanced patient care.
- **Agriculture**: agricultural robots perform tasks such as planting, harvesting, weeding, and monitoring crop health, improving efficiency, reducing labor costs, and optimizing agricultural practices.
- **Logistics and warehousing**: robots are used for automated material handling, order picking, sorting, and inventory management in warehouses and distribution centers, increasing productivity and reducing errors.
- **Space exploration**: robots are important for space exploration missions, performing tasks such as planetary exploration, satellite maintenance, and extraterrestrial research.
- **Search and rescue**: robots can be deployed in hazardous or inaccessible environments for search and rescue operations, including disaster response and exploring dangerous terrains.
- **Education and research**: robots are used in educational settings to teach programming, engineering concepts, and problem-solving skills. They also serve as research platforms for studying human-robot interaction and developing new robotics technologies.
- **Entertainment and service**: robots are employed in entertainment venues, theme parks, and interactive exhibits, as well as in service industries, such as hotels, restaurants, and customer service centers.

The field of robotics continues to advance rapidly, driven by developments in artificial intelligence, machine learning, and sensor technologies. As robots become more capable, versatile, and intelligent, they are increasingly integrated into various aspects of our lives, contributing to automation, efficiency, and new possibilities for human-robot collaboration.

14.2 ROLE OF HCI AND XAI IN ITS

We can find several roles of HCI and XAI in ITS.

14.2.1 Public Transportation Operations Using HCI and XAI

Public transportation operations can benefit from the integration of human-computer interaction (HCI) and explainable artificial intelligence (XAI) techniques [10, 11 and 12]. HCI focuses on designing systems and interfaces that are intuitive, user-friendly, and enhance the interaction between humans and computers. XAI aims to provide transparency and understandability in AI systems' decision-making processes. When applied to public transportation operations, HCI and XAI can improve the user experience, decision-making, and overall efficiency of the system.

- **User-friendly interfaces**: HCI principles can be applied to the design of interfaces used by both transportation operators and passengers. Intuitive and user-friendly interfaces can enhance the efficiency of tasks such as route planning, ticket purchasing, and real-time information access. Clear visualizations and interactive features can help passengers navigate the transportation system and provide operators with a comprehensive view of the network.
- **Passenger information systems**: HCI techniques can be used to design passenger information systems that provide real-time updates on schedules, delays, and route changes. These systems can be accessible through various platforms such as mobile applications, websites, and digital displays at stations. Clear and concise information presented in a user-friendly manner improves the passenger experience and reduces uncertainty.
- **Intelligent decision support**: XAI techniques can be employed to provide explanations and justifications for the decisions made by AI systems used in public transportation operations [13, 14]. For example, AI algorithms can suggest optimal schedules, routing, or fleet management strategies. By providing understandable explanations for these suggestions, operators can make informed decisions and have a better understanding of the underlying reasoning behind AI-generated recommendations.
- **Predictive maintenance**: XAI techniques can be used to provide insights and explanations for predictive maintenance systems in public transportation. AI algorithms can analyze sensor data from vehicles and infrastructure to predict maintenance needs and optimize maintenance schedules. Explanations of these predictions can help maintenance staff understand the underlying factors and take appropriate actions.

- **Incident management**: HCI and XAI can aid in incident management by providing clear and actionable information to operators and decision-makers during disruptions, emergencies, or incidents. User-friendly interfaces can display real-time incident information, suggested alternative routes, and instructions for passengers. XAI can provide explanations for decisions related to rerouting, rescheduling, and resource allocation during incidents.
- **Accessibility and inclusivity**: HCI principles should be applied to ensure that public transportation systems are accessible and inclusive for individuals with disabilities or special needs. User interfaces and assistive technologies should be designed with consideration for different abilities and should provide clear instructions, visual and auditory cues, and alternative communication methods.

By incorporating HCI and XAI in public transportation operations, the system can be more user-centric, efficient, and transparent. Passengers can have a better experience through user-friendly interfaces and real-time information, while operators can make informed decisions with the help of explanations and justifications provided by XAI. Ultimately, the integration of HCI and XAI can lead to improved public transportation services, increased user satisfaction, and more efficient operations.

14.2.2 Electronic Payment and Commercial Vehicle Operations Using HCI and XAI

The integration of human-computer interaction (HCI) and explainable artificial intelligence (XAI) techniques can enhance electronic payment systems and commercial vehicle operations, improving user experience, decision-making, and overall efficiency [15]. Here are some ways HCI and XAI can be applied in these domains.

Electronic payment systems:

- **User-friendly interfaces**: HCI principles can be applied to design intuitive and user-friendly interfaces for electronic payment systems. This includes mobile payment applications, point-of-sale terminals, and online payment platforms. Clear and easy-to-understand interfaces can facilitate seamless and efficient transactions for users.
- **Personalization and customization**: HCI techniques can enable personalized features in electronic payment systems, such as transaction history, spending insights, and tailored provides. Customizable interfaces and settings allow users to personalize their payment experience according to their preferences.
- **Secure and trustworthy design**: HCI can play a important role in designing interfaces that instill trust and confidence in users. Clear security measures, authentication processes, and visual cues can help users feel secure during transactions.
- **Error prevention and recovery**: HCI principles can help design interfaces that minimize errors during payment processes. Clear instructions, error messages, and intuitive design can prevent mistakes and guide users through error recovery steps.

Commercial vehicle operations:

- **Fleet management interfaces**: HCI can be used to design user-friendly interfaces for fleet management systems used by commercial vehicle operators. Intuitive dashboards and visualizations can provide operators with real-time data on vehicle locations, fuel consumption, maintenance schedules, and driver performance.
- **Intelligent routing and optimization**: XAI techniques can be applied to explain the rationale behind routing and optimization decisions made by AI algorithms in commercial vehicle operations. Operators can have a better understanding of suggested routes, load distribution, and scheduling recommendations, leading to more informed decision-making.
- **Driver assistance and feedback**: HCI can enhance interfaces for driver assistance systems in commercial vehicles. Intuitive displays, alerts, and feedback mechanisms can help drivers navigate routes, monitor vehicle performance, and improve driving behavior.
- **Safety and compliance**: HCI can contribute to the design of interfaces that promote safety and regulatory compliance in commercial vehicle operations. Clear guidelines, reminders, and visual cues can help drivers adhere to safety protocols, hours-of-service regulations, and vehicle maintenance requirements.

Integration of HCI and XAI:

- **Transparency and explanations**: XAI techniques can provide explanations for AI-driven decisions made in electronic payment and commercial vehicle operations. This helps users and operators understand the reasoning behind system recommendations, such as fraud detection, transaction approvals, or route optimizations.
- **User feedback and adaptation**: HCI principles can be used to gather user feedback and preferences in electronic payment systems and commercial vehicle operations. User input can be incorporated to enhance system recommendations and customize the user experience.
- **Continuous Improvement**: the integration of HCI and XAI allows for iterative improvements based on user feedback and system performance. User-centered design and user testing can provide insights into usability issues and areas for enhancement in electronic payment and commercial vehicle systems.

By using HCI and XAI in electronic payment systems and commercial vehicle operations, organizations can improve user satisfaction, streamline operations, and enhance decision-making. User-friendly interfaces, transparent explanations, personalized features, and safety measures contribute to a more efficient and trustworthy user experience in these domains.

14.2.3 Advanced Vehicle Control and Safety Systems Using HCI and XAI

Advanced vehicle control and safety systems (AVCSS) can greatly benefit from the integration of human-computer interaction (HCI) and explainable artificial intelligence (XAI) techniques. HCI focuses on designing interfaces that enable effective interaction between humans and machines [16], while XAI aims to provide transparency and understandability in AI systems' decision-making processes. When applied to AVCSS, HCI and XAI can enhance user experience, safety, and overall system performance. Here are some ways HCI and XAI can be utilized in AVCSS.

Driver assistance systems:

- **Intuitive interfaces**: HCI principles can be applied to design user-friendly interfaces for driver assistance systems, such as adaptive cruise control, lane-keeping assist, and collision warning systems. Clear visual displays, intuitive controls, and well-designed alerts can improve the driver's understanding and interaction with these systems.
- **Contextual information**: HCI techniques can provide drivers with real-time, context-sensitive information about the vehicle's surroundings, including traffic conditions, road hazards, and pedestrian presence. This information can be displayed through heads-up displays, instrument clusters, or augmented reality interfaces, enhancing situational awareness.
- **Explainable alerts and recommendations**: XAI can provide explanations for alerts and recommendations generated by AI algorithms in driver assistance systems. For example, if the system advises a lane change, it can explain the reasons behind the recommendation, such as a detected hazard or upcoming road conditions, allowing the driver to make informed decisions.

Autonomous driving systems:

- **Human-machine interfaces (HMIs)**: HCI plays a important role in designing intuitive and effective HMIs for autonomous vehicles. These interfaces should enable clear communication between the vehicle and the passengers, displaying relevant information about the driving mode, system status, and handover requests.
- **Trust and user comfort**: HCI techniques can be used to build trust and enhance user comfort in autonomous driving systems. Visualizations of the system's perception, reasoning, and decision-making processes can help passengers understand and trust the vehicle's actions. Additionally, providing passengers with control over certain aspects of the driving experience, such as speed or following distance, can increase their sense of comfort and confidence.
- **Explanations for system decisions**: XAI can provide explanations for the decisions made by autonomous driving systems. This includes justifications for route planning, object detection and classification, and responses to unexpected situations. Explanations enable passengers and regulators to understand the reasoning behind the system's actions, enhancing trust and safety.

Vehicle safety:

- **Collision avoidance systems**: HCI techniques can be employed to design interfaces that effectively communicate collision warnings and interventions

to the driver. Clear visual and auditory alerts, along with haptic feedback, can ensure that drivers promptly react to potential collisions.
- **Predictive safety systems**: XAI can provide explanations for predictive safety systems that anticipate and prevent potential accidents. By explaining the factors considered, such as driver behavior, vehicle conditions, and environmental factors, these systems can enable drivers to better understand and adapt their driving accordingly.

By integrating HCI and XAI in AVCSS, the user experience can be enhanced, safety can be improved, and trust in these systems can be established. User-friendly interfaces, contextual information, explainable alerts, and continuous system learning contribute to effective human-machine interaction.

14.2.4 Emergency Situation Management by Future Vehicles Using HCI and XAI

Emergency situation management by future vehicles can greatly benefit from the integration of human-computer interaction (HCI) and explainable artificial intelligence (XAI) techniques [17]. HCI focuses on designing interfaces and systems that facilitate effective communication and interaction between humans and machines, while XAI aims to provide transparency and understandability in AI systems' decision-making processes. When applied to emergency situation management in vehicles, HCI and XAI can enhance user safety, decision-making, and overall system performance. Here are some ways HCI and XAI can be utilized in this context.

Emergency communication and assistance:
- **Intuitive interfaces**: HCI principles can be applied to design user-friendly interfaces that enable quick and efficient communication with emergency services. These interfaces can provide clear options for initiating emergency calls, requesting assistance, and sharing important information, such as the vehicle's location and the nature of the emergency.
- **Contextual information sharing**: HCI techniques can facilitate the collection and transmission of relevant contextual information to emergency responders. This includes data about the vehicle's condition, occupants, and any hazardous materials present. Interfaces can ensure that this information is shared in a structured and easily understandable format.

Autonomous emergency maneuvers:
- **Human-machine interfaces (HMIs)**: HCI plays a important role in designing interfaces that communicate and coordinate emergency maneuvers in autonomous vehicles. Clear visual displays, audible alerts, and haptic feedback can effectively inform passengers about emergency situations and the actions being taken by the vehicle.
- **Explainable emergency actions**: XAI can provide explanations for the emergency actions taken by autonomous vehicles. This includes justifications

for evasive maneuvers, emergency braking, or emergency route deviations. Explanations enable passengers and emergency responders to understand the reasoning behind the vehicle's actions.

Predictive emergency systems:

- **Hazard detection and prediction**: HCI techniques can be employed to design interfaces that present real-time hazard detection and prediction information to the driver or passengers. Visualizations, alerts, and warnings can effectively communicate potential emergency situations, such as collisions, road hazards, or adverse weather conditions.
- **Explanations for system predictions**: XAI can provide explanations for the predictions made by emergency systems, such as collision detection and avoidance algorithms. By explaining the factors considered, such as vehicle dynamics, sensor data, and environment conditions, passengers can better understand the system's reasoning and respond appropriately.

14.2.5 XAI – HCI-Enabled Traffic Management Using IoTs in Near Future

In the near future, the integration of explainable artificial intelligence (XAI) and human-computer interaction (HCI) techniques can play a significant role in traffic management systems that use the internet of things (IoT) [18]. By combining these technologies, traffic management can become more efficient, responsive, and user-centric. Here's how XAI and HCI can enable traffic management using IoT in the near future.

Real-time traffic monitoring and prediction:

- **IoT data integration**: traffic management systems can use IoT devices, such as smart sensors, cameras, and connected vehicles, to collect real-time data on traffic conditions, including vehicle density, speed, and congestion. HCI techniques can be applied to design intuitive interfaces that present this data to traffic managers and users in a meaningful and actionable way.
- **Explainable traffic predictions**: XAI algorithms can be used to analyze historical and real-time data to predict traffic patterns, identify potential congestion areas, and estimate travel times. The explanations provided by XAI can help traffic managers understand the factors influencing these predictions and make informed decisions regarding traffic management strategies.

Intelligent traffic control and optimization:

- **Adaptive traffic signal control**: XAI can explain the decisions made by AI algorithms that control traffic signal timings based on real-time traffic conditions. HCI techniques can be used to design interfaces that display the current signal status and provide feedback on the timing changes to drivers, cyclists, and pedestrians. This promotes transparency and enables users to understand and adapt to the signal patterns.

- **Dynamic route guidance**: XAI-enabled algorithms can suggest optimal routes to drivers, taking into account real-time traffic conditions, road incidents, and user preferences. HCI techniques can be used to design interfaces that present these route recommendations in a clear and understandable manner, allowing users to make informed decisions based on the provided explanations.

User-centric travel information and assistance:

- **Personalized travel recommendations**: HCI principles can be applied to design interfaces that provide personalized travel recommendations to users based on their preferences, historical data, and real-time information. XAI can explain the reasoning behind these recommendations, enhancing user trust and confidence in the system.
- **Multi-modal journey planning**: HCI and XAI can enable interfaces that support multi-modal journey planning, considering various transportation modes such as cars, public transit, bicycles, and walking. The system can provide recommendations based on user preferences, real-time data, and explanations of the suggested options.

Incident management and emergency response:

- **Real-time incident detection**: IoT sensors and cameras can detect incidents such as accidents, road hazards, or extreme weather conditions. HCI techniques can be used to design interfaces that display real-time incident information to traffic managers and users, enabling prompt response and alternative route suggestions.
- **Explanations for response actions**: XAI can provide explanations for the actions taken by the traffic management system in response to incidents, such as rerouting traffic or dispatching emergency services. These explanations help traffic managers understand the underlying reasoning and communicate the actions to the affected users.

Note that by integrating XAI and HCI in traffic management systems using IoT in the near future, traffic flow can be optimized, user experience can be enhanced, and congestion can be reduced. Transparent explanations, intuitive interfaces, personalized recommendations, and real-time incident management contribute to more efficient and user-centric.

14.2.6 XAI – HCI Enabled Road Accident Detection Using IoTs in Near Future

Explainable artificial intelligence (XAI) and human-computer interaction (HCI) enabled road accident detection using IoTs (internet of things) is a promising application that can enhance road safety and emergency response systems [18, 19]. In the near future, advancements in technology are expected to bring about significant improvements in this area. XAI refers to the development of AI systems that can provide transparent and interpretable explanations for their actions and decisions. HCI focuses on the interaction between humans and computers, aiming to create user-friendly and intuitive interfaces. Combining XAI and HCI in the context of road accident detection using IoTs can result in a system that not only detects accidents but also provides understandable and actionable insights to both drivers and emergency responders.

Here's how such a system might work in the near future:

- **IoT Sensors**: roadside IoT sensors are deployed along roads, highways, and intersections. These sensors can include cameras, radar systems, lidar sensors, and vehicle-to-infrastructure communication devices. They continuously monitor the road conditions, traffic patterns, and vehicle movements.
- **Data collection and processing**: the IoT sensors capture data in real-time and transmit it to a central processing unit. The data includes images, videos, speed, location, and other relevant information. The processing unit analyzes the data using AI algorithms to identify potential accidents or hazardous situations.
- **Accident detection**: using computer vision and machine learning techniques, the system can detect important events such as collisions, sudden changes in speed, or unexpected vehicle movements. It can also take into account environmental factors like weather conditions and road surface conditions.
- **Explainable alerts**: once an accident or hazardous situation is detected, the system generates alerts for both drivers and emergency responders. The alerts are designed to be easily understandable by humans, providing clear and concise information about the nature of the incident, location, and severity.
- **Driver assistance**: the system can also provide real-time assistance to drivers in the vicinity of an accident. It can recommend alternative routes, warn about potential dangers ahead, or suggest actions to avoid further incidents. The explanations provided by the XAI component help drivers make informed decisions.
- **Emergency response integration**: emergency response systems can be seamlessly integrated into the accident detection system. When an accident is detected, the system can automatically alert nearby emergency services, providing them with detailed information to facilitate a quick response. This integration can help reduce response times and potentially save lives.

By combining XAI, HCI, and IoT technologies, road accident detection systems of the near future have the potential to significantly improve road safety and emergency response. These systems not only detect accidents but also provide transparent and understandable insights to drivers and emergency responders, ultimately working towards the goal of reducing accidents and minimizing their impact.

14.3 APPLICATIONS OF HCI AND XAI

The combination of human-computer interaction (HCI) and explainable artificial intelligence (XAI) can play a significant role in enhancing supply chain and logistics operations [20, 21, 22, 23 and 24]. HCI focuses on designing intuitive and user-friendly interfaces, while XAI ensures transparent and interpretable AI-driven decision-making. Here are some key areas where HCI and XAI can contribute to supply chain and logistics:

- **Inventory management**: HCI principles can be applied to design interfaces that enable inventory managers to monitor and control inventory levels effectively. These interfaces can provide real-time visibility into inventory data, including stock levels, locations, and demand patterns. XAI algorithms can analyze historical data, demand forecasts, and other relevant factors to provide transparent explanations for inventory management decisions, such as replenishment recommendations or stock allocation strategies.
- **Demand forecasting**: HCI and XAI can work together to improve demand forecasting accuracy. Interfaces can be designed to allow demand planners to input and analyze relevant data, such as historical sales, market trends, and external factors. XAI algorithms can then generate demand forecasts and provide explanations for the underlying factors and reasoning behind the forecasts. HCI principles can be employed to design interfaces that present the forecasts in an easily understandable format for decision-makers.
- **Route optimization**: HCI can contribute to designing interfaces that allow logistics managers to optimize routes for efficient transportation of goods. These interfaces can consider various factors like distance, traffic conditions, delivery time windows, and cost constraints. XAI algorithms can analyze real-time data and historical patterns to generate optimized route recommendations. Explanations for the route choices can be provided, helping logistics managers understand the considerations taken into account by the AI system.
- **Supply chain visibility**: HCI principles can be utilized to design interfaces that provide supply chain stakeholders with real-time visibility into the movement of goods and assets. These interfaces can integrate data from various sources, such as GPS trackers, RFID tags, and inventory management systems. XAI algorithms can analyze the data and provide explanations for any deviations, delays, or anomalies in the supply chain. The explanations can help stakeholders understand the causes and take necessary actions.
- **Risk management**: HCI and XAI can contribute to risk management in supply chain operations. Interfaces can be designed to allow supply chain managers to identify, assess, and mitigate risks, such as disruptions in transportation, supplier reliability, or demand volatility. XAI algorithms can analyze historical data, external factors, and risk models to provide explanations for risk assessments and recommend mitigation strategies. HCI principles can ensure that the interfaces facilitate effective risk management decision-making.
- **Supplier collaboration**: HCI can facilitate the design of interfaces that enable seamless collaboration and communication among supply chain stakeholders, including suppliers, manufacturers, and distributors. These interfaces can support activities such as order management, shipment tracking, and information sharing. XAI can provide insights and explanations for supply chain decisions, fostering transparency and trust among stakeholders during collaborative processes.

Note that by integrating HCI and XAI in supply chain and logistics operations, organizations can benefit from user-friendly interfaces, transparent decision-making, improved forecasting accuracy, optimized routes, enhanced supply chain visibility, effective risk management, and continuous improvement [25, 26]. These technologies enable more efficient and resilient supply chains, leading to cost savings, better customer service, and overall operational excellence.

14.3.1 Role of HCI and XAI in Banking and Finance

Human-computer interaction (HCI) and explainable artificial intelligence (XAI) can play significant roles in enhancing banking and finance services [27]. By combining HCI principles with XAI techniques, banks and financial institutions can provide user-friendly interfaces, transparent decision-making, personalized services, and improved customer experiences. Here are some key areas where HCI and XAI can contribute to banking and finance:

- **Personalized financial advice**: HCI can be used to design interfaces that enable customers to input their financial goals, risk tolerance, and preferences. XAI algorithms can analyze customer data, market trends, and investment models to generate personalized financial advice and recommendations. Explanations for the recommendations can be provided, allowing customers to understand the reasoning behind the advice and make informed decisions.
- **Fraud detection and prevention**: HCI and XAI can work together to improve fraud detection and prevention in banking and finance. Interfaces can be designed to allow fraud analysts to monitor and analyze transactional data, customer behavior, and external risk factors. XAI algorithms can analyze the data to identify patterns, anomalies, and potential fraudulent activities. Explanations for the fraud detection can be provided, helping analysts understand the factors considered by the AI system and facilitating better decision-making.
- **Risk assessment and management**: HCI principles can be employed to design interfaces that support risk assessment and management in banking and finance. These interfaces can allow risk managers to input and analyze relevant data, such as market trends, financial indicators, and customer profiles. XAI algorithms can analyze the data to assess risks, provide transparent explanations for risk assessments, and recommend risk mitigation strategies. HCI can ensure that the interfaces present risk information in a clear and understandable format.
- **Customer service and support**: HCI and XAI can enhance customer service and support in banking and finance. Interfaces, such as chatbots or virtual assistants, can provide personalized assistance to customers, answer inquiries, and guide them through various banking processes. XAI algorithms can analyze customer queries, transactional data, and knowledge bases to provide accurate and contextually relevant responses. Explanations for the responses can be provided, ensuring transparency and building customer trust.

- **Investment portfolio management**: HCI can contribute to designing interfaces that allow customers to manage their investment portfolios effectively. These interfaces can provide real-time portfolio tracking, performance analytics, and investment recommendations. XAI algorithms can analyze market data, customer preferences, and investment models to generate portfolio recommendations. Explanations for the recommendations can be provided, helping customers understand the factors considered by the AI system and make informed investment decisions.
- **Credit assessment and lending**: HCI and XAI can assist in credit assessment and lending processes. Interfaces can be designed to support loan applications, credit scoring, and risk assessment. XAI algorithms can analyze customer data, credit histories, and financial indicators to assess creditworthiness and provide transparent explanations for credit decisions. HCI can ensure that the interfaces guide customers through the lending process and present information in a clear and user-friendly manner.
- **Financial education and literacy**: HCI and XAI can contribute to improving financial education and literacy. Interfaces can be designed to provide educational resources, interactive tools, and personalized financial insights to customers. XAI algorithms can analyze customer financial data and provide explanations for financial insights and recommendations. HCI principles can be applied to design interfaces that facilitate learning and engagement, empowering customers to make sound financial decisions.

By integrating HCI and XAI in banking and finance, organizations can benefit from user-friendly interfaces, transparent decision-making, personalized services, fraud prevention, risk management, enhanced customer experiences, and improved financial literacy. These technologies enable banks and financial institutions to provide tailored solutions, mitigate risks, ensure regulatory compliance, and foster trust and loyalty among customers.

14.3.2 Role of HCI and XAI in Defence

Human-computer interaction (HCI) and explainable artificial intelligence (XAI) can play important roles in the defense sector, enabling efficient and effective decision-making, improved situational awareness, and enhanced human-machine collaboration [27, 28, and 29]. By combining HCI principles with XAI techniques, defense organizations can use the power of AI while ensuring transparency, interpretability, and usability. Here are some key areas where HCI and XAI can contribute to defense:

- **Command and control systems**: HCI can be applied to design interfaces that facilitate effective command and control of military operations. These interfaces should be intuitive, enabling military personnel to access and analyze real-time data, make informed decisions, and coordinate forces efficiently. XAI algorithms can provide explanations for the recommendations or predictions generated by AI systems, helping commanders understand the underlying reasoning and build trust in the AI's capabilities.

- **Decision support systems**: HCI and XAI can assist in developing decision support systems that provide accurate and transparent insights to military decision-makers. Interfaces can present information in a user-friendly manner, utilizing visualizations and interactive tools. XAI algorithms can analyze complex data, such as sensor inputs, satellite imagery, and intelligence reports, to generate interpretable recommendations or assessments. Explanations for the system's conclusions can be provided, enhancing decision-maker understanding and confidence.
- **Threat detection and analysis**: HCI and XAI can enhance threat detection and analysis capabilities in defense. Interfaces can be designed to display real-time sensor data, such as radar, sonar, or satellite imagery, in a way that allows operators to effectively identify and analyze potential threats. XAI algorithms can assist in interpreting the data, providing explanations for the detection and classification of threats. This can help operators understand the AI system's reasoning and improve their own situational awareness.
- **Training and simulation**: HCI principles can be utilized to design immersive and interactive training and simulation interfaces for defense personnel. These interfaces can enable realistic training scenarios and virtual simulations of complex operational environments. XAI can enhance the training experience by providing explanations for the AI system's behavior and decisions during the simulation. This fosters a deeper understanding of the AI's capabilities and limitations, facilitating more effective training outcomes.
- **Autonomous systems and robotics**: HCI and XAI play a important role in designing interfaces and interactions with autonomous systems and robotics used in defense. Interfaces should enable operators to monitor and control autonomous platforms effectively, receive real-time feedback, and understand the autonomous system's decision-making processes. XAI algorithms can explain the reasoning behind autonomous system decisions, ensuring operators comprehend the system's actions and trust its capabilities.
- **Cybersecurity and intrusion detection**: HCI and XAI can contribute to enhancing cybersecurity and intrusion detection in defense systems [29, 30]. Interfaces can be designed to provide real-time visualization of network traffic, system vulnerabilities, and potential cyber threats. XAI algorithms can analyze network data, log files, and anomaly detection models to identify suspicious activities and provide explanations for the detected anomalies. HCI can ensure that the interfaces present the information in a manner that facilitates rapid response and decision-making.
- **Collaboration and communication**: HCI principles can be applied to design interfaces that support seamless collaboration and communication among defense personnel. These interfaces should enable secure and efficient information sharing, real-time communication, and situational awareness among distributed teams. XAI can assist in interpreting and summarizing large volumes of data, providing concise and actionable insights for effective collaboration.

By integrating HCI and XAI in defense systems, organizations can benefit from enhanced decision-making, improved situational awareness, efficient training, effective threat detection, and better human-machine collaboration. These technologies empower defense personnel with transparent and interpretable AI systems, allowing them to make informed decisions, optimize resource allocation, and respond effectively to complex operational challenges.

14.3.3 Role of HCI and XAI in Agriculture

Human-computer interaction (HCI) and explainable artificial intelligence (XAI) can play significant roles in improving agricultural practices and enhancing productivity [31, 32]. By combining HCI principles with XAI techniques, farmers and agricultural stakeholders can access user-friendly interfaces, transparent decision-making, and actionable insights from AI systems. Here are some key areas where HCI and XAI can contribute to agriculture:

- **Precision farming**: HCI and XAI can assist in designing interfaces and decision support systems for precision farming practices. Interfaces can provide real-time data visualization of crop health, soil conditions, and weather patterns. XAI algorithms can analyze the data to generate actionable insights and recommendations for optimized irrigation, fertilizer application, and pest control. Explanations for the recommendations can be provided, helping farmers understand the underlying factors considered by the AI system.
- **Crop monitoring and disease detection**: HCI principles can be applied to design interfaces that allow farmers to monitor crop growth and detect diseases or nutrient deficiencies. These interfaces can integrate data from various sources, such as satellite imagery, drones, and sensors. XAI algorithms can analyze the data to identify patterns, anomalies, or disease indicators. Explanations for the detection results can be provided, assisting farmers in making informed decisions about crop management and treatment strategies.
- **Yield prediction and forecasting**: HCI and XAI can contribute to yield prediction and forecasting models in agriculture. Interfaces can be designed to input historical data, weather forecasts, and other relevant parameters. XAI algorithms can analyze the data to generate yield predictions and forecasts. Explanations for the predictions can be provided, helping farmers understand the factors influencing the yield outcomes and enabling them to optimize production planning and resource allocation.
- **Farm management and resource optimization**: HCI principles can be utilized to design interfaces that support farm management and resource optimization. These interfaces can integrate data on field conditions, equipment availability, labor resources, and market trends. XAI algorithms can analyze the data to provide recommendations for efficient resource allocation, scheduling of tasks, and optimal crop rotation strategies. HCI can ensure that the interfaces facilitate intuitive and actionable decision-making.

- **Supply chain and logistics**: HCI and XAI can assist in improving supply chain and logistics operations in agriculture. Interfaces can be designed to enable farmers and stakeholders to track and manage the transportation, storage, and distribution of agricultural products. XAI algorithms can analyze supply chain data to provide recommendations for optimized routing, inventory management, and delivery scheduling. Explanations for the recommendations can be provided, facilitating better decision-making in the supply chain.
- **Agricultural education and extension services**: HCI and XAI can contribute to agricultural education and extension services. Interfaces can be designed to provide farmers with educational resources, personalized recommendations, and access to expert knowledge. XAI algorithms can analyze farmer data and generate tailored insights and guidance. HCI principles can be employed to design interfaces that are user-friendly, accessible, and facilitate knowledge transfer.
- **Environmental sustainability**: HCI and XAI can play a role in promoting environmental sustainability in agriculture. Interfaces can be designed to provide farmers with insights on sustainable farming practices, conservation techniques, and environmental impact assessments. XAI algorithms can analyze data on soil health, water usage, and climate conditions to provide recommendations for sustainable farming methods. Explanations for the recommendations can be provided, enabling farmers to understand the environmental benefits of adopting sustainable practices.

By integrating HCI and XAI in agriculture, farmers can benefit from user-friendly interfaces, transparent decision-making, and actionable insights. These technologies enable precision farming, efficient resource management, improved crop monitoring, optimized supply chain operations, and environmental sustainability. Ultimately, HCI and XAI empower farmers to make data-driven decisions, increase productivity, and promote sustainable agricultural practices.

14.3.4 Role of HCI and XAI in Medicare

Human-computer interaction (HCI) and explainable artificial intelligence (XAI) have significant roles in the field of healthcare and Medicare, improving patient care, facilitating medical decision-making, and enhancing the overall healthcare experience [33, 34, and 35]. Here's how HCI and XAI contribute to Medicare:

- **User-friendly interfaces**: HCI principles guide the design of user interfaces for healthcare systems, including electronic health records (EHRs), telemedicine platforms, and patient portals. HCI ensures that these interfaces are intuitive, easy to navigate, and user-friendly for both healthcare providers and patients. User-friendly interfaces improve efficiency, reduce errors, and enhance the overall experience of interacting with healthcare systems.
- **Patient engagement and empowerment**: HCI and XAI techniques are used to develop interactive tools and applications that empower patients to actively

participate in their healthcare. These tools can provide personalized health information, reminders for medication adherence, and access to educational resources. HCI principles ensure that these tools are engaging, accessible, and tailored to the needs and preferences of individual patients, promoting patient engagement and self-care.

- **Clinical decision support systems**: HCI and XAI contribute to the development of clinical decision support systems (CDSS) in Medicare. CDSS utilizes AI and machine learning algorithms to assist healthcare providers in making evidence-based clinical decisions. HCI principles guide the design of user interfaces that present relevant information and recommendations to healthcare providers in a clear and understandable manner. XAI techniques enable the explanation of AI-generated recommendations, helping providers understand the underlying reasoning behind the suggestions.
- **Medical imaging and diagnostics**: HCI and XAI techniques play a role in medical imaging and diagnostics. HCI principles guide the design of user interfaces for viewing and analyzing medical images, ensuring that radiologists and other healthcare professionals can efficiently interpret and diagnose conditions. XAI techniques can provide explanations for AI algorithms used in medical image analysis, aiding in the understanding and validation of AI-assisted diagnostic decisions.
- **Personalized treatment and care**: HCI and XAI enable personalized treatment and care in Medicare. HCI principles guide the design of interfaces that capture patient data, preferences, and medical history to provide personalized treatment recommendations. XAI techniques can provide explanations for treatment decisions, helping healthcare providers and patients understand the rationale behind personalized recommendations and facilitating shared decision-making.

By using HCI and XAI principles and techniques, Medicare can improve patient care, enhance medical decision-making, promote patient engagement, and address ethical considerations. The combination of user-friendly interfaces, explainable AI, and personalized healthcare experiences contributes to a more efficient and patient-centered Medicare system.

14.3.5 Role of HCI and XAI in Education

Human-computer interaction (HCI) and explainable artificial intelligence (XAI) play important roles in the field of education, enhancing learning experiences, personalizing instruction, and supporting teachers and students [36, 37]. Here's how HCI and XAI contribute to education:

- **User-centric learning experiences**: HCI principles are applied to design user interfaces and interactions in educational technology. This includes learning management systems, educational apps, and online platforms. HCI ensures that these tools are intuitive, easy to navigate, and user-friendly, enhancing the overall learning experience for students and teachers.

- **Personalized learning**: HCI and XAI techniques enable personalized learning experiences for students. HCI principles guide the design of interfaces that adapt to individual student needs, preferences, and learning styles. XAI techniques can provide explanations for personalized recommendations or adaptive learning algorithms, helping students understand the rationale behind content suggestions and learning pathways.
- **Intelligent tutoring systems**: HCI and XAI contribute to the development of intelligent tutoring systems (ITS). These systems use AI and machine learning algorithms to provide personalized instruction, feedback, and assessment. HCI principles ensure that the interface and interactions with the tutoring system are engaging, supportive, and responsive. XAI techniques can explain the reasoning behind the system's instructional decisions, helping students understand how they can improve and learn more effectively.
- **Augmented reality (AR) and Virtual Reality (VR)**: HCI techniques are applied to create immersive learning experiences using AR and VR technologies. These technologies provide students with interactive and engaging virtual environments for educational purposes. HCI principles guide the design of user interfaces and interactions that facilitate learning, exploration, and collaboration within these virtual environments.
- **Teacher support and professional development**: HCI and XAI can support teachers in their instructional practices and professional development. HCI principles are used to design interfaces for educational tools that assist teachers in planning lessons, managing student data, and analyzing learning outcomes. XAI techniques can provide insights and explanations to teachers about student progress, identifying areas where additional support or intervention may be needed.

In summary, HCI and XAI play important roles in improving education by enhancing user experiences, personalizing instruction, providing intelligent tutoring, creating immersive learning environments, supporting teachers, and addressing ethical considerations. By incorporating HCI and XAI principles and techniques, education can become more effective, engaging, and inclusive.

14.3.6 Role of HCI and XAI in Entertainment

Human-computer interaction (HCI) and explainable artificial intelligence (XAI) play important roles in the field of entertainment, enhancing user experiences and enabling greater understanding and trust in AI-powered entertainment systems [38]. Here's how HCI and XAI contribute to the entertainment industry:

- **Enhanced user experience**: HCI focuses on designing interactive systems that are intuitive, user-friendly, and engaging. In the context of entertainment, HCI principles help create interfaces and interactions that enhance the user experience. For example, HCI techniques can be employed to design intuitive and responsive interfaces for gaming consoles, virtual reality (VR) systems, and mobile applications, allowing users to navigate and interact with entertainment content more seamlessly.

- **Personalization and recommender systems**: HCI and AI techniques can be combined to develop personalized recommender systems in the entertainment domain. These systems use user preferences, behavior, and feedback to provide tailored content recommendations. By understanding user interactions and preferences through HCI methods, entertainment platforms can provide more relevant suggestions, leading to a more enjoyable and personalized experience for users.
- **Immersive technologies**: HCI plays a important role in developing immersive technologies like virtual reality (VR), augmented reality (AR), and mixed reality (MR). These technologies provide novel ways of experiencing entertainment content, creating virtual worlds, and interacting with digital characters. HCI principles guide the design of intuitive user interfaces and interaction techniques, ensuring users can seamlessly engage with the virtual environment or characters in a natural and intuitive manner.
- **Explainability and trust**: as AI technologies are increasingly utilized in entertainment systems, there is a growing need for transparency and understanding of AI decision-making processes. XAI aims to provide explanations for AI system outputs and behavior, promoting trust and enabling users to understand how recommendations or outcomes are generated. In the context of entertainment, XAI techniques can help users understand why certain movies, songs, or games are recommended, providing insights into the underlying algorithms and data used by AI systems.

In summary, HCI and XAI play important roles in shaping the future of entertainment by enhancing user experiences, personalizing content, creating immersive environments, promoting transparency, and addressing ethical concerns. These interdisciplinary fields collaborate to design user-centric, trustworthy, and engaging entertainment systems.

14.3.7 Role of HCI and XAI in Manufacturing

Human-computer interaction (HCI) and explainable artificial intelligence (XAI) have significant roles in the manufacturing industry, facilitating efficient and safe operations, improving decision-making processes, and enhancing overall productivity [39, 40, and 41]. Here are some ways HCI and XAI contribute to manufacturing:

- **User interface design**: HCI principles guide the design of user interfaces for manufacturing systems, including human-machine interfaces (HMIs), control panels, and dashboards. These interfaces need to be intuitive, user-friendly, and tailored to the specific needs of operators and workers on the shop floor. HCI considerations ensure that workers can easily interact with machines, access information, and control manufacturing processes effectively.

- **Augmented reality (AR) and virtual reality (VR)**: HCI techniques are important in developing AR and VR applications for manufacturing. AR and VR technologies provide workers with real-time information, virtual training simulations, and remote collaboration capabilities. HCI principles help create immersive and intuitive interfaces that allow workers to interact with virtual objects, visualize data, and perform tasks more efficiently and accurately.
- **Intelligent decision support systems**: XAI plays a significant role in manufacturing by providing explanations and insights into AI-powered decision-making processes. XAI techniques can help manufacturing systems explain why certain decisions were made, such as process optimization, predictive maintenance, or quality control. These explanations improve transparency, build trust, and enable human operators to understand and validate AI-generated recommendations or actions.
- **Predictive maintenance**: HCI and XAI contribute to predictive maintenance in manufacturing. HCI principles guide the design of user interfaces for monitoring equipment health, collecting sensor data, and providing alerts or recommendations to maintenance personnel. XAI techniques enable the explanation of predictive maintenance models, allowing operators to understand the factors influencing equipment failure and take appropriate actions to prevent breakdowns.
- **Safety and risk mitigation**: HCI and XAI help improve safety in manufacturing environments. HCI principles are applied to design interfaces that provide clear instructions, warnings, and feedback to operators, ensuring safe operation of machinery and equipment. XAI techniques can identify potential safety risks and provide explanations for safety-related decisions, helping operators understand the reasoning behind safety measures and guidelines.
- **Collaborative robotics (cobots)**: HCI and XAI play a role in designing interfaces and interactions between humans and collaborative robots (cobots) in manufacturing. HCI principles ensure that cobots are easy to program, operate, and collaborate with human workers. XAI techniques enable explanations of cobot actions, enabling human operators to understand and trust the robot's behavior and decision-making processes.

By using HCI principles and XAI techniques, the manufacturing industry can optimize processes, improve worker safety, increase productivity, and enable effective human-machine collaboration. The combination of intuitive interfaces, transparent AI decision-making, and immersive technologies contributes to a more efficient and productive manufacturing ecosystem.

14.3.8 Role of HCI and XAI in Cyber Security

Human-computer interaction (HCI) and explainable artificial intelligence (XAI) play important roles in the field of cybersecurity, enhancing user awareness, improving threat detection and response, and promoting trust and transparency in security systems [41, 42 and 43]. Here's how HCI and XAI contribute to cybersecurity:

- **User awareness and training**: HCI principles are applied to design user interfaces and interactions that promote cybersecurity awareness and provide user training. HCI ensures that security messages and warnings are clear, concise, and understandable, helping users make informed decisions and take appropriate actions to protect their digital assets. User-friendly interfaces for security training and education programs help users develop good cybersecurity practices.
- **Usable security interfaces**: HCI principles guide the design of security interfaces that balance usability and security requirements. Security interfaces, such as password managers, two-factor authentication systems, and secure file transfer tools, need to be intuitive and easy to use to encourage adoption and proper usage by users. HCI ensures that security controls and mechanisms are designed with the user's mental models and capabilities in mind.
- **Threat detection and response**: XAI techniques contribute to cybersecurity by providing explanations for the behavior and decisions of AI-based threat detection systems. XAI helps security analysts understand why a particular event or behavior is flagged as a potential threat, improving the accuracy and reliability of detection systems. Explanations provided by XAI can assist in incident response and facilitate the investigation and mitigation of security incidents.
- **Intrusion detection and prevention**: HCI and XAI play a role in designing interfaces for intrusion detection and prevention systems (IDPS). HCI principles ensure that the interfaces present alerts and information about detected threats in a clear and actionable manner, enabling security analysts to respond effectively. XAI techniques can explain the reasoning behind the IDPS's actions, helping analysts understand the basis for alerts and improving trust in the system.
- **Trust and transparency**: HCI and XAI contribute to building trust and transparency in cybersecurity systems. HCI principles guide the design of interfaces that provide feedback, notifications, and explanations about security measures and actions taken by security systems. XAI techniques help provide explanations for the behavior of AI algorithms used in security systems, promoting trust and enabling users and security analysts to understand and validate security decisions.

In summary, HCI and XAI play important roles in improving cybersecurity by enhancing user awareness, designing usable security interfaces, improving threat detection and response, promoting trust and transparency, and addressing ethical considerations. By incorporating HCI and XAI principles and techniques, cybersecurity systems can become more effective, user-friendly, and trustworthy in protecting digital assets and ensuring online safety.

14.3.9 Role of HCI and XAI in Software Development

Human-computer interaction (HCI) and explainable artificial intelligence (XAI) have significant roles in software development, enhancing user experiences, improving software quality, and enabling the development of more intelligent and trustworthy systems. Here's how HCI and XAI contribute to software development:

- **User-centered design**: HCI principles guide the design of user interfaces and interactions in software development. HCI focuses on understanding user needs, goals, and preferences, ensuring that software is designed with the user in mind. User-centered design techniques, such as user research, personas, and usability testing, help developers create intuitive and user-friendly interfaces that enhance the overall user experience.
- **Iterative design and prototyping**: HCI promotes an iterative design approach in software development, where early prototypes are created and evaluated with user feedback. This iterative process helps developers refine their designs, uncover usability issues, and make improvements based on user insights. HCI techniques, such as usability testing and user feedback collection, ensure that software interfaces are continuously refined and optimized for user satisfaction and efficiency.
- **Intelligent and adaptive interfaces**: HCI and XAI techniques enable the development of intelligent and adaptive interfaces in software. HCI principles guide the design of interfaces that can adapt to user preferences, context, and behavior. XAI techniques can provide explanations for adaptive interface decisions, helping users understand and trust the system's behavior. This combination allows software to deliver personalized experiences, optimize workflows, and improve user productivity.
- **Usability and accessibility testing**: HCI emphasizes the importance of usability and accessibility in software development. Usability testing techniques, such as user observation and interviews, help identify and address usability issues early in the development process. Accessibility considerations ensure that software is designed to be inclusive and usable by individuals with disabilities. HCI techniques ensure that software is tested and refined to meet usability and accessibility standards.
- **Quality assurance and bug detection**: HCI and XAI contribute to quality assurance in software development. HCI principles help identify usability issues, interface inconsistencies, and user experience problems that can impact software quality. XAI techniques can assist in identifying and diagnosing software bugs, providing explanations for unexpected behaviors or errors. This aids developers in troubleshooting and improving software reliability.

By incorporating HCI and XAI principles and techniques, software development can create user-centric, intelligent, and reliable software systems. HCI ensures that software interfaces are designed to meet user needs and preferences, while XAI provides transparency and understanding in AI-driven software. The combination of HCI and XAI improves the overall quality, usability, and ethical aspects of software development.

14.3.10 Role of HCI and XAI in Textile Industry

Human-computer interaction (HCI) and explainable artificial intelligence (XAI) play important roles in the textile industry, enhancing manufacturing processes, improving product design and customization, and enabling efficient decision-making. Here's how HCI and XAI contribute to the textile industry:

- **Design and visualization**: HCI principles are applied to develop user-friendly interfaces and tools for textile design and visualization. HCI techniques enable designers to create and modify textile patterns, colors, and textures using intuitive software interfaces. Visualization tools help designers preview and simulate the appearance of textiles in different contexts, facilitating the design process and reducing time and material waste.
- **Customization and personalization**: HCI and XAI enable textile customization and personalization. HCI principles guide the design of interfaces and interactions that allow customers to customize textiles based on their preferences, such as choosing patterns, colors, and sizes. XAI techniques can provide explanations for AI-based recommendations, assisting customers in making informed decisions about personalized textile choices.
- **Supply chain management**: HCI and XAI contribute to efficient supply chain management in the textile industry. HCI principles guide the design of user interfaces for tracking and managing textile inventory, logistics, and production schedules. XAI techniques can provide insights and explanations for AI-driven demand forecasting, helping stakeholders understand the reasoning behind production and procurement decisions.
- **Quality control**: HCI and XAI techniques play a role in textile quality control. HCI principles guide the design of interfaces and interactions for automated inspection systems, enabling operators to review and analyze images or sensor data to identify fabric defects or inconsistencies. XAI techniques can provide explanations for AI algorithms used in quality control, helping operators understand the reasoning behind defect detection or classification.
- **Sustainability and material optimization**: HCI and XAI contribute to sustainable practices in the textile industry. HCI principles guide the design of interfaces that enable textile manufacturers to monitor and optimize resource consumption, waste generation, and environmental impact. XAI techniques can provide insights and explanations for AI algorithms used in material optimization, helping manufacturers make informed decisions about reducing waste and improving sustainability.
- **Customer feedback and sentiment analysis**: HCI and XAI techniques enable the analysis of customer feedback and sentiment related to textiles. HCI principles guide the design of interfaces for collecting and analyzing customer feedback, enabling textile companies to understand customer preferences and improve their products. XAI techniques can assist in sentiment analysis, providing explanations for AI algorithms that analyze customer reviews and feedback.

By using HCI and XAI principles and techniques, the textile industry can enhance design processes, improve customization and personalization, optimize supply chain management, ensure quality control, promote sustainability, and analyze customer feedback. The combination of user-centric interfaces, explainable AI, and data-driven decision-making contributes to the efficiency, innovation, and sustainability of the textile industry.

14.4 CONVERSATION AI IMPORTANCE IN HCI AND XAI

Conversational AI is expected to play a significant role in the near future due to several reasons. Here are some of the key aspects that highlight its importance:

- **Enhanced customer experience**: conversational AI enables businesses to provide personalized and interactive experiences to their customers. Through chatbots, virtual assistants, or voice-enabled devices, customers can receive real-time assistance, resolve queries, and access information conveniently. This improves customer satisfaction and loyalty.
- **Automation and efficiency**: conversational AI can automate repetitive and mundane tasks, freeing up human resources to focus on more complex and value-added activities. By using natural language processing (NLP) and machine learning algorithms, conversational AI systems can understand and respond to user queries, process requests, and perform specific actions autonomously.
- **24/7 availability**: conversational AI systems can operate round the clock, providing continuous support and engagement to users. This is particularly valuable for businesses operating in different time zones or those dealing with high volumes of customer inquiries. Users can access information or assistance at any time, improving responsiveness and accessibility.
- **Scalability and cost efficiency**: conversational AI solutions can handle multiple conversations simultaneously, making them highly scalable. This enables businesses to handle a large number of customer interactions without requiring a proportional increase in human resources. It can also reduce operational costs associated with customer support, as automated systems can handle a significant portion of inquiries.
- **Data insights and analytics**: conversational AI platforms generate vast amounts of data through user interactions. By analyzing this data, businesses can gain valuable insights into customer preferences, behavior patterns, and pain points. This information can be used to refine products and services, personalize marketing campaigns, and improve overall business strategies.
- **Future growth potential**: as AI technologies continue to advance, conversational AI is expected to become even more sophisticated. Natural language understanding, sentiment analysis, and context awareness will improve, enabling more natural and meaningful interactions. This opens up opportunities for businesses to create more engaging and immersive experiences for their customers.

In summary, conversational AI holds great potential to transform customer service, streamline business operations, and drive innovation across industries. Its importance in the near future lies in its ability to enhance customer experiences, automate tasks, and use data-driven insights for better decision-making.

14.4.1 Conversational AI Importance in XAI and HCI

Conversational AI has significant importance in both explainable AI (XAI) and human-computer interaction (HCI). Here's how it contributes to these domains:

- **Explainable AI (XAI)**: XAI focuses on making AI systems more transparent and understandable to humans. Conversational AI plays a important role in enabling explainability by providing natural language interfaces for users to interact with AI models and understand their decisions. Conversational AI systems can explain the reasoning behind AI outputs, clarify uncertainties, and provide contextual information to users, enhancing the trust and interpretability of AI systems.
- **Human-computer interaction (HCI)**: conversational AI improves the interaction between humans and computers by enabling more natural and intuitive communication. Traditional graphical user interfaces (GUIs) often require users to learn specific commands or navigate complex menus, while conversational interfaces allow users to interact using everyday language. This reduces the cognitive load on users and makes technology more accessible, especially for individuals with limited technical skills or disabilities.
- **User engagement and personalization**: conversational AI can enhance user engagement by creating more interactive and personalized experiences. Through natural language understanding and context awareness, conversational AI systems can adapt to user preferences, provide tailored recommendations, and deliver relevant information. This level of personalization leads to more engaging and satisfying interactions, promoting positive user experiences.
- **Assistance and support**: conversational AI can act as a virtual assistant, providing real-time support and assistance to users. In HCI, conversational interfaces enable users to easily access information, perform tasks, and receive guidance or recommendations. This is particularly valuable for individuals who may have difficulty navigating complex interfaces or require hands-free interaction, such as people with disabilities or those using voice-enabled devices.
- **Feedback and iterative design**: conversational AI systems can collect valuable feedback from users, helping to improve the design and functionality of AI applications. By analyzing user interactions, sentiment, and preferences, HCI practitioners and AI developers can identify pain points, address usability issues, and iterate on the system's design to optimize the user experience.

In summary, conversational AI contributes to XAI and HCI by improving transparency, enabling natural and intuitive interactions, enhancing personalization and user engagement, facilitating assistance and support, promoting collaboration, and providing valuable feedback for iterative design and improvement. By integrating conversational AI into these domains, we can create more understandable, usable, and user-centric AI systems.

14.5 HCI AND XAI: CURRENT CHALLENGES

Conversational AI, HCI (human-computer interaction), and XAI (explainable AI) are related but distinct fields. Here's a brief overview of each:

- **Conversational AI**: Conversational AI focuses on enabling natural language interactions between humans and computers. It involves the development of systems like chatbots, virtual assistants, and voice-enabled devices that can understand and respond to human queries and commands. Conversational AI aims to create more intuitive and engaging user experiences by using techniques such as natural language processing (NLP), dialogue management, and machine learning algorithms.
- **HCI (human-computer interaction)**: HCI is an interdisciplinary field that deals with the design, evaluation, and implementation of interactive computing systems. It focuses on improving the interaction between humans and computers, making technology more usable, accessible, and user-centric. HCI involves studying user behaviors, cognitive processes, and interface design principles to create intuitive and efficient interfaces. It encompasses various aspects like graphical user interfaces (GUIs), user experience (UX) design, information architecture, and usability testing.
- **XAI (explainable AI)**: XAI is concerned with making AI systems transparent and understandable to humans. As AI models become more complex and sophisticated, it becomes important to explain their decision-making processes and provide insights into how they arrive at their outputs. XAI techniques aim to provide explanations, justifications, and insights into AI systems' behavior, making them more interpretable and trustworthy. This field involves developing methods like rule-based explanations, visualizations, and model-agnostic approaches to provide understandable explanations for AI decisions.

While there is overlap between these fields, they have distinct focuses:

- Conversational AI emphasizes creating natural language interfaces and enhancing user interactions through dialogue systems and virtual assistants.
- HCI concentrates on designing intuitive and user-friendly interfaces, studying user behavior, and improving the overall interaction between humans and computers.
- XAI aims to make AI models and systems more transparent and explainable, enabling users to understand and trust AI decisions.

However, these fields can intersect and benefit from each other. For instance, HCI principles can guide the design of conversational AI interfaces to make them more usable and engaging. XAI techniques can be applied within conversational AI systems to provide explanations for the recommendations or decisions made by virtual assistants. In summary, conversational AI, HCI, and XAI are complementary fields that contribute to creating better human-computer interactions, understandable AI systems, and more intuitive user experiences.

14.5.1 Important Challenges towards HCI

Human-computer interaction (HCI) faces several important challenges that impact the design, development, and implementation of interactive computing systems. Here are some key challenges in HCI:

- **Usability**: designing interfaces that are intuitive, easy to learn, and efficient to use remains a central challenge. HCI practitioners need to consider diverse user groups with varying abilities, preferences, and prior knowledge. Striking a balance between simplicity and functionality while accommodating different user needs is important.
- **Accessibility and inclusivity**: ensuring that interactive systems are accessible to individuals with disabilities is an ongoing challenge. HCI needs to address issues related to visual impairments, hearing impairments, motor disabilities, cognitive limitations, and other accessibility requirements. Designing interfaces that accommodate a wide range of users and provide equal access to information and functionalities is essential.
- **Cross-platform and device compatibility**: with the proliferation of various devices and platforms, creating consistent user experiences across different interfaces and screen sizes is challenging. HCI professionals need to consider the responsive design, adaptability, and usability of interfaces across multiple platforms, including desktops, mobile devices, wearables, and emerging technologies like virtual reality and augmented reality.
- **Privacy and security**: as technology becomes increasingly pervasive, privacy and security concerns are important. HCI practitioners must address the challenge of designing interfaces that allow users to understand and control the data they share. Ensuring secure interactions, safeguarding sensitive information, and maintaining user trust are important aspects of HCI in today's digital landscape.
- **Context awareness**: designing interfaces that understand and adapt to the user's context presents a significant challenge. HCI needs to consider factors such as the user's location, time, preferences, and social environment to provide personalized and relevant experiences. Incorporating context-awareness into interactive systems requires understanding user behaviors, using sensor data, and designing intelligent interfaces that dynamically adapt to changing contexts.

Addressing these challenges requires interdisciplinary collaboration, involving HCI researchers, designers, psychologists, engineers, and other stakeholders. Additionally, ongoing research, user-centered design approaches, iterative evaluation, and user feedback are important for advancing the field of HCI and overcoming these important challenges.

14.5.2 Important Challenges towards XAI

Explainable AI (XAI) presents several important challenges that need to be addressed to ensure the transparency, interpretability, and trustworthiness of AI systems. Here are some key challenges in XAI:

- **Complexity of AI models**: AI models, such as deep neural networks, can be highly complex and operate as black boxes, making it challenging to understand how they arrive at their decisions. XAI needs to develop techniques to explain the reasoning and internal workings of these complex models in a human-understandable manner.
- **Model-agnostic explanations**: XAI faces the challenge of providing explanations that are not tied to specific AI models or algorithms. Developing model-agnostic approaches allows explanations to be generated regardless of the underlying AI techniques, enabling broader applicability and interpretability across different types of models.
- **Balancing explanation accuracy and simplicity**: XAI aims to strike a balance between providing accurate explanations and keeping them simple enough for users to understand. The challenge lies in distilling complex AI processes into explanations that are comprehensible, yet accurate and meaningful enough for users to trust the AI system.
- **Handling uncertainty**: AI models often operate in uncertain and probabilistic domains, where explanations cannot be deterministic. XAI needs to tackle the challenge of effectively communicating uncertainty and conveying confidence levels in AI outputs to users, ensuring that they understand the limitations and potential risks associated with the decisions made by AI systems.
- **User-centric explanations**: XAI should focus on generating explanations that align with the users' mental models and cognitive capabilities. This challenge involves understanding users' information needs, tailoring explanations to their level of expertise, and adapting explanations to different user contexts and preferences.
- **Trade-off between accuracy and explainability**: AI models can achieve high accuracy by using complex, opaque techniques, but these models are often less interpretable. XAI faces the challenge of striking a balance between model accuracy and explainability, finding methods that provide sufficient transparency without significantly sacrificing predictive performance.

Addressing these challenges requires interdisciplinary collaboration involving experts from AI, machine learning, human-computer interaction, cognitive science, and ethics. Research and development efforts should focus on developing new XAI techniques, creating standards and guidelines for explainability, and incorporating user feedback and evaluations to improve the interpretability of AI systems. Additionally, educating users and stakeholders about the limitations and possibilities of XAI is important for building trust and acceptance of AI technologies.

14.5.3 Important Challenges towards Conversational AI

Conversational AI faces several important challenges that impact its development, deployment, and effectiveness. Here are some key challenges in conversational AI:

- **Natural language understanding**: accurately understanding and interpreting user input in natural language is a significant challenge. Conversational AI

systems need to handle various linguistic nuances, context dependencies, slang, abbreviations, and user intents to provide meaningful and accurate responses. Achieving robust natural language understanding (NLU) is important for effective conversational interactions.
- **Context and continuity**: maintaining context and continuity in conversations is essential for natural and engaging interactions. Conversational AI systems must be able to remember and refer back to previous exchanges, understand implicit references, and adapt responses based on the ongoing conversation. Managing context across multiple turns and providing coherent and contextually relevant responses are ongoing challenges.
- **Multimodal interactions**: conversational AI traditionally relies on text-based interactions, but there is an increasing demand for multimodal capabilities that incorporate voice, images, videos, and other modalities. Handling multimodal input and generating appropriate multimodal responses require advanced technologies to understand and generate content in different formats, adding complexity to conversational AI systems.
- **Personalization and user context**: delivering personalized conversational experiences tailored to individual users' preferences, needs, and context is a challenge. Conversational AI systems need to gather and utilize user data effectively while addressing privacy concerns. They should adapt to user preferences, provide recommendations, and maintain a consistent user profile across different interactions and channels.
- **Handling ambiguity and errors**: conversational AI must be able to handle ambiguity, errors, and misunderstandings that commonly occur in human conversations. Understanding and recovering from user errors, disambiguating unclear queries, and handling out-of-scope or unrelated inputs are challenges that require robust error handling and graceful fallback mechanisms.
- **Social and emotional intelligence**: conversational AI systems need to exhibit social and emotional intelligence to create more natural and empathetic interactions. Understanding and responding to emotions, detecting sarcasm, humor, or implicit meanings, and expressing appropriate social cues pose challenges in building conversational agents that can effectively engage and empathize with users.
- **Ethical and bias considerations**: conversational AI systems should be designed and deployed with ethical considerations in mind. Ensuring fairness, avoiding bias, and mitigating potential harm or misuse are important challenges. Addressing issues like biased language generation, privacy concerns, and responsible handling of sensitive information are important for maintaining user trust.

Addressing these challenges requires advancements in natural language processing, machine learning, dialogue management, and user-centered design. It necessitates ongoing research, development, and evaluation of conversational AI systems, incorporating user feedback and conducting rigorous testing to improve their accuracy, usability, and ethical implications. Additionally, ethical guidelines and regulations can play a role in ensuring responsible deployment and usage of conversational AI technologies.

14.6 HCI AND XAI: FUTURE RESEARCH OPPORTUNITIES

This section will discuss several future research opportunities of HCI and XAI.

14.6.1 Future Research Opportunities of HCI

Human-computer interaction (HCI) is a dynamic field that provides several exciting research opportunities for the future. Here are some potential areas for future research in HCI:

- **Augmented and virtual reality (AR/VR)**: with the growing popularity of AR/VR technologies, there is ample scope for research in designing intuitive and immersive user experiences. Future HCI research can focus on enhancing interaction techniques, designing realistic virtual environments, addressing motion sickness issues, and exploring novel applications in fields like education, healthcare, and entertainment.
- **Natural user interfaces**: HCI can discuss new interaction paradigms beyond traditional graphical user interfaces (GUIs). Research can focus on developing natural and intuitive interfaces based on touch, gesture, voice, gaze, and other modalities. Advances in machine learning and computer vision can enable more accurate and robust input recognition and interaction techniques.
- **Human-robot interaction**: as robots become increasingly integrated into our lives, HCI can investigate effective ways for humans to interact with robots. Research opportunities include designing social robots that exhibit empathy and emotional intelligence, understanding user expectations and mental models of robots, and exploring collaborative tasks and communication modalities between humans and robots.
- **Ubiquitous computing and internet of things (IoT)**: HCI can delve into designing seamless interactions and interfaces in the era of ubiquitous computing and IoT. Exploring novel ways to interact with smart devices, designing intelligent and context-aware interfaces, and addressing privacy and security challenges in IoT environments are potential research areas.
- **Adaptive and personalized interfaces**: future HCI research can focus on developing interfaces that adapt to individual users' preferences, abilities, and contexts. This includes investigating techniques for user modeling, personalization algorithms, and interfaces that can dynamically adjust to different user needs and situations.
- **Social computing and online communities**: HCI can discuss the design of online platforms and social computing systems that foster collaboration, information sharing, and community building. Research opportunities include understanding social dynamics in online communities, designing inclusive and accessible platforms, and addressing issues like online misinformation and digital well-being.

- **Explainable AI and trustworthy systems**: with the increasing adoption of AI systems, HCI can contribute to developing techniques and interfaces for explainable AI (XAI) to enhance transparency, interpretability, and user trust. Research can focus on designing interfaces that facilitate user understanding of AI decisions and building systems that address biases, fairness, and accountability.
- **Human-centered AI**: HCI research can focus on human-centric approaches to AI, where AI technologies are developed to augment human capabilities, facilitate decision-making, and improve user experiences. This involves investigating ways to integrate AI seamlessly into users' daily lives, understanding user expectations and mental models of AI systems, and designing interfaces that enable effective human-AI collaboration.
- **Sustainability and green computing**: HCI can contribute to designing energy-efficient and environmentally sustainable computing systems. Research can focus on reducing energy consumption in interactive systems, promoting sustainable user behaviors, and exploring interfaces that raise awareness and encourage eco-friendly practices.

These research opportunities highlight the breadth and potential impact of HCI in shaping the future of human-technology interactions. Embracing interdisciplinary collaboration, user-centered design approaches, and considering societal implications will be important for advancing HCI research in these areas.

14.6.2 Future Research Opportunities of XAI

Explainable artificial intelligence (XAI) is an evolving field with significant research opportunities for the future. Here are some potential areas for future research in XAI:

- **Interpretable deep learning**: deep learning models have achieved remarkable success but are often considered black boxes. Future research can focus on developing techniques to make deep learning models more interpretable without sacrificing their predictive performance. This includes investigating methods for extracting meaningful explanations from complex neural networks and understanding the internal workings of deep learning models.
- **Model-agnostic explanations**: XAI research can discuss techniques that provide explanations for a wide range of machine learning models, not limited to specific algorithms or architectures. Developing model-agnostic explanation methods allows users to understand and trust a variety of AI systems, fostering transparency and accountability in AI applications.
- **Causal reasoning and counterfactual explanations**: XAI can investigate methods that go beyond correlation-based explanations and delve into causal reasoning. Understanding the causal relationships between inputs and outputs can lead to more robust and reliable explanations. Additionally, generating counterfactual explanations, which illustrate how changes in inputs

would have led to different outcomes, can provide valuable insights into AI decision-making.

- **User-centric explanations**: future XAI research can focus on designing explanations that are tailored to specific user needs and preferences. This includes exploring interactive explanations that allow users to query the AI system, ask for clarification, or customize the level of detail in explanations. Adapting explanations to individual users' cognitive abilities and mental models is another area of potential research.
- **Visual and interactive explanations**: XAI can investigate the use of visualizations and interactive interfaces to convey explanations effectively. Visual explanations can use human perceptual capabilities to present complex information in a more intuitive and understandable manner. Developing interactive interfaces that allow users to discuss and interact with AI models' internals can enhance user engagement and understanding.
- **Trust and decision-making**: understanding how users perceive and trust AI systems' explanations is an important area of research. Future XAI studies can investigate the factors that influence users' trust in AI systems and how different types of explanations impact trust. Furthermore, exploring how users incorporate AI explanations into their decision-making processes can provide valuable insights for designing more effective and trustworthy AI systems.
- **Privacy and security in XAI**: XAI research can address the challenges of providing explanations while respecting privacy and ensuring the security of sensitive data. Developing techniques that provide meaningful explanations without revealing confidential or sensitive information is important. Additionally, investigating potential vulnerabilities and adversarial attacks on XAI systems is important for maintaining the security of AI deployments.
- **Human-AI collaboration**: XAI can discuss the design of collaborative systems where humans and AI work together to solve complex problems. Investigating how to effectively integrate human judgment and expertise with AI's analytical capabilities, and designing interfaces that facilitate seamless collaboration between humans and AI, are potential research directions.
- **Multimodal explanations**: XAI can extend its focus to include explanations that incorporate multiple modalities, such as text, images, and auditory cues. Exploring how to combine different modalities to provide more comprehensive and rich explanations can enhance users' understanding and engagement with AI systems.
- **Cross-domain and generalizable explanations**: XAI research can aim to develop explanation techniques that can be applied across different domains and tasks. Generalizable explanations would allow users to understand AI systems' behavior and decision-making processes across a wide range of applications, fostering trust and usability.

These research opportunities highlight the potential for advancing XAI and addressing the need for transparent, interpretable, and accountable AI systems.

14.6.3 Future Research Opportunities of Conversational AI

Here are some potential areas for future research in conversational AI:

- **Context-aware conversational systems**: future research can focus on developing conversational systems that can effectively understand and utilize contextual information. This includes understanding the context of previous conversations, user preferences, and environmental factors to provide more personalized and relevant responses.
- **Multimodal conversations**: conversational AI can discuss the integration of multiple modalities, such as text, speech, images, and gestures, to enable richer and more natural interactions. Research opportunities include designing systems that can seamlessly switch between different modalities and understanding the challenges and opportunities in multimodal input understanding and generation.
- **Emotional intelligence**: enhancing conversational agents' emotional intelligence can lead to more engaging and empathetic interactions. Future research can focus on developing techniques to detect and respond to users' emotions, expressions of empathy, and adapting the conversational style and tone to match users' emotional states.
- **Explainable and transparent conversations**: conversational AI systems can benefit from being more transparent and providing explanations for their responses. Future research can investigate methods for generating explanations in real-time during conversations, making the decision-making process of the conversational agent more interpretable and accountable to users.
- **Lifelong learning and adaptation**: conversational AI systems can benefit from continuously learning and adapting to users' changing preferences, language usage, and knowledge. Research opportunities include developing techniques for lifelong learning in conversational agents, exploring methods for user-driven adaptation, and addressing the challenge of handling concept drift and evolving language trends.
- **Socially intelligent conversational systems**: conversational AI can strive to create more socially intelligent systems that understand social norms, etiquette, and engage in natural and socially appropriate conversations. Research can focus on developing techniques to model and respond to conversational nuances like turn-taking, politeness, and cultural variations.
- **Collaboration and coordination**: conversational AI can play a role in supporting collaborative tasks and group conversations. Future research can investigate how conversational agents can facilitate coordination, negotiation, and information sharing in group settings, enabling more effective and efficient collaboration among users.
- **Ethical considerations**: conversational AI research should also consider ethical considerations, such as privacy, fairness, and bias. Exploring methods to mitigate biases and ensure fairness in conversational interactions, protecting user privacy, and adhering to ethical guidelines are important areas for future research.

- **User experience and engagement**: future research can focus on enhancing the user experience and engagement with conversational agents. This includes investigating methods for generating more natural and engaging dialogue, designing effective user interfaces for conversational interactions, and considering the impact of personalization on user satisfaction.
- **Real-world applications**: conversational AI has potential applications in various domains, including healthcare, customer service, education, and entertainment. Future research can discuss domain-specific challenges and opportunities, tailoring conversational AI systems to the specific needs and requirements of different application areas.

These research opportunities highlight the broad scope for advancements in conversational AI, encompassing technical, social, and ethical aspects. Interdisciplinary collaboration, incorporating user feedback, and addressing the practical challenges of deploying conversational systems in real-world settings will be important for advancing the field.

14.7 CONCLUSION

The applications of Human-Computer Interaction (HCI), explainable artificial intelligence (XAI), and conversational artificial intelligence (CAI) across various real-life sectors hold immense potential to enhance user experiences, improve decision-making processes, and drive innovation. Through HCI, users can interact with computers and digital interfaces in intuitive and efficient ways, leading to more engaging and user-friendly experiences in sectors such as education, healthcare, entertainment, and retail. By integrating principles of usability, accessibility, and user-centered design, HCI technologies enable individuals to interact with digital systems seamlessly, regardless of their technical expertise or physical capabilities. Explainable artificial intelligence (XAI) plays a important role in enhancing transparency, accountability, and trustworthiness in AI-driven decision-making processes across diverse domains, including finance, healthcare, cybersecurity, and autonomous systems. By providing interpretable explanations for AI-generated insights and recommendations, XAI algorithms enable users to understand the underlying reasoning behind AI-driven decisions, identify potential biases or errors, and make informed choices based on reliable information. Note that conversational artificial intelligence (CAI) technologies, such as virtual assistants, chatbots, and voice-activated systems, revolutionize human-computer interactions by enabling natural language communication and personalized assistance in real-time. From customer service and e-commerce to productivity tools and smart home devices, CAI systems facilitate seamless communication and task completion, enhancing user productivity, convenience, and satisfaction. In real-life sectors, the integration of HCI, XAI, and CAI technologies provides transformative benefits like enhanced user experience, transparent and trustworthy AI, personalized assistance and automation, empowering users and driving innovation. In summary, the integration of HCI, XAI, and CAI technologies in real-life sectors represents a paradigm shift in human-computer interactions, enabling more natural, transparent, and personalized interactions between humans and machines. As these technologies continue to evolve and mature, they have the potential to reshape industries, improve societal well-being, and unlock new opportunities for human creativity, productivity, and collaboration.

REFERENCES

1 Abd-Alsabour, N. M., El-Sayed, S. E., Ali, H. A. (2020). Using Explainable Artificial Intelligence to Diagnose Heart Disease. *In 2020 11th International Conference on Information Technology (ICIT)* (pp. 1–7). IEEE.

2 Ahmadi, A., Gamon, M. (2018). Explaining Machine Learning Models in Human-AI Interaction: Survey and Review. arXiv preprint arXiv:1810.02942.

3 Alhaddad, M., Wong, K. W., Fiaidhi, J., Lin, C. T. (2021). Intelligent and Explainable Human-Computer Interaction System for Personalized Education. *IEEE Access*, 9, 16941–16960.

4 Azodi, C. B., Furnas, G. W., Landauer, T. K. (1999). Adaptive incremental learning for user modeling. *User Modeling and User-Adapted Interaction*, 9(2-3), 147–189.

5 Bajaj, N., Pardasani, P. (2021). Conversational AI Systems: A Comparative Study. In *2021 4th International Conference on Computing, Communication and Security (ICCCS)* (pp. 1–6). IEEE.

6 Balachandran, K., Shanmugapriya, M. (2019). *A Survey on Explainable Artificial Intelligence. In 2019 3rd International Conference on Electronics, Communication and Aerospace Technology (ICECA)* (pp. 394–397). IEEE.

7 Dali, K., Jain, V., Akhtar, S. (2020). Conversational artificial intelligence in healthcare. In *Proceedings of the Second International Conference on Computer and Communication Systems* (pp. 442–446). Springer.

8 Dey, A., Jain, A., Guleria, A. (2021). Explainable Artificial Intelligence (XAI): A Comprehensive Review. In *2021 3rd International Conference on Inventive Research in Computing Applications (ICIRCA)* (pp. 1–7). IEEE.

9 Fiebrink, R., Cook, P. R., Trueman, D. (2011). Human model evaluation in interactive supervised learning. In *Proceedings of the SIGCHI Conference on Human Factors in Computing Systems* (pp. 147–156). Association for Computing Machinery.

10 Hupont, I., Khanna, P. (2020). Human-Computer Interaction in the Age of Artificial Intelligence. In *Design, User Experience, and Usability: Designing Interactions.* Springer (pp. 243–255).

11 Jia, Y., Li, C., Shen, Z. (2019). Explainable Artificial Intelligence in Human-Robot Interaction: A Review. In *2019 4th International Conference on Robotics and Automation Engineering (ICRAE)* (pp. 235–239). IEEE.

12 Liu, W., He, Y., Wang, Z., Qian, Y. (2021). Applications of Conversational Artificial Intelligence in Marketing. In *2021 International Conference on Intelligent Information Technologies (ICIIT)* (pp. 81–85). IEEE.

13 Maruyama, T., Sumi, Y. (2019). Explanation generation for human-computer interaction based on explainable artificial intelligence. In *Proceedings of the 24th International Conference on Intelligent User Interfaces* (pp. 398–408). Association for Computing Machinery.

14 Olanrewaju, R. F., Moh, S. (2020). Conversational Artificial Intelligence (CAI) and Its Applications in E-Commerce: A Review. *Journal of Theoretical and Applied Information Technology*, 98(7), 1439–1450.

15 Rader, E., Murphy, R., Peintner, B., Blaha, L. M. (2021). Interdisciplinary approaches to human-computer interaction and explainable artificial intelligence in healthcare. In *Explainable AI for Healthcare* (pp. 217–244). Springer.
16 Saha, A., Alam, M. J. E., Nasipuri, M. (2019). Conversational artificial intelligence (CAI) in retail business: An analytical study. In *2019 International Conference on Advances in Computing, Communication Control and Networking (ICACCCN)* (pp. 152–155). IEEE.
17 Shah, A., Pangarkar, N., Peddinti, S. (2018). Conversational AI: Next frontier in customer experience management. *Business Horizons*, 61(4), 557–566.
18 Susanto, R., Wicaksana, I. G., Eka Putri, A. K. (2021). Human-Computer Interaction based on Explainable Artificial Intelligence in Health Care. In *2021 2nd International Conference on Data and Information Science (ICoDIS)* (pp. 167–171). IEEE.
19 Wang, Z., Zhang, J., Xu, Y. (2021). Application of Conversational Artificial Intelligence (CAI) in Education Industry. In *2021 International Conference on Intelligent Information Technologies (ICIIT)* (pp. 107–110). IEEE.
20 Yang, C., Ritter, F.E. (2017). Integrating human-computer interaction and artificial intelligence: Opportunities and challenges. In *Proceedings of the 2017 CHI Conference Extended Abstracts on Human Factors in Computing Systems* (pp. 2772–2779). Association for Computing Machinery.
21 Singh, R., Tyagi, A.K., Arumugam, S. K. (2024) Imagining the Sustainable Future With Industry 6.0: A Smarter Pathway for Modern Society and Manufacturing Industries. In *Machine Learning Algorithms Using Scikit and TensorFlow Environments*. IGI Global, DOI: 10.4018/978-1-6684-8531-6.ch016
22 Tyagi, A.K., Richa. (2023). Digital Twin Technology: Opportunities and Challenges for Smart Era's Applications. In *Proceedings of the 2023 Fifteenth International Conference on Contemporary Computing (IC3-2023)*. Association for Computing Machinery, (pp. 328–336). https://doi.org/10.1145/3607947.3608015
23 Tyagi, A.K., Tiwari, S. (2024). The Future of Artificial Intelligence in Blockchain Applications. In *Machine Learning Algorithms Using Scikit and TensorFlow Environments*. IGI Global, DOI: 10.4018/978-1-6684-8531-6.ch018
24 Tyagi, A.K., Manoj Nair, M. (2022). Preserving Privacy using Distributed Ledger Technology in Intelligent Transportation System. In *Proceedings of the 2022 Fourteenth International Conference on Contemporary Computing (IC3-2022)*. Association for Computing Machinery, (pp. 582–590). https://doi.org/10.1145/3549206.3549306
25 Deshmukh, A., Patil, D., Tyagi, A.K., Arumugam, S.S., Arumugam. (2022). Recent Trends on Blockchain for Internet of Things based Applications: Open Issues and Future Trends. In *Proceedings of the 2022 Fourteenth International Conference on Contemporary Computing (IC3-2022)*. Association for Computing Machinery (pp. 484–492). https://doi.org/10.1145/3549206.3549289
26 Jayaprakash, V., Tyagi, A.K. (2022). Security Optimization of Resource-Constrained Internet of Healthcare Things (IoHT) Devices Using Asymmetric Cryptography for Blockchain Network. In *Proceedings of International Conference on Network Security and Blockchain Technology*, Giri, D., Mandal, J.K., Sakurai, K., De, D. (eds). ICNSBT 2021. Lecture Notes in Networks and Systems, vol 481. Springer. https://doi.org/10.1007/978-981-19-3182-6_18
27 Pandey, A.A., Fernandez, T.F., Bansal, R., Tyagi, A.K. (2022). Maintaining Scalability in Blockchain. In *Intelligent Systems Design and Applications*, Abraham, A., Gandhi, N., Hanne, T., Hong, TP., Nogueira Rios, T., Ding, W. (eds). ISDA 2021. Lecture Notes in Networks and Systems, vol 418. Springer. https://doi.org/10.1007/978-3-030-96308-8_4

28 Manoj Nair, M., Tyagi, A.K. (2023). 6G: Technology, Advancement, Barriers, and the Future. In *6G-Enabled IoT and AI for Smart Healthcare*, Kumar, A., Jain, R., Gupta, M., Islam, S. M. N. (eds). CRC Press.
29 Tyagi, A.K., Fernandez, T.F., Mishra, S., Kumari, S. (2021). Intelligent Automation Systems at the Core of Industry 4.0. In *Intelligent Systems Design and Applications*, Abraham, A., Piuri, V., Gandhi, N., Siarry, P., Kaklauskas, A., Madureira, A. (eds). ISDA 2020. Advances in Intelligent Systems and Computing, vol 1351. Springer. https://doi.org/10.1007/978-3-030-71187-0_1
30 Tyagi, A., Kukreja, S., Nair, M.M., Tyagi, A.K. (2022). Machine Learning: Past, Present and Future, *Neuroquantology*, 20(8), DOI: 10.14704/nq.2022.20.8.NQ44468
31 Manoj Nair, M., Tyagi, A.K. (2023). Blockchain Technology for Next-generation Society: Current Trends and Future Opportunities for Smart Era. In *Blockchain Technology for Secure Social Media Computing*. Institution of Engineering and Technology. DOI: 10.1049/PBSE019E_ch11
32 Tyagi, A.K., Lakshmi Priya, R., Mishra, A.K., Balamurugan, G. (2023). Industry 5.0: Potentials, Issues, Opportunities, and Challenges for Society 5.0. In *Privacy Preservation of Genomic and Medical Data*, Tyagi, A. K. (ed). Wiley (pp. 409–32).
33 Manoj Nair, M., Tyagi, A.K. (2023). AI, IoT, Blockchain, and Cloud Computing: The Necessity of the Future. In *Distributed Computing to Blockchain*, Pandey, R., Goundar, S., Fatima, S. (eds). Academic Press (pp. 189–206), ISBN 9780323961462, https://doi.org/10.1016/B978-0-323-96146-2.00001-2.
34 Deshmukh, A., Patil, D.S., Shyam Mohan, J.S., Balamurugan, G., Tyagi, A.K. (2023). Transforming Next Generation-Based Artificial Intelligence for Software Development: Current Status, Issues, Challenges, and Future Opportunities. In *Emerging Technologies and Digital Transformation in the Manufacturing Industry*, Tiburcio, A., Ortiz-Rodriguez, F., Hernández-González, L. M., Tiwar, S. (eds). IGI Global. DOI: 10.4018/978-1-6684-8088-5.ch003.
35 Deshmukh, A., Patil, D. S., Pawar, P. D., Kumari, S. (2023). Recent Trends for Smart Environments With AI and IoT-Based Technologies: A Comprehensive Review. In *Handbook of Research on Quantum Computing for Smart Environments*, Tyagi, A. (Ed.). IGI Global (pp. 435–452). https://doi.org/10.4018/978-1-6684-6697-1.ch023
36 Tyagi, A.K. (2023) Decentralized Everything: Practical Use of Blockchain Technology in Future Applications, In *Distributed Computing to Blockchain*, Pandey, R., Goundar, S., Fatima, S. (eds). Academic Press (pp. 19–38), ISBN 9780323961462, https://doi.org/10.1016/B978-0-323-96146-2.00010-3.
37 Deshmukh, A., Patil, D.S., Soni, G., Tyagi, A.K. (2023). Cyber Security: New Realities for Industry 4.0 and Society 5.0. In *Handbook of Research on Quantum Computing for Smart Environments*, Tyagi, A. (Ed.) IGI Global (pp. 299–325). https://doi.org/10.4018/978-1-6684-6697-1.ch017
38 Dangey, R., Tandon, A., Tyagi, A.K. (2023). Emerging Internet of Things (IoTs) Scenarios Using Machine Learning for 6G Over 5G-Based Communications. In *6G-Enabled IoT and AI for Smart Healthcare,* Kumar, A., Jain, R., Gupta, M., Islam, S. M. N. (eds). CRC Press (pp. 215–237).
39 Hariharan, R., Tyagi, A.K., Soni, G. (2023). A Survey on Blockchain-Internet of Things-Based Solutions. In *Privacy Preservation and Secured Data Storage in Cloud Computing*. IGI Global. DOI: 10.4018/979-8-3693-0593-5.ch005
40 Balamurugan, G., Tyagi, A.K., Richa, (2023). A Survey on Privacy Preserving and Trust Building Techniques of Blockchain-Based Systems. In *Privacy Preservation and Secured Data Storage in Cloud Computing.* IGI Global. DOI: 10.4018/979-8-3693-0593-5.ch019

41 Sai Dhakshan, Y., Tyagi, A.K. (2023) Introduction to Smart Healthcare: Healthcare Digitization. In *6G-Enabled IoT and AI for Smart Healthcare,* Kumar, A., Jain, R., Gupta, M., Islam, S. M. N. (eds). CRC Press. DOI: 10.1201/9781003321668-1

42 Tyagi, A.K., Nair, M.M. (2021) Deep Learning for Clinical and Health Informatics. In *Computational Analysis and Deep Learning for Medical Care: Principles, Methods, and Applications*, Tyagi, A.K. (ed). Wiley-Scrivener. DOI: 10.1002/9781119785750.ch5

43 Tyagi, A.K., Gupta, M., Aswathy, S.U., Chetanya, V. (2021). Healthcare Solutions for Smart Era: An Useful Explanation from User's Perspective. In *Recent Trends in Blockchain for Information Systems Security and Privacy*, Tyagi, A. K., Abraham, A. (eds). CRC Press. DOI: 10.1201/9781003139737-13

15 AI for Industry 4.0 with Real-world Problems

Bireshwar Dass Mazumdar, Manmohan Mishra, Anindya Ghatak, and Virendra Kumar Verma

15.1 INTRODUCTION

The advent of Industry 4.0, or the fourth industrial revolution, has ushered in a profound transformation within the manufacturing industry. This revolution is characterized by the integration of cutting-edge technologies such as artificial intelligence (AI), the internet of things (IoT), and robotics, effectively reshaping the landscape of manufacturing operations (Kiciński et al., 2021). AI has been a crucial driver in the development of Industry 4.0, enabling machines to learn from data, adapt to new situations, and make autonomous decisions (Dopico et al., 2016).In this research, we will explore the potential of AI to address real-world problems in Industry 4.0. The manufacturing industry can benefit from AI-powered solutions for predictive maintenance, quality control, supply chain optimization, autonomous robots, and energy efficiency, among others (Maddikunta et al., 2022). These applications can improve efficiency, productivity, and sustainability, leading to a more innovative and prosperous industry. However, it is essential to consider the ethical implications of AI on workers, society, and the environment (Di Vaio et al., 2020). In this article, we will examine successful implementations of AI in addressing real-world problems in the manufacturing industry and strategies for responsible and ethical purpose of AI in Industry 4.0.

15.1.1 The Role of AI in Industry 4.0 Revolution

Industry 4.0 is the fourth industrial revolution characterized by the integration of advanced technologies such as AI, IoT, and robotics into manufacturing processes (Tjahjono et al., 2017). It aims to create a highly linked and automated environment where machines can communicate with each other and make autonomous decisions, resulting in increased efficiency and productivity (Oztemel, E., & Gursev, S. 2020). The role of AI in Industry 4.0 is crucial as it enables machines to learn from data, analyze it, and make decisions based on the information gathered (Zhong et al., 2017). AI-powered solutions can be used optimize the industrial characteristics for win-win situations. By leveraging the power of AI, the manufacturing industry can improve performance, reduce costs, and enhance sustainability (Stoica et al., 2017). The integration of AI in

DOI: 10.1201/9781003480860-15

Industry 4.0 represents a significant shift in how the manufacturing industry operates and has the potential to revolutionize it for the better (Zhong et al., 2017).

15.1.2 Problems in the Manufacturing Industry

Real-world problems of the manufacturing industry are critical for several reasons. First, the manufacturing industry is a significant contributor to the global economy, and any inefficiencies or issues can have a ripple effect on other industries and the economy as a whole (Ivanov, D., Dolgui, A., & Sokolov, B. 2019). Therefore, resolving real-world problems can improve the industry's efficiency and productivity, contributing to economic growth.

The real-world problems such as energy consumption, waste generation, and environmental impact are crucial to address, and AI-based solutions can improve the manufacturing industry's sustainability and reduce these negative impacts (Ahmad et al., 2021). Sustainable manufacturing practices can lead to cost savings, improved brand reputation, and a more environmentally conscious industry.

Finally, addressing real-world problems in the manufacturing industry can uplift the safety, security and well-being of workers (Peters, S. E., et al., 2020). AI-powered solutions such as predictive maintenance and autonomous robots can reduce the risk of accidents and injuries, leading to a safer work environment (Pan, Y., & Zhang, L. 2021). The manufacturing industry can become more efficient, sustainable, and safe, benefitting not only the industry but also society as a whole.

This chapter explores the potential for AI to solve real-world problems in Industry 4.0, the fourth industrial revolution, with a focus on the manufacturing industry. The article starts with an introduction to Industry 4.0 and the role of AI in this revolution. It then discusses the importance of addressing problems in the industry, highlighting the economic, environmental, and social benefits.

The chapter then delves into various AI-powered solutions for real-world problems in the manufacturing industry (Ivanov, D., Dolgui, A., & Sokolov, B. 2019). Each solution is discussed in detail, providing examples of successful implementations and the benefits they bring to the industry.

The chapter also addresses the ethical implications of AI on workers, society, and the environment. It discusses the importance of responsible and ethical use of AI in Industry 4.0 and provides strategies for achieving this goal.

In the conclusion, the key points discussed and highlight the potential of AI to revolutionize the manufacturing industry by solving real-world problems while promoting sustainability and ethical practices.

15.2 PREDICTIVE MAINTENANCE FOR MANUFACTURING INDUTRIES

Predictive maintenance is an AI-driven solution that leverages machine learning algorithms to anticipate equipment failure prior to its actual occurrence. By enabling maintenance teams to proactively address issues, this approach empowers them to undertake repairs and maintenance tasks in a timely manner, effectively minimizing downtime and optimizing operational efficiency. It involves analyzing real-time data from sensors to detect anomalies, identify patterns, and generate alerts when maintenance is required. Predictive maintenance can help reduce downtime, increase productivity, and prolong equipment life, resulting in cost savings and improved efficiency(Ahmad et al., 2017).

One example of predictive maintenance in the manufacturing industry is General Electric's (GE) Predix platform (Zonta, et al., 2020). Predix platform use the sensor data for analysing the machine performance and predicts equipment failure before it occurs. By using Predix, GE was able to reduce unplanned downtime by up to 5%, resulting in significant cost savings for their customers (Kwon et al., 2016).

Another example is SKF, a Swedish bearings manufacturer, which implemented an AI-based predictive maintenance system to monitor the health of its production machinery (Gao et al., 2015). The system uses various machine learning techniques to predict the machine health and when maintenance is required. This has resulted in a 25% reduction in maintenance costs and a 70% reduction in downtime (Machen et al., 2017).

Predictive maintenance is an excellent example of how AI can help solve real-world problems in the manufacturing industry, improving efficiency, reducing costs, and increasing sustainability (Sarker, 2021). By detecting equipment failures before they occur, manufacturers can perform maintenance proactively, reducing the risk of accidents and injuries, and prolonging equipment life.

15.2.1 Benefits of Predictive Maintenance

Predictive maintenance is an AI-based maintenance techniques that uses ML algorithms to analyse equipment health (Chen, J., et al., 2021). It involves collecting and analyzing real-time data from sensors, equipment, and other sources to identify patterns and generate alerts when maintenance is required (Çınar et al., 2020). By using predictive maintenance, manufacturers can avoid unexpected failure. By this way manufacturers can reduce maintenance costs, and improve efficiency, sustainability, and reliability.

The benefits of predictive maintenance include:

- **Reduced downtime**: by predicting equipment failure before it occurs, maintenance teams can perform repairs proactively, avoiding unexpected downtime that can disrupt production schedules (Lee & Kim, 2019).
- **Increased efficiency**: predictive maintenance can help manufacturers optimize equipment performance, resulting in increased efficiency and productivity (Sakib & Wuest, 2018).
- **Cost savings**: by reducing unexpected downtime and optimizing equipment performance, manufacturers can save on maintenance costs and avoid expensive repairs (Ramere & Laseinde, 2021).
- **Improved safety**: predictive maintenance can help reduce the risk of accidents and injuries by detecting potential equipment failures before they occur (Pech, Vrchota & Bednář, 2021).
- **Prolonged equipment life**: by performing maintenance proactively, manufacturers can extend the life of their equipment, resulting in long-term cost savings (Lerch & Gotsch, 2015).

Hence, predictive maintenance is a powerful tool for manufacturers looking to improve their maintenance strategies and solve real-world problems in the manufacturing industry.

15.2.2 Techniques for AI-based Predictive Maintenance

According to Figure 15.1 following are phases of AI that can be used for predictive maintenance in the manufacturing industry:

- **Sensor data analysis**: AI algorithms can analyze data from sensors installed on equipment to detect anomalies and identify patterns (Himeur, Y., et al., 2021). By detecting changes in the equipment's behavior, the AI system can predict when maintenance is required (Ambika, 2020).
- **Machine learning**: AI-powered ML techniques can be trained to analyze historical data from equipment to identify patterns that lead to equipment failure. Based on this analysis, the system can predict when maintenance is required and generate alerts for maintenance teams (Ghazal et al., 2021).
- **Predictive analytics**: AI-powered predictive analytics can be used to analyse the data to predict equipment health. By using predictive analytics, manufacturers can optimize their maintenance schedules, reducing downtime and maintenance costs (Javaid et al., 2022).
- **Digital twins**: digital twins are virtual models of physical equipment that can be used to simulate equipment behavior and predict when maintenance is required. By creating digital twins, manufacturers can perform predictive maintenance in a virtual environment, reducing the risk of downtime and increasing efficiency (Qi et al., 2021).

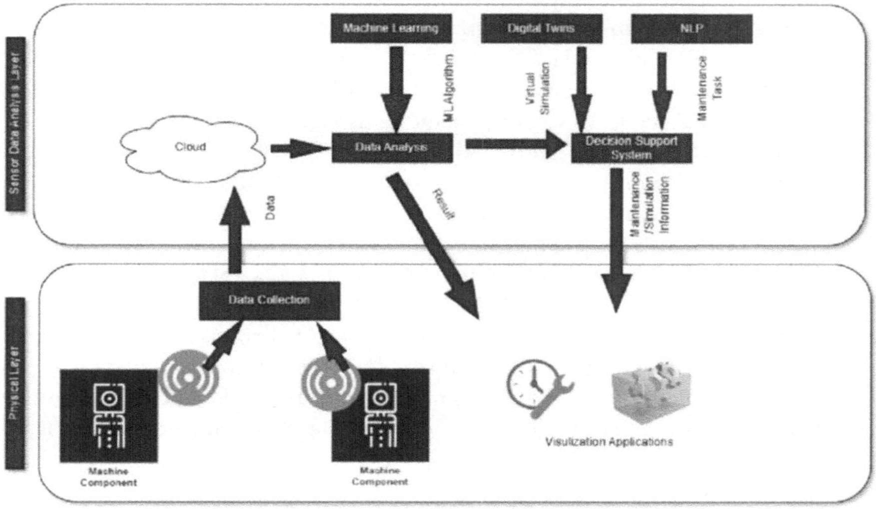

FIGURE 15.1 Phases of Predictive Maintenance in the Manufacturing Industry.

- **Natural language processing**: AI-powered natural language processing can be used to analyze maintenance logs and other textual data to identify patterns that lead to equipment failure. By using natural language processing, manufacturers can automate the process of analyzing maintenance logs, reducing the time and cost required for maintenance (Mah, Skalna & Muzam, 2022).

Overall, AI can be a powerful tool for predictive maintenance in the manufacturing industry, helping manufacturers optimize equipment performance, reduce downtime, and improve efficiency and sustainability. By using AI-powered algorithms to predict equipment failure, manufacturers can proactively address real-world problems in the manufacturing industry.

15.2.3 Successful Case Studies of AI-based Predictive Maintenance

There are several successful implementations of AI for predictive maintenance in the manufacturing industry. The following are some examples:

- **General Electric's (GE) Predix platform**: GE's Predix platform uses AI algorithms to analyze sensor data from machines and predict equipment failure before it occurs. By using Predix, GE was able to reduce unplanned downtime by up to 5%, resulting in significant cost savings for their customers (De Vasconcelos Batalha & Parli, 2017d).
- **SKF's AI-based predictive maintenance system**: SKF, a Swedish bearings manufacturer, implemented an AI-based predictive maintenance system to monitor the health of its production machinery. The system uses machine learning to analyze data from sensors installed on the equipment. This has resulted in a 25% reduction in maintenance costs and a 70% reduction in downtime (Cheng et al., 2020; Pule et al., 2023).
- **Siemens' MindSphere platform**: Siemens' MindSphere platform uses AI algorithms to analyze data from sensors installed on equipment to predict equipment failure before it occurs. By using MindSphere, Siemens was able to reduce downtime by up to 30%, resulting in significant cost savings for their customers (Petrik & Herzwurm, 2019).
- **BMW's AI-based predictive maintenance system**: BMW implemented an AI-based predictive maintenance system that analyzes data from sensors installed on its production machinery to predict when maintenance is required. This has resulted in a 5% reduction in downtime and a 15% reduction in maintenance costs (Ahmed et al., 2022).
- **Ford's AI-based predictive maintenance system**: Ford uses an AI-based predictive maintenance system that analyzes data from sensors installed on its production machinery to predict when maintenance is required. This has resulted in a 35% reduction in maintenance costs and a 50% reduction in downtime (Sharanya, 2022).

These successful implementations of AI for predictive maintenance demonstrate the significant benefits that can be achieved by using AI to optimize maintenance strategies in the manufacturing industry. By detecting equipment failures before they occur, manufacturers can perform maintenance proactively, reducing the risk of accidents and injuries, and prolonging equipment life, resulting in cost savings and improved efficiency.

15.3 AI-BASED TECHNIQUES FOR QUALITY CONTROL

Quality control is the process of ensuring that products or services meet specified quality standards. In the manufacturing industry, quality control is essential to ensure that products are safe, reliable, and meet customer expectations (Helo & Hao, 2022). AI can be used for quality control in the manufacturing industry by analyzing data from sensors, cameras, and other sources to detect defects and other quality issues. The following are some of the AI based techniques for quality control in the manufacturing industry:

- **Vision systems**: AI-based vision systems can be used to analyze images of products to detect defects and other quality issues. By using machine learning algorithms, these systems can be trained to identify specific defects, such as cracks, scratches, and color variations, resulting in improved accuracy and efficiency (Javaid et al., 2023).
- **Statistical process control**: AI based statistical process control can be used to predict data from sensors to monitor product quality in real-time. By detecting deviations from established quality standards, the system can alert production teams, allowing them to take corrective action before the production process is compromised (Montgomery & Friedman, 2020).
- **Natural language processing (NLP)**: AI integrated NLP can be used to predict customer feedback, warranty claims, and other textual data to identify patterns that indicate quality issues. By using natural language processing, manufacturers can automate the process of analyzing customer feedback, resulting in improved responsiveness and better customer satisfaction (Kang et al., 2020).
- **Predictive analytics**: AI-powered predictive analytics can be used to analyze data from multiple sources, including sensors, cameras, and maintenance logs, to predict quality issues before they occur (Dutta et al., 2021). By using predictive analytics, manufacturers can optimize their quality control processes, reducing the risk of defects and other quality issues (Soori et al., 2023).
- **Robotics**: AI-powered robotics can be used to automate quality control processes, reducing the risk of human error and improving efficiency. By using robotics, manufacturers can perform quality control inspections at a higher frequency, resulting in improved accuracy and reduced variability (Babu, Franciosa & Ceglarek, 2019).

AI can be a powerful tool for quality control in the manufacturing industry, helping manufacturers improve product quality, reduce waste, and improve customer satisfaction. By using AI-powered algorithms to detect defects and other quality issues, manufacturers can proactively address real-world problems in the manufacturing industry.

15.3.1 Importance of AI-based Quality Control in the Manufacturing Industry

Quality control is the process of ensuring that products or services meet specified quality standards. In the manufacturing industry, quality control is essential to ensure that products are safe, reliable, and meet customer expectations (Xu et al., 2018). It involves a range of activities, including inspection, testing, and analysis, to ensure that products are manufactured to a consistent standard (Okpala & Korzeniowska, 2021).

The importance of quality control in the manufacturing industry cannot be overstated (Adam et al., 2021). Poor quality products can lead to safety issues, customer dissatisfaction, and increased costs (Sahoo & Lo, 2022). Quality control processes help manufacturers identify defects and other quality issues, allowing them to take corrective action before products are shipped to customers. These results in improved product quality, reduced waste, and improved customer satisfaction (Lee et al., 2018).

In addition to improving product quality, quality control also helps manufacturers identify opportunities for process improvement (Zheng et al., 2018). By analyzing quality control data, manufacturers can identify trends and patterns that indicate areas for improvement, allowing them to optimize their production processes and reduce costs (Tao et al., 2018).

Hence, quality control is a major issue of the manufacturing industry for ensuring the product quality as per safe, reliable, and customer expectations. By implementing effective AI based quality control techniques, manufacturers can optimize the manufacturing process.

15.3.2 AI Techniques for Detection of Defects and Anomalies to Maintain Quality

AI can be used for quality control in the manufacturing industry with number of AI techniques, including the detection of defects and anomalies. Here are some of the ways AI can be used:

- **Image recognition**: AI can help to analyze images of products to detect defects and other quality issues. Machine learning algorithms can be trained to identify specific defects, such as cracks, scratches, and color variations, resulting in improved accuracy and efficiency (Benbarrad et al., 2021).
- **Anomaly detection**: AI can be used to detect anomalies in manufacturing processes, which could indicate potential quality issues (Belhadi et al., 2022). Machine learning algorithms can be trained to detect patterns in data that are outside the norm, allowing manufacturers to take corrective action before defects occur (Cao et al., 2022).

Overall, AI can be a powerful tool for quality control in the manufacturing industry. By using AI-powered algorithms to detect defects and anomalies, manufacturers can proactively address quality issues, resulting in improved product quality, reduced waste, and improved customer satisfaction (van Wynsberghe, 2021)

15.3.3 Case Studies on AI for Quality Control

Here are some case studies and examples of successful implementations of AI for quality control in the manufacturing industry:

- **Foxconn**: Foxconn, the world's largest contract electronics manufacturer. It uses AI to detect defects in its production processes (Wong et al., 2022) . The company uses computer vision algorithms to analyze images of products, identifying defects such as scratches, dents, and other imperfections. By using AI-based quality control processes, Foxconn has been able to reduce defects and improve product quality (Ding et al., 2020).
- **Bosch**: Bosch, a leading manufacturer of automotive components, uses AI to improve its quality control processes. The company uses machine learning algorithms to analyze for detecting anomalies in real-time (Soori et al., 2023). By using AI-powered quality control, Bosch has been able to optimize its production processes, reducing defects and improving product quality (Bilgeri et al., 2019).
- **BMW**: BMW uses AI-powered quality control to detect defects in its production processes. The company uses machine learning for obtaining the predictive analysis using the data from cameras and sensors installed on production lines, identifying defects such as misaligned parts and color variations. By using AI-powered quality control, BMW has been able to reduce defects and improve product quality (Khalifa et al., 2021).
- **Siemens**: Siemens, a leading manufacturer of industrial equipment, uses AI to improve its quality control processes. The company uses ML algorithms to predict from sensors data on production lines, detecting anomalies in real-time. By using AI-powered quality control, Siemens has been able to optimize its production processes, reducing defects and improving product quality (Trakadas et al., 2020).
- **Panasonic**: Panasonic, a leading manufacturer of consumer electronics, uses AI-powered quality control to improve its production processes (Attaran, 2020). The company uses computer vision algorithms to analyze images of products, identifying defects such as scratches, dents, and other imperfections. By using AI-powered quality control, Panasonic has been able to reduce defects and improve product quality (He, 2022).

These case studies show about AI-based quality control in the manufacturing industry. By using machine learning algorithms manufacturers can detect defects and anomalies in real-time, improving product quality and reducing waste.

15.4 SUPPLY CHAIN OPTIMIZATION

Supply chain optimization involves using technology and data to streamline and optimize the flow of goods and services from suppliers to customers. This involves improving processes.

AI can be used to optimize supply chain process of manufacturing industries by identifying sectors (Dash et al., 2019). AI based patterns and trends can help to optimize operations (Belhadi et al., 2022). Here are some ways AI can be used for supply chain optimization in the manufacturing industry:

15.4.1 Benefits of Supply Chain Optimization

Supply chain optimization is the process of using technology, data analysis, and other tools to streamline and optimize the flow of goods and services from suppliers to customers. The goal of supply chain optimization is to improve efficiency, reduce costs, and minimize waste in the supply chain.

By optimizing the supply chain, manufacturers can improve their competitiveness and profitability. Some of the benefits of supply chain optimization include:

- **Reduced costs**: by optimizing the supply chain, manufacturers can reduce costs associated with procurement, inventory management, logistics, and distribution (Moons et al., 2019).
- **Improved efficiency**: by streamlining processes and reducing waste, supply chain optimization can improve the overall efficiency of the manufacturing process (Mastos et al., 2020).
- **Increased flexibility**: optimized supply chains are more flexible, allowing manufacturers to respond quickly to changes in demand or supply chain disruptions (Fragapane et al., 2022).
- **Improved customer satisfaction**: by improving delivery times and reducing the risk of stockouts, supply chain optimization can improve customer satisfaction (Gallego-García et al., 2021).
- **Enhanced competitiveness**: By optimizing the supply chain, manufacturers can improve their competitiveness, allowing them to compete more effectively in the global marketplace (Christopher, 2017).

Supply chain optimization is a critical process for manufacturers looking to improve their operations and remain competitive in the modern business landscape. By leveraging AI algorithms to predict patterns, manufacturers can streamline their operations with minimum costs. It will be helpful to improve customer satisfaction.

15.4.2 Techniques for AI-based Supply Chain Optimization

AI can be used for supply chain optimization in several ways, including demand prediction and logistics optimization. Here are some examples:

- **Demand prediction**: AI algorithms can analyze historical sales data, market trends, weather patterns, and other relevant factors to predict future demand for products. By accurately predicting demand, manufacturers can optimize their production processes, reduce waste, and minimize inventory costs (Liu et al., 2022).
- **Logistics optimization**: AI algorithms can predict the transportation routes, shipping schedules, and inventory levels to optimize logistics processes. By optimizing transportation routes and scheduling, manufacturers can reduce transportation costs and improve delivery times. AI can also be used to optimize

inventory levels, ensuring that manufacturers have the right products in stock at the right time (Abosuliman & Almagrabi, 2021).

- **Supplier selection**: AI algorithms can predict the supplier performance, such as delivery times, quality, and reliability, to select the best suppliers for a manufacturer's needs. By selecting the best suppliers, manufacturers can reduce the risk of delays, quality issues, and other supply chain disruptions (Brintrup et al., 2020).
- **Predictive maintenance**: AI algorithms can predict equipment failures and maintenance needs using the data from sensors and other sources. By predicting maintenance needs, manufacturers can reduce downtime and improve overall equipment efficiency (Sayyad et al.,2021).
- **Risk management**: AI algorithms can predict the potential risks, such as weather events or political unrest, to identify and mitigate potential supply chain disruptions. By managing risk, manufacturers can ensure that their supply chains remain resilient and efficient (Ivanov, 2021).

AI can be a powerful tool for optimizing supply chain processes in the manufacturing industry. AI algorithms help to predict patterns. Hence manufacturers can streamline their operations with minimum costs. It will be helpful to improve customer satisfaction.

15.4.3 Case Studies of AI for Supply Chain Optimization

Here are some examples of successful implementations of AI for supply chain optimization:

- **Walmart**: AI algorithms are leveraged by Walmart to enhance its supply chain processes, encompassing tasks like demand prediction and inventory management, with the aim of optimization. By accurately predicting demand and optimizing inventory levels, Walmart has been able to reduce costs and improve customer satisfaction (Bahramimianrood & Bathaei, 2021).
- **Amazon**: Amazon uses AI algorithms to optimize its logistics processes, including route optimization and delivery scheduling. By optimizing logistics processes, Amazon has been able to reduce transportation costs and improve delivery times (Muñoz-Villamizar, A., Velázquez-Martínez et al., 2021).
- **Zara**: Zara uses AI algorithms to optimize its production processes, including demand prediction and manufacturing scheduling. By accurately predicting demand and optimizing production schedules, Zara has been able to reduce waste and improve efficiency (Ross, 2022).
- **BMW**: BMW uses AI algorithms to optimize its supply chain processes, including logistics and supplier selection. By optimizing logistics and selecting the best suppliers, BMW has been able to reduce costs and improve quality (Khan et al., 2023).
- **Maersk**: Maersk uses AI algorithms to optimize its shipping routes and scheduling. By optimizing shipping routes, Maersk has been able to reduce transportation costs and improve delivery times (Mirxalilovich & Sabirbayevich, 2023).

Overall, these companies demonstrate the potential for AI to revolutionize supply chain optimization in the manufacturing industry. By leveraging AI algorithms to predict patterns, manufacturers can streamline their operations with minimum costs. It will be helpful to improve customer satisfaction.

15.5 AUTONOMOUS ROBOTS

Autonomous robots can perform tasks without human intervention (Rysz & Mehta, 2021). These autonomous robots navigate their surroundings and cary out tasks through advanced sensors, cameras, and other technologies with minimal human intervention. In the manufacturing industry, autonomous robots can be used for a variety of tasks, including assembly, material handling, and inspection.

There are several benefits to using autonomous robots in the manufacturing industry:

- **Increased efficiency**: autonomous robots have the capability to operate continuously without the need for breaks or downtime, resulting in heightened efficiency and productivity levels (Javaid et al., 2021).
- **Improved safety**: by replacing humans in dangerous or hazardous tasks, autonomous robots can improve safety in the manufacturing environment (Galin & Meshcheryakov, 2019).
- **Lower costs**: by reducing the need for human labor, autonomous robots can reduce labor costs and improve profitability (Rysz & Mehta, 2021).
- **Improved quality**: autonomous robots exhibit exceptional precision and accuracy when executing tasks, consequently yielding enhanced quality in manufacturing processes (Soori et al., 2023).
- **Flexibility**: autonomous robots can be reprogrammed or reconfigured to perform different tasks, allowing for greater flexibility in manufacturing processes (Karagiannis et al., 2019).

In recent years, there have been several successful implementations of autonomous robots in the manufacturing industry. For example, BMW uses autonomous robots to transport materials and perform other tasks in its factories. Tesla uses autonomous robots to perform assembly tasks on its electric vehicles. Overall, autonomous robots have the potential to revolutionize manufacturing processes and improve efficiency, safety, and quality in the industry.

15.5.1 Application of Autonomous Robots in the Manufacturing Industry

Autonomous robots are robots that can operate and make decisions without human intervention. These systems are outfitted with a range of sensors, cameras, and other advanced technologies enabling them to perceive and traverse their surroundings. By leveraging algorithms and artificial intelligence, they possess the capability to make informed decisions and execute various tasks effectively. In the manufacturing industry, autonomous robots have a wide range of applications. Some common applications of autonomous robots in manufacturing include:

- **Material handling**: autonomous robots can be used to transport materials and products throughout a manufacturing facility, decreasing reliance on human labor while enhancing effectiveness (Kusiak, 2023).
- **Assembly**: autonomous robots can be used to perform assembly tasks, such as welding, fastening, and painting, with a high degree of precision and accuracy (Javaid et al., 2021).
- **Inspection**: autonomous robots can be used to inspect products and parts, identifying defects and anomalies and ensuring quality control (Czimmermann et al., 2021).
- **Packaging**: autonomous robots have the potential to streamline product packaging processes, diminishing the reliance on human labor and significantly enhancing operational efficiency (Grobbelaar et al., 2021).
- **Maintenance**: autonomous robots can be used to perform maintenance tasks, such as cleaning and equipment inspections, reducing the need for human labor and improving safety (Bengtsson & Lundström, 2018).

Overall, autonomous robots have the potential to improve efficiency, quality and safety in the manufacturing industry, and they are becoming increasingly popular as manufacturers look to automate their operations and reduce costs.

15.5.2 Autonomous Robots for Industries

Artificial intelligence (AI) plays an important role for enabling autonomous robots to perform tasks such as assembly, material handling, and inspection. Here are some examples of how AI can be used to power autonomous robots in these tasks:

- **Assembly**: AI can be used to provide robots with the ability to identify and manipulate objects. Computer vision algorithms can enable robots to recognize objects and determine how to grasp and manipulate them. Machine learning algorithms can be used to teach robots how to perform specific assembly tasks and refine their techniques over time (Tussyadiah et al., 2020).
- **Material handling**: AI robots can navigate and interact with their environment. Robotics algorithms can be used to enable robots to plan paths, avoid obstacles, and manipulate objects. Reinforcement learning algorithms can be used to help robots learn how to move efficiently and effectively through their environment (Soori et al., 2023).
- **Inspection**: AI can be used to enable robots to identify and classify defects and anomalies. Computer vision algorithms can be used to enable robots to recognize and analyze images of products or parts, identifying any issues or defects. Machine learning algorithms can be used to help robots learn how to identify defects and distinguish between normal and abnormal products or parts (He et al., 2023).

AI is critical for enabling autonomous robots to perform complex tasks in manufacturing environments. By providing robots with the ability to perceive and interact with their environment, and by using machine learning algorithms to help robots learn and improve over time, AI can help manufacturers improve efficiency, quality, and safety in their operations.

15.5.3 Case Studies of Autonomous Robots

There are several successful examples of AI-powered autonomous robots being used in the manufacturing industry. Here are some examples:

- **FANUC**: FANUC is a Japanese company that specializes in industrial robots and CNC systems. FANUC has developed a range of AI-powered robots, including the FANUC CR-35iA, which is capable of handling heavy payloads and performing a range of tasks, including material handling, assembly, and inspection. The CR-35iA uses advanced sensors and machine learning algorithms to enable it to navigate complex environments and adapt to changing conditions (Zhu et al., 2017).
- **BMW**: BMW has implemented a range of AI-powered robots in its manufacturing plants. These robots perform a range of tasks, including welding, painting, and assembly. The robots are equipped with sensors and actuators that enable them to perceive their environment and adapt to changing conditions. BMW has reported that the use of these robots has improved efficiency and quality in its manufacturing operations (Michalos et al., 2022).
- **Amazon**: Amazon has implemented a range of AI-powered robots in its fulfilment centres. These robots are used to transport and sort packages, improving efficiency and reducing the need for human labor. The robots use sensors and machine learning algorithms to navigate through the warehouse and avoid obstacles (Robbins, 2020).
- **Boston Dynamics** (Gupta et al., 2021): Boston Dynamics is a robotics company that has developed a range of advanced robots, including the Spot robot. The Spot robot is a four-legged robot (Mohamed et al., 2019) that is capable of navigating through complex environments and performing a range of tasks, including inspection and material handling. The Spot robot is equipped with cameras and sensors that enable it to perceive its environment, and it uses machine learning algorithms to adapt to changing conditions (Gupta et al., 2021).

Overall, these examples demonstrate the potential of AI-powered autonomous robots to improve efficiency, quality, and safety in the manufacturing industry. As the development of technological component, we can expect to see more advanced and capable robots being deployed in manufacturing environments.

15.6 ENERGY EFFICIENCY

Energy efficiency refers to the use of technology and practices to reduce the amount of energy required to produce goods and services. In the manufacturing industry, energy efficiency is important for several reasons. First, energy costs can be a significant expense for manufacturers, and improving energy efficiency can help to reduce these costs (Mohamed et al., 2019). Second, improving energy efficiency can help to reduce the environmental impact of manufacturing operations, by reducing the amount of greenhouse gas emissions and other pollutants that are released into the atmosphere (Min et al., 2022). Finally, improving energy efficiency can help to ensure a reliable and sustainable supply of energy for manufacturing operations.

15.6.1 AI for Energy Efficiency in the Manufacturing Industry

AI can be used to improve energy efficiency in the manufacturing industry in several ways. As an illustration, artificial intelligence (AI) algorithms have the potential to enhance energy utilization in manufacturing operations. By scrutinizing data gathered from sensors and various channels, these algorithms can pinpoint areas where energy consumption can be minimized, thereby optimizing the overall efficiency of the processes.AI can also be used to predict energy usage and demand, enabling manufacturers to adjust their operations to minimize energy consumption during periods of high demand or high prices. Finally, AI can be used to identify and prioritize energy efficiency projects, by analyzing data on energy usage and identifying areas where improvements can be made.

15.6.2 Case Studies on AI for Energy Efficiency

There are several successful examples of AI being used to improve energy efficiency in the manufacturing industry. Here are a few examples:

- **Siemens**: Siemens has developed an AI-powered energy management system that uses machine learning algorithms to optimize energy consumption in manufacturing processes (Trakadas et al., 2020). By leveraging data derived from sensors and diverse sources, the system assesses and detects prospects for diminishing energy consumption and enhancing operational efficiency.
- **General Motors**: General Motors has implemented an AI-powered energy management system in several of its manufacturing plants. The system uses machine learning algorithms to predict energy usage and demand, enabling the company to adjust its operations to minimize energy consumption during periods of high demand or high prices (Shah et al., 2023).
- **Schneider Electric**: Schneider Electric has innovated an energy management platform infused with AI capabilities, harnessing the power of machine learning algorithms to optimize energy utilization within manufacturing processes. This advanced platform carefully scrutinizes data acquired from sensors and various sources, intelligently identifying opportunities to curtail energy consumption and enhance overall operational efficiency. (Khan et al., 2023).
- **BMW**: BMW has implemented an AI-powered energy management system in its production facilities. The system analyzes real-time data on energy consumption and production schedules to optimize energy usage and reduce costs.

BMW has reported a 15% reduction in energy consumption and associated CO2 emissions since implementing the system (Khan et al., 2023).

- **Schneider Electric**: Schneider Electric has developed an AI-powered energy management system called EcoStruxure. The system uses machine learning (ML) algorithms to predict energy usage data from manufacturing plants and identify sectors where energy consumption can be reduced. Schneider Electric has reported that the system has helped its customers reduce energy consumption by up to 30% (Choudary, 2021).
- **Fujitsu**: Fujitsu has implemented an AI-powered energy management system in its semiconductor manufacturing plant in Japan. The system uses machine learning algorithms to analyze data on energy usage and production schedules to optimize energy consumption. Fujitsu has reported a 10% reduction in energy consumption since implementing the system (Kodama et al., 2020,).
- **Honeywell**: Honeywell has developed an AI-powered energy management system called Energy Vision. The system uses ML algorithms to predict energy usage data from manufacturing plants and identify sectors where energy consumption can be reduced. Honeywell has reported that the system has helped its customers reduce energy consumption by up to 20% (Choy et al., 2005). These examples demonstrate the potential of AI to optimize consumption of energy in manufacturing and help companies reduce their carbon footprint while improving their bottom line.

These examples show the potential of AI to improve energy efficiency in the manufacturing industry. Hence, we can expect to see more advanced and capable AI systems being deployed to optimize use of energy resources and hence it reduces costs in manufacturing operations.

15.6.3 Importance of Energy Efficiency in the Manufacturing Industry

Energy efficiency is the ability to achieve the same or better results while using less energy. In the manufacturing industry, energy efficiency is essential for reducing energy costs, increasing productivity, and reducing environmental impact (Dell'Anna, 2021). It involves the use of technology and practices that minimize energy consumption without affecting the quality or quantity of the manufactured products. Energy efficiency measures can range from upgrading equipment and lighting to changing manufacturing processes and implementing renewable energy sources (Fresner et al., 2017). By improving energy efficiency, manufacturers can reduce operating costs and increase profits while contributing to a more sustainable future.

Energy efficiency is crucial in the manufacturing industry for several reasons:

- **Cost savings**: energy costs can account for a significant portion of a manufacturer's expenses. By improving energy efficiency, manufacturers can reduce their energy consumption and save money on their energy bills (Elia et al., 2020).

- **Increased productivity**: energy-efficient equipment and processes can help manufacturers produce goods faster and more efficiently, increasing productivity and output (Chau et al.,2021).
- **Environmental impact**: the manufacturing industry plays a pivotal role in the emission of greenhouse gases, thereby contributing to the phenomenon of climate change. By improving energy efficiency, manufacturers can reduce their carbon footprint and environmental impact (Liu et al., 2016).
- **Regulatory compliance**: many countries have regulations and standards aimed at reducing energy consumption and greenhouse gas emissions. Manufacturers that fail to comply with these regulations may face fines or other penalties (Tsai & Tsai, 2022).
- **Competitive advantage**: manufacturers that prioritize energy efficiency can differentiate themselves from their competitors and attract environmentally-conscious consumers who value sustainability (Rusinko, 2007).

Overall, energy efficiency is essential for manufacturers to remain competitive, reduce costs, and meet environmental regulations while contributing to a sustainable future.

15.6.4 AI-Based Optimizing Energy Consumption in Manufacturing Plants

AI can be used to optimize energy consumption in manufacturing plants in several ways:

- **Predictive maintenance**: by using machine learning algorithms, AI can predict when equipment is likely to fail, allowing maintenance teams to fix issues before they cause significant energy wastage (Soori et al., 2023).
- **Energy management systems**: utilizing AI, it becomes feasible to create energy management systems that effectively monitor energy consumption throughout the entirety of a plant, enabling the identification of specific areas where energy usage can be optimized and reduced. (Elia et al., 2020).
- **Process optimization**: AI can be used to optimize manufacturing processes by identifying inefficiencies and recommending changes that reduce energy consumption without affecting product quality (Montgomery et al., 2020).
- **Predictive analytics**: AI can predict future energy usage based on historical energy consumption data factors such as production schedules and weather conditions. This information can be used to adjust energy usage in real-time to avoid peak demand charges and reduce energy waste (Sayyad et al., 2021).
- **Renewable energy integration**: AI can be harnessed to effectively oversee the integration of renewable energy sources like solar and wind power into manufacturing plants (Choudary, 2021). This integration aims to ensure the optimal utilization of such renewable sources while simultaneously reducing reliance on conventional fossil fuels (Rusinko, 2007).

By using AI to optimize energy consumption in manufacturing plants, manufacturers can reduce their energy costs, improve environmental sustainability, and meet regulatory requirements.

15.7 ETHICAL CONSIDERATIONS AND IMPLICATIONS OF AI IN THE MANUFACTURING INDUSTRY

As with any technology, AI in the manufacturing industry raises important ethical considerations that must be addressed (Tang et al., 2020; Martin et al., 2022). The use of AI in the manufacturing industry raises important ethical implications that need to be considered (Coeckelbergh, 2020; Berrah et al., 2021). These include:

- **Job displacement**: the adoption of AI and automation technologies in manufacturing can lead to the displacement of human workers, potentially resulting in job losses and economic disruption. Companies must consider the impact of AI on their workforce and implement measures to mitigate negative effects, such as retraining and reskilling programs.
- **Bias and discrimination**: AI systems can perpetuate existing biases and discrimination if they are not designed and tested properly. This can lead to unfair treatment of certain groups, such as women and minorities. To prevent bias and discrimination, AI systems must be designed to ensure fairness and equity for all users.
- **Privacy and security**: AI systems that collect and work on large amounts of data can pose privacy and security risks if the data is not protected properly. Companies must be able to determine that their AI systems are compliant with relevant data protection laws and have appropriate safeguards in place to protect sensitive data.
- **Transparency and accountability**: AI systems can be opaque and difficult to understand, which can make it hard to hold companies accountable for their decisions. To promote transparency and accountability, companies must be transparent about how their AI systems work and ensure that they can be audited and explained.
- **Environmental impact**: While AI can help reduce energy consumption and improve sustainability in manufacturing, the production and disposal of AI hardware can also have a significant environmental impact. Companies must consider the environmental impact of their AI systems throughout their entire lifecycle, from production to disposal.
- **Safety and liability**: AI-powered autonomous robots and machines can pose safety risks if they malfunction or are not properly designed. Companies must ensure that their AI systems are designed with safety in mind and that they take responsibility for any accidents or incidents caused by their AI systems.

By addressing these ethical considerations and implications, companies can ensure that their AI systems are not only effective but also ethical and responsible. This will help to promote acceptance of AI technology in the manufacturing industry.

15.7.1 Discussion of Potential Negative Impacts on Workers, Society, and the Environment

The adoption of AI in the manufacturing industry can have potential negative impacts on workers, society, and the environment.

One of the main concerns is the displacement of human workers. The increased automation brought by AI and robotics can lead to job losses and economic disruption, particularly in areas where manufacturing is a significant employer. The consequences of such biases and discrimination within AI systems can extend beyond individuals directly affected, impacting not only workers and their families but also the broader community as a whole (Bag & Pretorius, 2022). In addition, there is a risk of exacerbating inequalities and discrimination (Berrah et al., 2021). AI systems can perpetuate existing biases and discrimination if they are not designed and tested properly (Mikalef et al., 2022). This can lead to unfair treatment of certain groups, such as women and minorities. This can have negative consequences for the affected individuals, as well as for society as a whole.

The use of AI in manufacturing can also have environmental impacts. While AI can help reduce energy consumption and improve sustainability in manufacturing, the production and disposal of AI hardware can also have a significant environmental impact. This includes the use of rare earth metals, energy consumption during production, and the disposal of AI hardware after use.

Moreover, there is a risk that AI could be used for nefarious purposes, such as the development of autonomous weapons or the creation of fake news and propaganda. This could have serious consequences for society and could exacerbate existing geopolitical tensions.

To mitigate these potential negative impacts, it is important that AI is developed and used in an ethical and responsible way. This includes ensuring that workers are not unfairly impacted by automation that AI systems are designed to be fair and unbiased, and that environmental impacts are considered throughout the entire lifecycle of AI hardware. It is also important that there is transparency and accountability around the use of AI, and those ethical considerations are integrated into the development and use of AI in manufacturing (Mikalef et al., 2022).

15.7.2 Strategies for Responsible and Ethical Use of AI in Industry 4.0

To ensure responsible and ethical use of AI in Industry 4.0, several strategies can be implemented (Berrah et al., 2021):

- **Develop ethical guidelines and principles**: companies should establish ethical guidelines based principles for use of AI systems in the manufacturing industry. These guidelines should be based on established ethical frameworks and should be regularly reviewed and updated to ensure they remain relevant.
- **Involve diverse stakeholders**: AI development teams should involve diverse stakeholders, including workers, community members, and ethical experts, in the development and testing of AI systems. This can help to ensure that the systems will be designed and implemented in a fair and unbiased method.

- **Ensure transparency and accountability**: companies should ensure that AI systems are transparent and accountable. This includes providing clear explanations of how the systems work and how decisions are made, as well as establishing mechanisms for accountability and redress in case of errors or harms caused by the systems.
- **Address bias and discrimination**: companies should take steps to address bias and discrimination in AI systems. This includes using diverse and representative data sets, testing for bias, and ensuring that AI systems are regularly audited for fairness.
- **Prioritize worker well-being**: companies should prioritize worker well-being in the development and use of AI systems. This includes ensuring that workers are not unfairly impacted by automation, providing retraining and up-skilling opportunities, and ensuring that workers are able to maintain their dignity and autonomy.
- **Consider environmental impacts**: companies should consider the environmental impacts of AI hardware throughout its entire lifecycle, from production to disposal. This includes minimizing the use of rare earth metals, reducing energy consumption during production, and ensuring that hardware is recycled or disposed of in an environmentally responsible way.

By implementing these strategies, companies can ensure that AI is developed and used in a responsible and ethical way in the manufacturing industry. This can help mitigate potential negative impacts on workers, society, and the environment, and ensure that AI is used to drive positive outcomes for all stakeholders.

15.8 CONCLUSION

In conclusion, AI is playing a significant role in Industry 4.0, offering numerous benefits for the manufacturing industry, including predictive maintenance, quality control, supply chain optimization, autonomous robots, and energy efficiency. However, the use of AI in the manufacturing industry also poses ethical and social challenges, such as potential negative impacts on workers, society, and the environment. Therefore, it is essential to develop responsible and ethical AI systems that prioritize transparency, accountability, fairness, and worker well-being. By doing so, we can maximize the benefits of AI in the manufacturing industry while minimizing potential negative impacts, leading to a more sustainable and socially responsible future.

This chapter discusses the role of AI in Industry 4.0 and its applications in the manufacturing industry. The main points covered in this chapter include:

- Industry 4.0 is a technological revolution that involves the integration of digital technologies into the manufacturing industry.
- Predictive maintenance using AI can help manufacturers to reduce downtime, lower maintenance costs, and increase productivity.
- AI can be used for quality control in manufacturing to detect defects and anomalies, which will be helpful for improving product quality and reduce industrial waste.

- Supply chain optimization using AI can improve demand prediction, logistics optimization, and inventory management, leading to reduced costs and increased efficiency.
- Autonomous robots powered by AI can perform a variety of tasks in the manufacturing industry, such as assembly, material handling, and inspection, leading to increased productivity and efficiency.
- AI can be used to optimize energy consumption in manufacturing plants, leading to reduced costs and increased sustainability.
- The use of AI in the manufacturing industry also poses ethical and social challenges, such as potential negative impacts on workers, society, and the environment.
- To ensure the responsible and ethical use of AI in Industry 4.0, it is essential to prioritize transparency, accountability, fairness, and worker well-being.

Overall, this chapter highlights the potential benefits of AI in the manufacturing industry and the importance of responsible and ethical use of these technologies to achieve a sustainable and socially responsible future.

15.8.1 Reflection on the Potential of AI Integrated Manufacturing Industry

As an AI language model, I can say that AI has immense potential to address real-world problems in the manufacturing industry. The integration of AI technologies such as predictive maintenance, quality control, supply chain optimization, autonomous robots, and energy efficiency can lead to significant improvements in productivity, efficiency, and sustainability (Liu et al., 2022). These technologies will be helpful for manufacturers to reduce costs, improve product quality, and increase worker safety.

AI can also help to address some of the most pressing challenges facing the manufacturing industry, such as reducing downtime, improving inventory management, and optimizing energy consumption. Furthermore, AI can support a shift towards more sustainable and socially responsible manufacturing practices by enabling manufacturers to minimize waste, reduce environmental impact, and improve worker well-being.

However, it is important to recognize that the use of AI in the manufacturing industry also poses ethical and social challenges, such as potential job displacement, worker safety, and the responsible use of data. Therefore, it is crucial that the ethical use of AI is prioritized, with a focus on transparency, accountability, fairness, and worker well-being.

Overall, I believe that AI has enormous potential to address real-world problems in the manufacturing industry and socially responsible future.

15.8.2 Call to Action for Responsible and Ethical use of AI in Industry 4.0

As AI continues to transform the manufacturing industry, it is essential that we prioritize the responsible and ethical use of this technology. This requires collaboration between industry leaders, policy-makers, and stakeholders to establish clear guidelines and best practices for the development, deployment, and governance of AI technologies in the manufacturing sector.

As individuals, we can also play a role in promoting the responsible and ethical use of AI in Industry 4.0. We can educate ourselves on the potential benefits and risks of AI and advocate for transparent and accountable use of this technology. We can also support companies that prioritize ethical and socially responsible practices in their use of AI and hold those who do not accountable for their actions.

Ultimately, the responsible and ethical use of AI in Industry 4.0 is essential to ensure the well-being of workers, protecting the environment, and creating a sustainable and equitable future. By working together to promote ethical AI practices, we can harness the full potential of this technology to address real-world problems and drive progress towards a better future.

REFERENCES

Abosuliman, S. S., & Almagrabi, A. O. (2021). Routing and scheduling of intelligent autonomous vehicles in industrial logistics systems. *Soft Computing*, *25*, 11975–11988.

Adam, M., Wessel, M., & Benlian, A. (2021). AI-based chatbots in customer service and their effects on user compliance. *Electronic Markets*, *31*(2), 427–445.

Ahmad, S., Lavin, A., Purdy, S., & Agha, Z. (2017). Unsupervised real-time anomaly detection for streaming data. *Neurocomputing*, *262*, 134–147.

Ahmad, T., Zhang, D., Huang, C., Zhang, H., Dai, N., Song, Y., & Chen, H. (2021). Artificial intelligence in sustainable energy industry: Status Quo, challenges and opportunities. *Journal of Cleaner Production*, *289*, 125834.

Ahmed, A. A. A., Agarwal, S., Kurniawan, I. G. A., Anantadjaya, S. P., & Krishnan, C. (2022). Business boosting through sentiment analysis using Artificial Intelligence approach. *International Journal of System Assurance Engineering and Management*, *13*(Suppl 1), 699–709.

Ambika, P. (2020). Machine learning and deep learning algorithms on the Industrial Internet of Things (IIoT). *Advances in Computers*, *117*(1), 321–338.

Attaran, M. (2020, July). Digital technology enablers and their implications for supply chain management. In *Supply Chain Forum: An International Journal* (Vol. 21, No. 3, pp. 158–172). Taylor & Francis.

Babu, M., Franciosa, P., & Ceglarek, D. (2019). Spatio-Temporal Adaptive Sampling for effective coverage measurement planning during quality inspection of free form surfaces using robotic 3D optical scanner. *Journal of Manufacturing Systems*, *53*, 93–108.

Bag, S., & Pretorius, J. H. C. (2022). Relationships between industry 4.0, sustainable manufacturing and circular economy: proposal of a research framework. *International Journal of Organizational Analysis*, *30*(4), 864–898.

Bahramimianrood, B., & Bathaei, M. (2021). The impact of information technology on knowledge management in the supply chain. *Journal of Social, Management and Tourism Letter*, *2021*, 1–11.

Belhadi, A., Kamble, S., Fosso Wamba, S., & Queiroz, M. M. (2022). Building supply-chain resilience: an artificial intelligence-based technique and decision-making framework. *International Journal of Production Research*, *60*(14), 4487–4507.

Benbarrad, T., Salhaoui, M., Kenitar, S. B., & Arioua, M. (2021). Intelligent machine vision model for defective product inspection based on machine learning. *Journal of Sensor and Actuator Networks*, *10*(1), 7.

Bengtsson, M., & Lundström, G. (2018). On the importance of combining "the new" with "the old"–One important prerequisite for maintenance in Industry 4.0. *Procedia manufacturing*, *25*, 118–125.

Berrah, L., Cliville, V., Trentesaux, D., & Chapel, C. (2021). Industrial performance: an evolution incorporating ethics in the context of industry 4.0. *Sustainability*, *13*(16), 9209.

Bilgeri, D., Gebauer, H., Fleisch, E., & Wortmann, F. (2019). Driving process innovation with IoT field data. *MIS Quarterly Executive*, *18*, 191–207.

Brintrup, A., Pak, J., Ratiney, D., Pearce, T., Wichmann, P., Woodall, P., & McFarlane, D. (2020). Supply chain data analytics for predicting supplier disruptions: a case study in complex asset manufacturing. *International Journal of Production Research*, *58*(11), 3330–3341.

Cao, Q., Zanni-Merk, C., Samet, A., Reich, C., De Beuvron, F. D. B., Beckmann, A., & Giannetti, C. (2022). KSPMI: a knowledge-based system for predictive maintenance in industry 4.0. *Robotics and Computer-Integrated Manufacturing*, *74*, 102281.

Chau, M. Q., Nguyen, X. P., Huynh, T. T., Chu, V. D., Le, T. H., Nguyen, T. P., & Nguyen, D. T. (2021). Prospects of application of IoT-based advanced technologies in remanufacturing process towards sustainable development and energy-efficient use. *Energy Sources, Part A: Recovery, Utilization, and Environmental Effects*, *43*, 1–25.

Chen, J., Lim, C. P., Tan, K. H., Govindan, K., & Kumar, A. (2021). Artificial intelligence-based human-centric decision support framework: an application to predictive maintenance in asset management under pandemic environments. *Annals of Operations Research*, *306*, 1–24.

Cheng, J. C., Chen, W., Chen, K., & Wang, Q. (2020). Data-driven predictive maintenance planning framework for MEP components based on BIM and IoT using machine learning algorithms. *Automation in Construction*, *112*, 103087.

Choudary, S. P. (2021). *Platform Scale: For A Post-Pandemic World*. Penguin Random House India Private Limited.

Choy, K. L., Lee, W. B., Lau, H. C., & Choy, L. C. (2005). A knowledge-based supplier intelligence retrieval system for outsource manufacturing. *Knowledge-based systems*, *18*(1), 1–17.

Christopher, M. (2017). Relationships and alliances Embracing the era of network competition. In *Strategic Supply Chain Alignment* (pp. 286–351). Routledge.

Çınar, Z. M., Abdussalam Nuhu, A., Zeeshan, Q., Korhan, O., Asmael, M., & Safaei, B. (2020). Machine learning in predictive maintenance towards sustainable smart manufacturing in industry 4.0. *Sustainability*, *12*(19), 8211.

Coeckelbergh, M. (2020). *AI Ethics*. Mit Press.

Czimmermann, T., Chiurazzi, M., Milazzo, M., Roccella, S., Barbieri, M., Dario, P., ... & Ciuti, G. (2021). An autonomous robotic platform for manipulation and inspection of metallic Surfaces in Industry 4.0. *IEEE Transactions on Automation Science and Engineering*, *19*(3), 1691–1706.

Dash, R., McMurtrey, M., Rebman, C., & Kar, U. K. (2019). Application of artificial intelligence in automation of supply chain management. *Journal of Strategic Innovation and Sustainability*, *14*(3), 43–53.

De Vasconcelos Batalha, A., & Parli, A. L. (2017). Industry 4.0 with a Lean perspective-Investigating IIoT platforms' possible influences on data driven *Lean*. Student Paper. www.diva-portal.org/smash/get/diva2:1118830/FULLTEXT01.pdf

Dell'Anna, F. (2021). Green jobs and energy efficiency as strategies for economic growth and the reduction of environmental impacts. *Energy Policy*, *149*, 112031.

Di Vaio, A., Palladino, R., Hassan, R., & Escobar, O. (2020). Artificial intelligence and business models in the sustainable development goals perspective: A systematic literature review. *Journal of Business Research*, *121*, 283–314.

Ding, H., Gao, R. X., Isaksson, A. J., Landers, R. G., Parisini, T., & Yuan, Y. (2020). State of AI-based monitoring in smart manufacturing and introduction to focused section. *IEEE/ASME Transactions on Mechatronics*, *25*(5), 2143–2154.

Dopico, M., Gómez, A., De la Fuente, D., García, N., Rosillo, R., & Puche, J. (2016). A vision of industry 4.0 from an artificial intelligence point of view. In *Proceedings on the international conference on artificial intelligence (ICAI)* (p. 407). The Steering Committee of The World Congress in Computer Science, Computer Engineering and Applied Computing (WorldComp).

Dutta, G., Kumar, R., Sindhwani, R., & Singh, R. K. (2021). Digitalization priorities of quality control processes for SMEs: A conceptual study in perspective of Industry 4.0 adoption. *Journal of Intelligent Manufacturing*, *32*(6), 1679–1698.

Elia, A., Taylor, M., Gallachóir, B. Ó., & Rogan, F. (2020). Wind turbine cost reduction: A detailed bottom-up analysis of innovation drivers. *Energy Policy*, *147*, 111912.

Fragapane, G., Ivanov, D., Peron, M., Sgarbossa, F., & Strandhagen, J. O. (2022). Increasing flexibility and productivity in Industry 4.0 production networks with autonomous mobile robots and smart intralogistics. *Annals of Operations Research*, *308*(1–2), 125–143.

Fresner, J., Morea, F., Krenn, C., Uson, J. A., & Tomasi, F. (2017). Energy efficiency in small and medium enterprises: Lessons learned from 280 energy audits across Europe. *Journal of Cleaner Production*, 142, 1650–1660.

Galin, R., & Meshcheryakov, R. (2019). Review on human–robot interaction during collaboration in a shared workspace. In *Interactive Collaborative Robotics: 4th International Conference, ICR 2019, Istanbul, Turkey, August 20–25, 2019, Proceedings 4* (pp. 63–74). Springer International Publishing.

Gallego-García, D., Gallego-García, S., & García-García, M. (2021). An optimized system to reduce procurement risks and stock-outs: a simulation case study for a component manufacturer. *Applied Sciences*, *11*(21), 10374.

Gao, R., Wang, L., Teti, R., Dornfeld, D., Kumara, S., Mori, M., & Helu, M. (2015). Cloud-enabled prognosis for manufacturing. *CIRP Annals*, *64*(2), 749–772.

Ghazal, T. M., Hasan, M. K., Alshurideh, M. T., Alzoubi, H. M., Ahmad, M., Akbar, S. S., & Akour, I. A. (2021). IoT for smart cities: Machine learning approaches in smart healthcare—A review. *Future Internet*, *13*(8), 218.

Grobbelaar, W., Verma, A., & Shukla, V. K. (2021). Analyzing human robotic interaction in the food industry. *Journal of Physics: Conference Series*, 1714(1), 012032.

Gupta, A., Anpalagan, A., Guan, L., & Khwaja, A. S. (2021). Deep learning for object detection and scene perception in self-driving cars: Survey, challenges, and open issues. *Array*, *10*, 100057.

He, F., Yuan, L., Mu, H., Ros, M., Ding, D., Pan, Z., & Li, H. (2023). Research and application of artificial intelligence techniques for wire arc additive manufacturing: a state-of-the-art review. *Robotics and Computer-Integrated Manufacturing*, *82*, 102525.

He, J. (2022). *An AI-Based Pick-and-Place Control for Quality Enhancement in Surface Mount Technology* (Doctoral dissertation, State University of New York at Binghamton).

Helo, P., & Hao, Y. (2022). Artificial intelligence in operations management and supply chain management: An exploratory case study. *Production Planning & Control*, *33*(16), 1573–1590.

Himeur, Y., Ghanem, K., Alsalemi, A., Bensaali, F., & Amira, A. (2021). Artificial intelligence based anomaly detection of energy consumption in buildings: A review, current trends and new perspectives. *Applied Energy*, 287, 116601.

Ivanov, D. (2021). Lean resilience: AURA (Active Usage of Resilience Assets) framework for post-COVID-19 supply chain management. *The International Journal of Logistics Management*, *33*(4), 1196–1217.

Ivanov, D., Dolgui, A., & Sokolov, B. (2019). The impact of digital technology and Industry 4.0 on the ripple effect and supply chain risk analytics. *International Journal of Production Research*, *57*(3), 829–846.

Javaid, M., Haleem, A., Khan, I. H., & Suman, R. (2023). Understanding the potential applications of Artificial Intelligence in Agriculture Sector. *Advanced Agrochem*, *2*(1), 15–30.

Javaid, M., Haleem, A., Singh, R. P., & Suman, R. (2021). Substantial capabilities of robotics in enhancing industry 4.0 implementation. *Cognitive Robotics*, *1*, 58–75.

Javaid, M., Haleem, A., Singh, R. P., & Suman, R. (2022). Artificial intelligence applications for industry 4.0: A literature-based study. *Journal of Industrial Integration and Management*, *7*(01), 83–111.

Kang, Y., Cai, Z., Tan, C. W., Huang, Q., & Liu, H. (2020). Natural language processing (NLP) in management research: A literature review. *Journal of Management Analytics*, *7*(2), 139–172.

Karagiannis, P., Matthaiakis, S. A., Andronas, D., Filis, K., Giannoulis, C., Michalos, G., & Makris, S. (2019). Reconfigurable assembly station: a consumer goods industry paradigm. *Procedia CIRP*, *81*, 1406–1411.

Khalifa, N., Abd Elghany, M., & Abd Elghany, M. (2021). Exploratory research on digitalization transformation practices within supply chain management context in developing countries specifically Egypt in the MENA region. *Cogent Business & Management*, *8*(1), 1965459.

Khan, H., Kushwah, K. K., Singh, S., Thakur, J. S., & Sadasivuni, K. K. (2023). Machine Learning in Additive Manufacturing. *Nanotechnology-Based Additive Manufacturing: Product Design, Properties and Applications*, *2*, 601–636.

Kiciński, J., Chaja, P., Kiciński, J., & Chaja, P. (2021). Industry 4.0—The Fourth Industrial Revolution. In *Climate Change, Human Impact and Green Energy Transformation* (pp. 115–140). Springer.

Kodama, Y., Odajima, T., Arima, E., & Sato, M. (2020, September). Evaluation of power management control on the supercomputer Fugaku. In *2020 IEEE International Conference on Cluster Computing (CLUSTER)* (pp. 484–493). IEEE.

Kusiak, A. (2023). Smart manufacturing. In *Springer Handbook of Automation* (pp. 973–985). Springer International Publishing.

Kwon, D., Hodkiewicz, M. R., Fan, J., Shibutani, T., & Pecht, M. G. (2016). IoT-based prognostics and systems health management for industrial applications. *IEEE Access*, *4*, 3659–3670.

Lee, J., Davari, H., Singh, J., & Pandhare, V. (2018). Industrial Artificial Intelligence for industry 4.0-based manufacturing systems. *Manufacturing Letters*, *18*, 20–

Lee, S. M., Lee, D., & Kim, Y. S. (2019). The quality management ecosystem for predictive maintenance in the Industry 4.0 era. *International Journal of Quality Innovation*, *5*, 1–11.

Lerch, C., & Gotsch, M. (2015). Digitalized product-service systems in manufacturing firms: A case study analysis. *Research-Technology Management*, *58*(5), 45–52.

Liu, P., Hendalianpour, A., Hamzehlou, M., & Feylizadeh, M. (2022). Cost reduction of inventory-production-system in multi-echelon supply chain using game theory and fuzzy demand forecasting. *International Journal of Fuzzy Systems*, *24*(4), 1793–1813.

Liu, Z., Feng, K., Davis, S. J., Guan, D., Chen, B., Hubacek, K., & Yan, J. (2016). Understanding the energy consumption and greenhouse gas emissions and the implication for achieving climate change mitigation targets. *Applied Energy*, *184*, 737–741.

Machen, A., Wang, S., Leung, K. K., Ko, B. J., & Salonidis, T. (2017). Live service migration in mobile edge clouds. *IEEE Wireless Communications*, *25*(1), 140–147.

Maddikunta, Praveen Kumar Reddy, et al. "Industry 5.0: A survey on enabling technologies and potential applications." *Journal of Industrial Information Integration* 26 (2022): 100257.

Mah, P. M., Skalna, I., & Muzam, J. (2022). Natural Language Processing and Artificial Intelligence for Enterprise Management in the Era of Industry 4.0. *Applied Sciences*, *12*(18), 9207.

Martin, K., Shilton, K., & Smith, J. E. (2022). Business and the ethical implications of technology: Introduction to the symposium. In *Business and the Ethical Implications of Technology* (pp. 1–11). Springer Nature Switzerland.

Mastos, T. D., Nizamis, A., Vafeiadis, T., Alexopoulos, N., Ntinas, C., Gkortzis, D., ... & Tzovaras, D. (2020). Industry 4.0 sustainable supply chains: An application of an IoT enabled scrap metal management solution. *Journal of Cleaner Production*, *269*, 122377.

Michalos, G., Karagiannis, P., Dimitropoulos, N., Andronas, D., & Makris, S. (2022). Human robot collaboration in industrial environments. In *The 21st Century Industrial Robot: When Tools Become Collaborators* (pp. 17–39). Springer.

Mikalef, P., Conboy, K., Lundström, J. E., & Popovič, A. (2022). Thinking responsibly about responsible AI and 'the dark side'of AI. *European Journal of Information Systems*, *31*(3), 257–268.

Min, J., Yan, G., Abed, A. M., Elattar, S., Khadimallah, M. A., Jan, A., & Ali, H. E. (2022). The effect of carbon dioxide emissions on the building energy efficiency. *Fuel*, *326*, 124842.

Mirxalilovich, I. O., & Sabirbayevich, S. D. (2023). Streamlining operations: how intellectualization algorithms improve efficiency in transport and logistics. *Science and Innovation*, *2*(Special Issue 3), 832–835.

Mohamed, N., Al-Jaroodi, J., & Lazarova-Molnar, S. (2019). Leveraging the capabilities of industry 4.0 for improving energy efficiency in smart factories. *IEEE Access*, *7*, 18008–18020.

Montgomery, D. C., & Friedman, D. J. (2020). Statistical process control in a computer-integrated manufacturing environment. In *Statistical process control in automated manufacturing* (pp. 67–87). CRC Press.

Moons, K., Waeyenbergh, G., & Pintelon, L. (2019). Measuring the logistics performance of internal hospital supply chains–a literature study. *Omega*, *82*, 205–217.

Muñoz-Villamizar, A., Velázquez-Martínez, J. C., Haro, P., Ferrer, A., & Mariño, R. (2021). The environmental impact of fast shipping ecommerce in inbound logistics operations: A case study in Mexico. *Journal of Cleaner Production*, *283*, 125400.

Okpala, C. O. R., & Korzeniowska, M. (2021). Understanding the relevance of quality management in agro-food product industry: From ethical considerations to assuring food hygiene quality safety standards and its associated processes. *Food Reviews International*, *39*, 1–74.

Oztemel, E., & Gursev, S. (2020). Literature review of Industry 4.0 and related technologies. *Journal of Intelligent Manufacturing*, *31*, 127–182.

Pan, Y., & Zhang, L. (2021). Roles of artificial intelligence in construction engineering and management: A critical review and future trends. *Automation in Construction*, *122*, 103517.

Pech, M., Vrchota, J., & Bednář, J. (2021). Predictive maintenance and intelligent sensors in smart factory. *Sensors*, *21*(4), 1470.

Peters, S. E., Trieu, H. D., Manjourides, J., Katz, J. N., & Dennerlein, J. T. (2020). Designing a Participatory Total Worker Health® organizational intervention for commercial construction subcontractors to improve worker safety, health, and well-being: the "ARM for Subs" trial. *International Journal of Environmental Research and Public Health, 17*(14), 5093.

Petrik, D., & Herzwurm, G. (2019, August). IoT ecosystem development through boundary resources: a Siemens MindSphere case study. In *Proceedings of the 2nd ACM SIGSOFT International Workshop on Software-Intensive Business: Start-Ups, Platforms, and Ecosystems* (pp. 1–6). ACM. https://doi.org/10.1145/3340481.334273

Pule, M., Matsebe, O., & Samikannu, R. (2023, June). Investigation of rotating equipment condition monitoring techniques in Botswana's big industries. In *AIP Conference Proceedings* (Vol. 2581, No. 1). AIP Publishing.

Qi, Q., Tao, F., Hu, T., Anwer, N., Liu, A., Wei, Y., ... & Nee, A. Y. C. (2021). Enabling technologies and tools for digital twin. *Journal of Manufacturing Systems, 58*, 3–21.

Ramere, M. D., & Laseinde, O. T. (2021). Optimization of condition-based maintenance strategy prediction for aging automotive industrial equipment using FMEA. *Procedia Computer Science, 180*, 229–238.

Robbins, S. (2020). AI and the path to envelopment: knowledge as a first step towards the responsible regulation and use of AI-powered machines. *AI & SOCIETY, 35*, 391–400.

Ross, K. (2022). Artificial Intelligence in Fashion Manufacturing: From Factory Operation to Advisory Role. In *Leading Edge Technologies in Fashion Innovation: Product Design and Development Process from Materials to the End Products to Consumers* (pp. 95–116). Cham: Springer International Publishing.

Rusinko, C. (2007). Green manufacturing: an evaluation of environmentally sustainable manufacturing practices and their impact on competitive outcomes. *IEEE Transactions on Engineering Management, 54*(3), 445–454.

Rysz, M. W., & Mehta, S. S. (2021). A risk-averse optimization approach to human-robot collaboration in robotic fruit harvesting. *Computers and Electronics in Agriculture, 182*, 106018.

Sahoo, S., & Lo, C. Y. (2022). Smart manufacturing powered by recent technological advancements: A review. *Journal of Manufacturing Systems, 64*, 236–250.

Sakib, N., & Wuest, T. (2018). Challenges and opportunities of condition-based predictive maintenance: a review. *Procedia Cirp, 78*, 267–272.

Sarker, I. H. (2021). Machine learning: Algorithms, real-world applications and research directions. *SN Computer Science, 2*(3), 160.

Sayyad, S., Kumar, S., Bongale, A., Kamat, P., Patil, S., & Kotecha, K. (2021). Data-driven remaining useful life estimation for milling process: sensors, algorithms, datasets, and future directions. *IEEE Access, 9*, 110255–110286.

Shah, J. M., Natraj, N. A., Hallur, G. G., & Aslekar, A. (2023, March). Artificial Intelligence (AI) in the Automotive Industry and the use of Exoskeletons in the Manufacturing Sector of the Automotive Industry. In *2023 International Conference on Sustainable Computing and Data Communication Systems (ICSCDS)* (pp. 428–432). IEEE.

Sharanya, S. (2022). A Cyber Physical System Framework for Industrial Predictive Maintenance Using Machine Learning. In *Real-Time Applications of Machine Learning in Cyber-Physical Systems* (pp. 241–269). IGI Global.

Soori, M., Arezoo, B., & Dastres, R. (2023). Internet of things for smart factories in industry 4.0, a review. *Internet of Things and Cyber-Physical Systems, 3*, 192–204.

Stoica, I., Song, D., Popa, R. A., Patterson, D., Mahoney, M. W., Katz, R., ... & Abbeel, P. (2017). A berkeley view of systems challenges for ai. arXiv preprint arXiv:1712.05855.

Tang, Y., Xiong, J., Becerril-Arreola, R., & Iyer, L. (2020). Ethics of blockchain: A framework of technology, applications, impacts, and research directions. *Information Technology & People*, 33(2), 602–632.

Tao, F., Qi, Q., Liu, A., & Kusiak, A. (2018). Data-driven smart manufacturing. *Journal of Manufacturing Systems*, 48, 157–169.

Tjahjono, B., Esplugues, C., Ares, E., & Pelaez, G. (2017). What does industry 4.0 mean to supply chain?. *Procedia Manufacturing*, 13, 1175–1182.

Trakadas, P., Simoens, P., Gkonis, P., Sarakis, L., Angelopoulos, A., Ramallo-González, A. P., ... & Karkazis, P. (2020). An artificial intelligence-based collaboration approach in industrial iot manufacturing: Key concepts, architectural extensions and potential applications. *Sensors*, 20(19), 5480.

Trakadas, P., Simoens, P., Gkonis, P., Sarakis, L., Angelopoulos, A., Ramallo-González, A. P., ... & Karkazis, P. (2020). An artificial intelligence-based collaboration approach in industrial iot manufacturing: Key concepts, architectural extensions and potential applications. *Sensors*, 20(19), 5480.

Tsai, W. T., & Tsai, C. H. (2022). Interactive analysis of green building materials promotion with relevance to energy consumption and greenhouse gas emissions from Taiwan's building sector. *Energy and Buildings*, 261, 111959.

Tussyadiah, I. (2020). A review of research into automation in tourism: Launching the Annals of Tourism Research Curated Collection on Artificial Intelligence and Robotics in Tourism. *Annals of Tourism Research*, 81, 102883.

van Wynsberghe, A. Sustainable AI: AI for sustainability and the sustainability of AI. AI Ethics1, 213–218 (2021)

Wong, L. W., Tan, G. W. H., Ooi, K. B., Lin, B., & Dwivedi, Y. K. (2022). Artificial intelligence-driven risk management for enhancing supply chain agility: A deep-learning-based dual-stage PLS-SEM-ANN analysis. *International Journal of Production Research*, *62*, 1–21.

Xu, L. D., Xu, E. L., & Li, L. (2018). Industry 4.0: state of the art and future trends. *International journal of production research*, *56*(8), 2941–2962.

Zheng, P., Wang, H., Sang, Z., Zhong, R. Y., Liu, Y., Liu, C., ... & Xu, X. (2018). Smart manufacturing systems for Industry 4.0: Conceptual framework, scenarios, and future perspectives. *Frontiers of Mechanical Engineering*, *13*, 137–150.

Zhong, R. Y., Xu, X., Klotz, E., & Newman, S. T. (2017). Intelligent manufacturing in the context of industry 4.0: a review. *Engineering*, *3*(5), 616–630.

Zhu, S., Chen, J., Li, J., Zhang, X., & Huang, Z. (2017, April). Current Patent Trends Analysis and Research for Intelligent Industrial Robots Industry – based on deburring robot. In *2017 7th International Conference on Manufacturing Science and Engineering (ICMSE 2017)* (pp. 451–457). Atlantis Press.

Zonta, T., Da Costa, C. A., da Rosa Righi, R., de Lima, M. J., da Trindade, E. S., & Li, G. P. (2020). Predictive maintenance in the Industry 4.0: A systematic literature review. *Computers & Industrial Engineering*, *150*, 106889.

16 Industry 4.0 in Manufacturing, Communication, Transportation, Healthcare

Virendra Kumar Verma, Shrikant Tiwari, Bireshwar Dass Mazumdar, and Yadav Krishna Kumar Rajnath

16.1 INTRODUCTION TO INDUSTRY 4.0

Industry 4.0, often denoted as the fourth industrial revolution, is seen as a transformative era for businesses with the integration of technologies such as the internet of things (IoT), artificial intelligence (AI), cloud computing, and analytics (Badri, A., et al., 2018).

Industry 4.0 is regarded as the continuation of the industrial revolution, which commenced with the incorporation of the production process into the textile industry in the 18th century (Industry 1.0), followed by the introduction of mass production in the early 20th century (Industry 2.0), and the advent of automation in the 1970s (Industry 3.0). Industry 4.0 is considered to herald a new era where the integration of technology and automation is taken to the next level, traditional factories are transformed into smart factories, and new levels of efficiency, quality control, and customization are enabled (Zhong, R. Y. et al., 2017). Industry 4.0 is expected to change many sectors, from manufacturing to communications, transportation and healthcare. It enables the advancement of new business models, improved customer experiences, and increased competitiveness. (Vaidya, S., et al., 2018). Figure 16.1 shows the Industry 4.0 key areas and their application field..

Key concepts that define Industry 4.0 includes:

- **Cyber-physical systems (CPS)**: these are physical systems integrated with digital systems, enabling communication and cooperation among them. CPS forms the backbone of Industry 4.0 and facilitates the automation of various industrial processes (Hozdić, E. 2015).

DOI: 10.1201/9781003480860-16

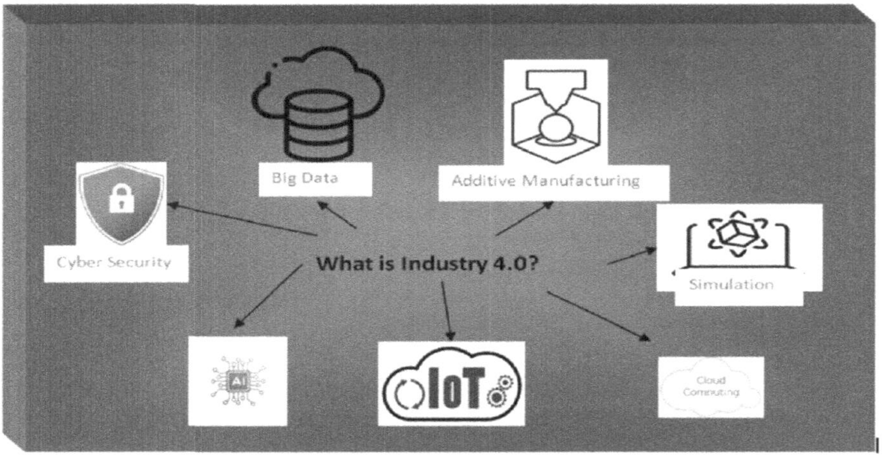

FIGURE 16.1 Introduction of Industry 4.0.

- **Internet of things (IoT)**: the internet of things (IoT) constitutes a network of physical devices, vehicles, and other items embedded with software, sensors, and network connectivity, allowing them to collect and exchange data (Munirathinam, S. 2020).
- **Artificial intelligence (AI)**: AI demonstrates the capacity of machines to simulate human intelligence, enabling them to perform tasks typically requiring human intelligence, such as pattern recognition, future predictions, and problem-solving (Dellermann, D. et al., 2019).
- **Cloud computing**: cloud computing signifies the provision of computing services over the internet, permitting organizations to remotely store and access data and applications (Jadeja, Y., & Modi, K. 2012).
- **Additive manufacturing**: a 3D printing manufacturing process that genrates objects layer by layer using digital models (Gardan, J. 2017).

Together, these concepts enable the creation of smart factories and other industrial systems that are more efficient, productive, and adaptable to changing demands.

16.1.1 Historical Context and Evolution of Industry 4.0

Industry 4.0 represents the latest evolution in a long chronicle of industrial revolutions that began in the late 18th era with the introduction of mechanization in textile manufacturing (Industry 1.0). The 2nd industrial revolution (Industry 2.0) ensued in the early 20th era, characterized by the introduction of production lines and assembly. The 3rd industrial innovation (Industry 3.0) occurred in the 1960s, marked by the introduction of automation and the extensive use of computers and digital technologies (Zhong, R. Y., et al., 2017).

Industry 4.0 emerged in the early 2010s as a response to the need for increased efficiency, productivity, and competitiveness in manufacturing and other industries. The term "Industry 4.0" was first utilized in Germany in 2011 and was coined to describe the 4th industrial revolution that was unfolding. The concept was subsequently embraced by other countries and industries (Piccarozzi, M., et al., 2018).

Industry 4.0 is anticipated to revolutionize various industries, including manufacturing, transportation, healthcare, and communication.

16.1.2 Importance and Impact on Various Industries

Industry 4.0 transforms various industries by enabling new levels of efficiency, productivity, and innovation. Figure 16.2 shows Industry 4.0 application in manufacturing, communication, transportation, healthcare. Here are some examples of the effect of Industry 4.0 on various industrial fields:

- **Manufacturing**: the manufacturing sector is expected to be revolutionized by Industry 4.0, with traditional factories being transformed into smart factories. Smart factories are highly automated and digitized, utilizing technologies such as IoT, AI, and big data analysis processes, which result in cost reduction and improved quality control (Chen, B. et al., 2017).

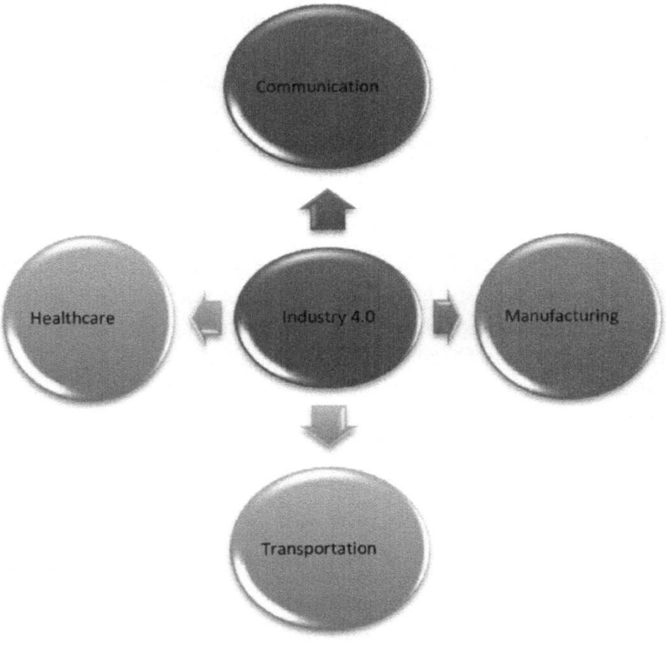

FIGURE 16.2 Different Sectors of Industry 4.0.

- **Transportation**: the transportation industry is indeed being transformed by Industry 4.0, facilitating the advancement of connected and autonomous vehicles. Connected vehicles are equipped with sensors and various communication technologies that enable interaction with other vehicles, infrastructure, and the cloud (Barreto, L., et al., 2017).
- **Healthcare**: Industry 4.0 is reshaping the healthcare industry by enabling the development of personalized and data-driven healthcare solutions. Healthcare providers can utilize data from wearable devices, sensors, and other sources to monitor patients in real-time and offer personalized treatments (Tseng, M. L., et al., 2021).
- **Communication**: the communication industry is undergoing changes due to Industry 4.0, enabling the development of 5G networks, which offer faster speeds and lower latency compared to previous network generations (Wollschlaeger, M. et al., 2017).

16.2 INDUSTRY 4.0 IN MANUFACTURING

Industry 4.0 is indeed bringing about a revolution in manufacturing by facilitating the establishment of smart factories characterized by high levels of automation, digitization, and connectivity. Here are some examples of the impact of Industry 4.0 on manufacturing:

- **Predictive maintenance**: predictive maintenance solutions, facilitated by Industry 4.0, are developed to identify the necessity of maintenance through the utilization of sensors and machine learning algorithms (Dalzochio, J. et al., 2020).
- **Digital twins**: Industry 4.0 enables the development of digital twin solutions that generate virtual models of physical assets. These models can simulate different scenarios and enhance the performance of machines and processes (Kenett, R. S., & Bortman, J. 2022).
- **Autonomous production**: autonomous production solutions, facilitated by Industry 4.0, are developed to utilize robots and other autonomous devices for task execution, reducing the requirement for human intervention and enabling continuous production (Madakam, S., et al., 2019).
- **Data analytics**: Industry 4.0 enables the collection and analysis of vast amounts of data from sensors, machines, and various sources within manufacturing operations. This data can be utilized to optimize the manufacturing process, improve product quality, and minimize waste (Pilloni, V. 2018).

16.2.1 Overview of Traditional Manufacturing Processes and Challenges

Traditional manufacturing processes typically involve a series of steps that are performed in a linear fashion. These steps can include design, raw material acquisition, production, assembly, quality control, and shipping. While this approach has been effective for many years, it is often slow and inflexible, making it difficult to adapt to changing market conditions or customer needs (Pilloni, V. 2018).

There are several challenges associated with traditional manufacturing processes, including:

- **Lack of flexibility**: traditional manufacturing processes are often inflexible and cannot easily modify to changes in customer demand or fair conditions. This can indicate to excess inventory, stockouts, and missed opportunities.
- **Low efficiency**: traditional manufacturing processes can be slow and inefficient, resulting in high labor and production costs.
- **Poor quality control**: traditional manufacturing processes may not have robust quality control measures in place, which can result in defects, recalls, and customer dissatisfaction.
- **Limited data analysis**: traditional manufacturing processes may not capture and analyze data effectively, making it difficult to optimize the manufacturing process and improve efficiency.
- **Environmental impact**: traditional manufacturing processes may generate significant waste and generate large amount of adverse impact on the environment.

16.2.2 Smart Factories and Advanced Manufacturing Technologies

Smart factories are a mainelement of Industry 4.0, and they represent a significant advancement over traditional manufacturing processes. Smart factories use advanced manufacturing technologies to create highly automated, digitized, and connected production processes (Chen, B., et al. 2017). Figure 16.3 shows some examples of advanced manufacturing technologies that are used in intelligent industrial units. Here are some examples of advanced manufacturing technologies that are used in intelligentindustrial unit:

- **Robotics and automation**: smart factories use robots and automation to perform tasks such as assembly, packaging, and material handling.
- **Additive manufacturing**: 3D printing, also seen as additive manufacturing, is a technology that involves creating objects layer by layer from digital designs. This innovative process allows manufacturers to produce complex geometries and reduce waste, making it particularly suitable for prototyping and low-volume production.
- **Augmented reality**: augmented reality (AR) technologies overlays digital communication onto the physical globe. In smart factories, AR provides workers with real-time instructions and feedback, to perform the tasks more efficiently and accurately.
- **Internet of things (IoT)**: the connectivity of physical devices to the internet, enabling communication and data exchange, is encompassed by the internet of things (IoT). In smart factories, IoT devices such as sensors and smart machines play a crucial role in providing real-time data on manufacturing processes. This data facilitates the optimization of manufacturing operations and reduction of downtime.

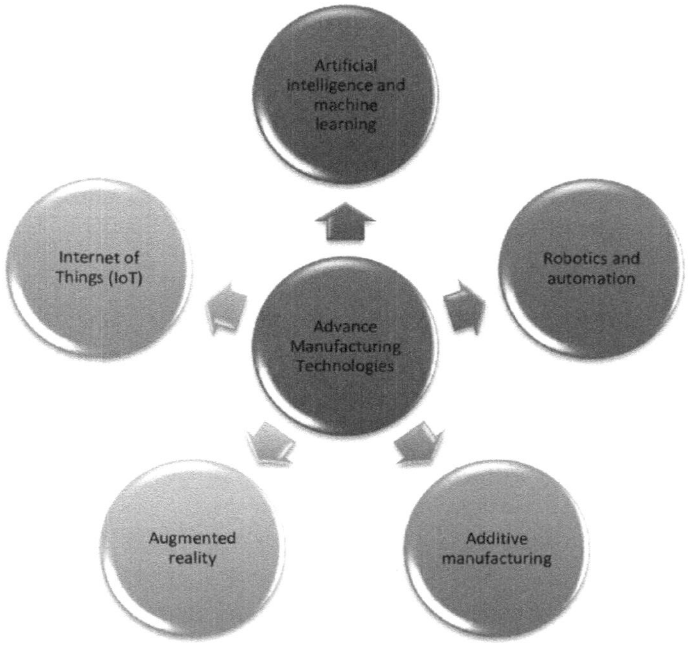

FIGURE 16.3 Advanced Manufacturing Technologies used in Smart Factories.

16.2.3 Examples of Industry 4.0 Implementation in Manufacturing

Numerous examples illustrate the execution of Industry 4.0 in manufacturing. Here are some specific instances:

- **Siemens**: Siemens, a global electronics and electrical engineering company, has implemented Industry 4.0 in its Amberg factory in Germany. The factory uses autonomous robots, intelligent transportation systems, and real-time analytics for optimization. (Li, B. H., et al., 2017).
- **General Electric**: General Electric (GE) has implemented Industry 4.0 in its Brilliant Factory in Pune, India. The factory uses advanced analytics, IoT sensors, and AI to optimize production methods and improve product quality. GE estimates that the factory has increased productivity by 20%, reduced downtime by 50%, and improved product quality by 40% (Bhat, T. P. 2020).
- **Airbus**: Airbus, is a global aerospace manufacturer company, has also implemented Industry 4.0 in its Factory of the Future in Hamburg, Germany. The factory uses autonomous robots, additive manufacturing, and digital twin technology to create highly efficient and flexible production processes (Ernst, F., & Frische, P. 2015).
- **BMW**: BMW, a global automotive manufacturer, has implemented Industry 4.0 in its Dingolfing plant in Germany. The plant uses IoT sensors, data

analytics, and autonomous robots to optimize production processes (Rosa, D. I. D. S. 2022).

- **Haier**: Haier, a global home appliance manufacturer, has implemented Industry 4.0 in its Qingdao factory in China. The factory uses digital twin technology, IoT sensors, and advanced analytics to optimize production methods and improve product quality (Huang, P. Y., et al., 2012).

16.2.4 Benefits of Industry 4.0 in Manufacturing

Figure 16.4 shows some benefits of Industry 4.0 in manufacturing.

Benefits of Industry 4.0 in manufacturing:

- **Improved efficiency**: improved efficiency is achieved through the utilization of Industry 4.0 technologies such as automation, the internet of things (IoT), and artificial intelligence (AI), which play a crucial role in optimizing manufacturing processes and reducing production time, ultimately leading to increased efficiency.
- **Improved quality control**: advanced manufacturing technologies such as 3D printing and digital twin technology can improve product quality by enabling more precise and accurate manufacturing.
- **Flexibility**: Industry 4.0 technologies provides more flexible manufacturing processes, allowing manufacturers to quickly respond to varying customer demands and market conditions.
- **Reduced costs**: Industry 4.0 technologies can reduce manufacturing costs by reducing waste, improving supply chain efficiency, and optimizing production processes.

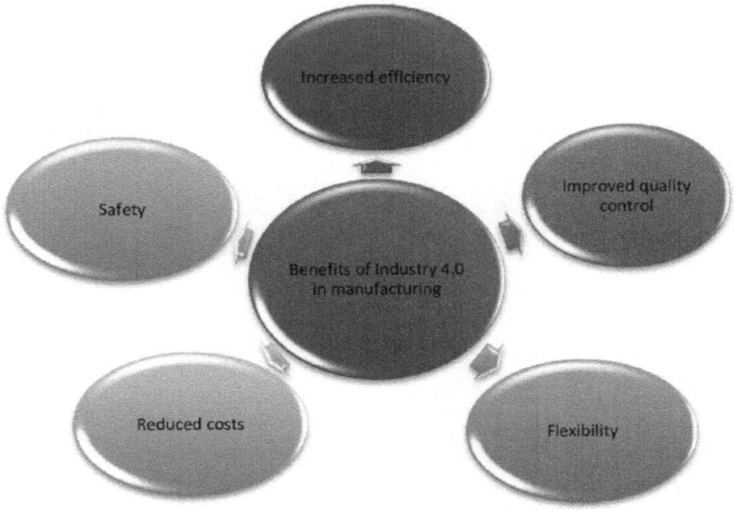

FIGURE 16.4 Benefits of Industry 4.0 in Manufacturing.

- **Safety**: safety is significantly improved through the utilization of automation and robotics, as they reduce the requirement for human intervention in hazardous manufacturing processes. By automating tasks that are unsafe, repetitive, or involve exposure to dangerous conditions, workplace safety is enhanced, and workers are protected.

16.2.5 Drawbacks of Industry 4.0 in Manufacturing

Figure 16.5 shows some drawbacks of Industry 4.0 in manufacturing.

- **Cost**: Industry 4.0 technologies operation can be costly, requiring more investment in installing new hardware, software, and infrastructure.
- **Cybersecurity risks**: connected devices and networks can be vulnerable to cyberattacks, and manufacturers need to take steps to ensure the security of their systems.
- **Job displacement**: automation and robotics can reduce the need for human labor, leading to job displacement and potential social and economic impacts.
- **Complexity**: Industry 4.0 technologies can be complex and require significant expertise to implement and maintain, which may be challenging for smaller manufacturers.

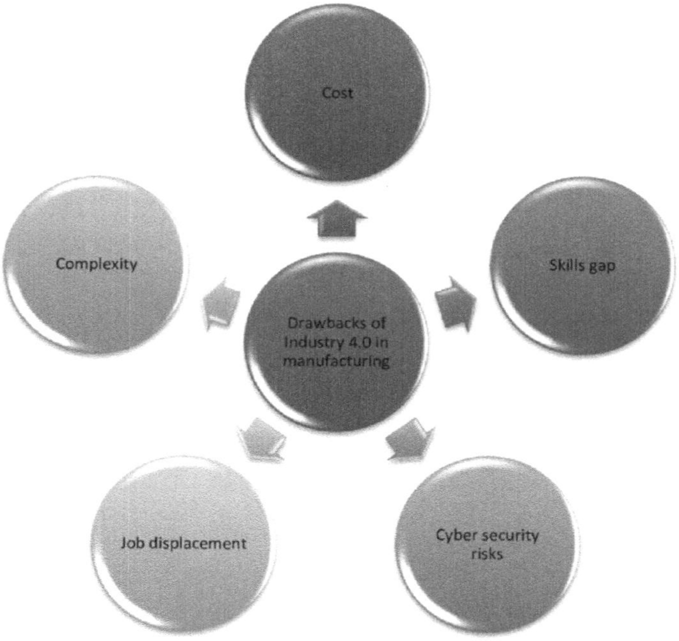

FIGURE 16.5 Drawbacks of Industry 4.0 in Manufacturing.

16.2.6 Future Trends and Potential Developments

The future trends and potential developments in Industry 4.0 for manufacturing are vast and exciting (Xu, L. D., et al., 2018). Here are some of the most significant trends and developments to watch for:

- **Advanced analytics and artificial intelligence (AI)**: these are increasingly in demand to analyze the substantial amount of data generated by Industry 4.0 technologies, deriving insights to further optimize production.
- **Digital twins**: used to generate a digital replica of a physical object or system, these are becoming more sophisticated and widely used in manufacturing.
- **5G connectivity**: 5G connectivity offers the potential for faster, more reliable, and more secure wireless communication between Industry 4.0 devices and systems, enabling new manufacturing applications and opportunities.
- **Autonomous robots and vehicles**: autonomous robots and vehicles are becoming more advanced and capable, enabling new possibilities for manufacturing, such as fully automated production lines and autonomous warehouse operations.
- **Sustainable manufacturing**: as sustainability gains increasing importance, Industry 4.0 technologies are being developed to facilitate more sustainable manufacturing processes, including the utilization of renewable energy sources and waste reduction.

16.3 INDUSTRY 4.0 IN COMMUNICATION

Industry 4.0 is reshaping the manufacturing industry and influencing other sectors, including communication. The integration of Industry 4.0 technologies with communication infrastructure is leading to the development of more intelligent communication systems. Here are some ways in which Industry 4.0 is impacting communication:

- **5G technology**: 5G technology plays a pivotal role as a backbone for the success of Industry 4.0, providing high-speed, low-latency, and reliable wireless connectivity that enables seamless communication between different devices and systems (Rao, S. K., & Prasad, R. 2018).
- **Smart cities**: Industry 4.0 technologies are integrated into city infrastructure to establish smart cities. These cities utilize communication networks to connect various devices and systems, such as transportation, energy, and waste management, to enhance efficiency and sustainability (Lom, M., et al., 2016).
- **IoT**: the IoT serves as a crucial component of Industry 4.0 in communication, involving the connection of various devices and systems to exchange information and communicate with each other over the internet. This opens up new opportunities for automation, remote monitoring, and control (Wollschlaeger, M., Sauter, T., & Jasperneite, J. 2017).
- **AI**: AI also plays a significant role in Industry 4.0 within communication, enabling intelligent decision-making and automation in communication systems,

such as chatbots, voice assistants, and predictive maintenance systems (Lee, J., et al., 2018).

- **Cybersecurity**: as communication networks become more interconnected, cybersecurity emerges as a critical concern. Industry 4.0 technologies, such as AI and blockchain, can be utilized to bolster communication security and mitigate cyber attacks (Culot, G., et al., 2019).

16.4 OVERVIEW OF TRADITIONAL COMMUNICATION NETWORKS AND TECHNOLOGIES

Traditional communication networks and technologies refer to the methods used to transfer data, voice, and video between different devices and systems (Akyildiz, I. F., et al, 2008). Here are some of the most common traditional communication networks and technologies:

- **Wired networks**: wired networks use physical cables to move data, voice, and video between devices. Wired networks, like Ethernet, use fiber optic cables and coaxial cables.
- **Wireless networks**: wireless networks use electromagnetic waves to transfer data, voice, and video amongmechanisms. Examples of wireless networks include Wi-Fi, Bluetooth, and cellular networks.
- **Voice communication**: voice communication technologies include analog and digital telephone networks, as well as voice over internet protocol (VoIP) systems, which use the internet to transmit voice data.
- **Video communication**: video communication technologies include video conferencing systems, which enable remote users to participate in meetings, and streaming services, which deliver video content over the internet.
- **Messaging and collaboration**: messaging and collaboration technologies include email, instant messaging, and file sharing systems, which enable users to exchange data and collaborate on projects.
- **Security**: security technologies include firewalls, encryption, and intrusion detection systems, which protect communication networks from cyber threats.

16.4.1 Advanced Communication Technologies Enabled by Industry 4.0

Industry 4.0 is enabling the stage and implementation of advanced transmission technologies that are transforming the way we communicate and exchange information (Aceto, G., Persico, V., & Pescapé, A. 2019). Figure 16.6 shows some examples of advanced communication technologies that are enabled by Industry 4.0.

Here are some examples of advanced communication technologies that are enabled by Industry 4.0:

- **5G networks**: 5G networks represent the current generation of wireless networks and are highly beneficial for the success of Industry 4.0. They offer

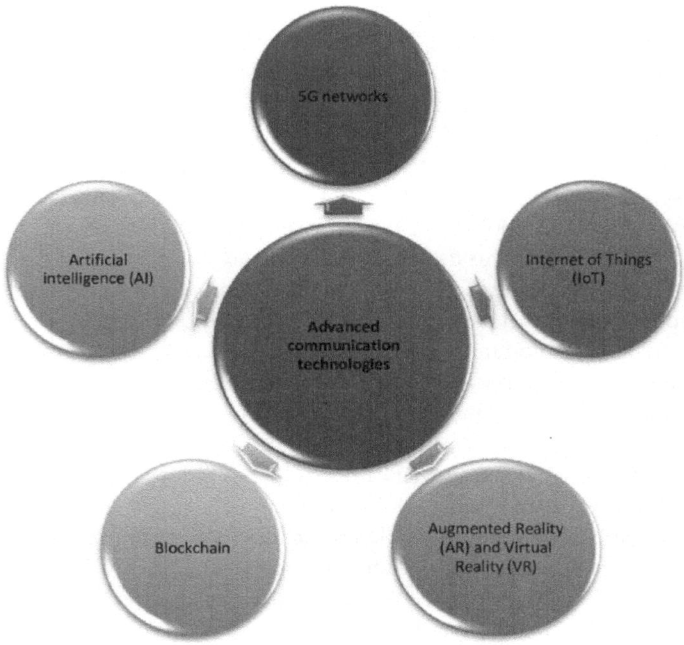

FIGURE 16.6 Advanced Communication Technologies Enabled by Industry 4.0.

faster data transfer, reduced latency, and heightened reliability, facilitating more efficient and dependable communication between devices and systems.

- **IoT**: the internet of things (IoT) constitutes a network of intersected devices and systems that connect and exchange data with each other, facilitated by Industry 4.0.
- **AI**: AI is utilized to empower intelligent communication systems, including chatbots, voice assistants, and predictive maintenance systems. These systems can analyse data and make intelligent decisions to enhance communication efficiency.
- **Blockchain**: blockchain serves as a dispersed ledger technology for secure and transparent communication and data swap among machines and systems. It can enhance communication security and mitigate cyber threats.
- **Augmented reality (AR) & virtual reality (VR)**: AR and VR technologies are employed to create envelop communication experiences. For instance, in manufacturing, AR can provide remote support and training, while in healthcare, VR can simulate medical procedures and surgeries.

16.4.2 Examples of Industry 4.0 Implementation in Communication

Industry 4.0 is transforming the communication industry in many ways, and there are numerous examples of its implementation. Here are some examples of Industry 4.0 implementation in communication (Thoben, K. D., et al., 2017):

- **Smart cities**: smart cities are enabled by Industry 4.0 technologies and communication systems that use IoT, AI, and big data to develop the effectiveness of city services, such as traffic administration, waste executives, and public safety.
- **Telecommunications**: telecommunications companies are implementing Industry 4.0 technologies, like 5G networks and AI, to improve communication speed, reliability, and security. For example, 5G networks can support new communication technologies, such as augmented and virtual reality.
- **Chatbots and voice assistants**: chatbots and voice assistants are becomes more popular in customer service, and they are powered by AI and natural language processing. These communication technologies enable more efficient and personalized communication with customers.
- **Predictive maintenance**: predictive maintenance systems are being implemented in communication networks to improve efficiency and reduce downtime. These organisations apply artificial intelligence and machine learning to recognize maintenance needs before failure.
- **Collaborative workspaces**: collaborative workspaces are becoming more popular, and they are being enabled by advanced communication technologies, such as video conferencing, file sharing, and virtual reality. These technologies enable remote collaboration and communication between teams, improving productivity and reducing travel costs.

16.4.3 Benefits and Drawbacks of Industry 4.0 in Communication

Industry 4.0 is bringing major benefits to the communication industry, but it also has some drawbacks (Liu, Y., & Xu, X. 2017). Figure 16.7 shows some of the benefits of Industry 4.0 in communication.

Benefits:

- **Improved efficiency**: it is indeed achieved through the utilization of Industry 4.0 technologies, including the internet of things (IoT), artificial intelligence (AI), and big data, which play a crucial role in enhancing communication and decision-making processes within industries, ultimately resulting in improved productivity and reduced costs.
- **Higher customer experience**: advanced communication technologies, such as chatbots and voice assistants, enable more personalized and efficient communication with customers, improving their satisfaction.
- **Innovation**: Industry 4.0 is driving innovation in the communication industry by enabling new and advanced communication technologies.

Figure 16.8 shows some of the drawbacks of Industry 4.0 in communication.

Drawbacks:

- **Security concerns**: advanced communication technologies, such as IoT and AI, create new security risks, such as cyber attacks and data breaches.

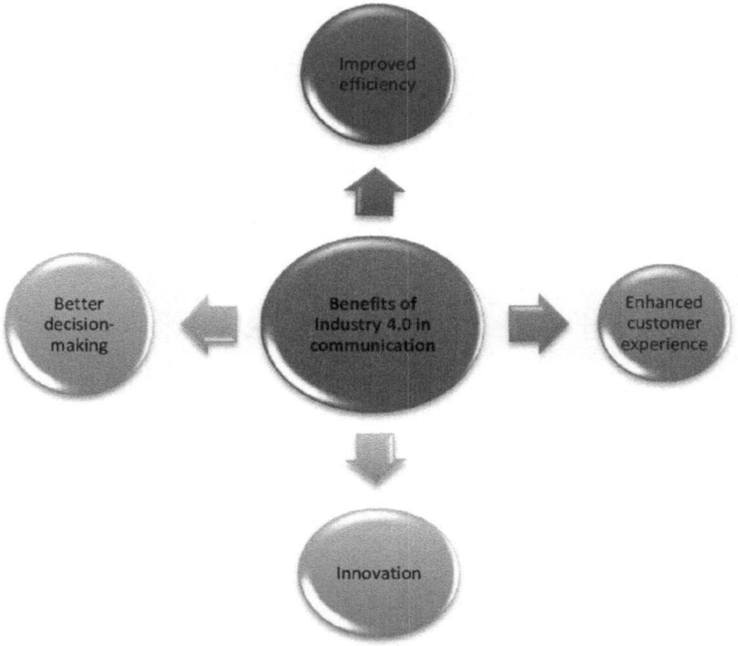

FIGURE 16.7 Benefits of Industry 4.0 in Communication.

- **High installation costs**: indeed, increased investment in infrastructure, training, and maintenance is required for the successful integration and operation of Industry 4.0 technologies.
- **Job displacement**: automation facilitated by Industry 4.0 technologies may result in job displacement, necessitating new skills and training for workers.
- **Data overload**: the adoption of Industry 4.0 technologies leads to the generation of vast amounts of data, which can be overwhelming and challenging for businesses to manage effectively.

16.4.4 Future Trends and Potential Developments

The outlook of Industry 4.0 in the communication industry is thrilling, with many ability developments and trends on the horizon (Xu, L. D., et al., 2018).These are the some of the future trends and potential developments of Industry 4.0 in communication:

- Increased adoption of 5G networks
- Advancements in AI and ML
- Expansion of the IoT
- Increased use of virtual and augmented reality
- Focus on sustainability.

FIGURE 16.8 Drawbacks of Industry 4.0 in Communication.

16.5 INDUSTRY 4.0 IN TRANSPORTATION

Industry 4.0 is bringing substantial changes to the transportation industry, with new technologies and processes transforming the way goods and people move around the world (Barreto, L., Amaral, A., & Pereira, T. 2017). Here's an overview of Industry 4.0 in transportation:

- **Intelligent transportation systems**: connected and autonomous vehicles, advanced data analytics, and smart infrastructure are leveraged by intelligent transportation systems (ITS) to enhance the efficiency, safety, and sustainability of transportation systems. By integrating these technologies, ITS aims to minimize traffic congestion, improve fuel efficiency, and enhance the overall travel experience for individuals and businesses.
- **Supply chain optimization**: supply chains are optimized through the utilization of Industry 4.0 technologies, enhancing efficiency and reducing costs. Advanced data analytics can assist in optimizing routes, reducing delivery times, and ensuring the timely delivery of goods.
- **Smart logistics**: Industry 4.0 technologies such as IoT and AI are utilized in smart logistics systems to improve the planning, execution, and monitoring of logistics operations.

- **Predictive maintenance**: predictive maintenance can be enabled by Industry 4.0 technologies, allowing transportation companies to monitor the health of their vehicles and equipment in real-time, predict failures before they occur, and reduce downtime.

16.5.1 Overview of Traditional Transportation Systems and Challenges

Traditional transportation systems typically rely on manual processes, outdated technology, and limited visibility into operations, resulting in a range of challenges for transportation companies (Rodrigue, J. P. 2020). Here's an overview of some of the challenges faced by traditional transportation systems:

- **Inefficient processes**: traditional transportation systems often rely on manual processes, such as paper-based tracking and data entry, which are time-onerous and prone to faults. This can result in inefficiencies and delays in the transportation process.
- **Lack of visibility**: traditional transportation systems often lack visibility into the movement of goods, making it difficult to track shipments and ensure timely delivery. This lack of visibility can also result in higher transportation costs and increased risk of theft or damage.
- **Limited data analytics**: traditional transportation systems often lack the capability to save and examine data, which can hinder the ability to optimize transportation routes, reduce fuel costs, and improve overall efficiency.
- **Safety concerns**: traditional transportation systems can be susceptible to safety concerns, such as accidents and theft. These concerns can result in higher insurance costs, legal liabilities, and negative brand reputation.
- **Environmental impact**: traditional transportation systems often rely on fossil fuels, contributing to pollution and carbon emissions. This can result in negative environmental impacts and increased regulatory pressure.

16.5.2 Autonomous Vehicles and Smart Traffic Management Systems

Autonomous vehicles and smart traffic management systems are two cases of Industry 4.0 tools that are transforming the transportation industry (Khayyam, H., et al., 2020). Here's an overview of these technologies:

- **Autonomous vehicles**: vehicles that can use a range of sensing devices, cameras, and advanced software to navigate routes without human input. These vehicles can potentially reduce accidents, have better efficiency, and lower transportation costs.
- **Smart traffic management systems**: the optimization of traffic flow, congestion reduction, and safety enhancement are achieved through the utilization of real-time data and advanced technologies by intelligent traffic management systems.

16.5.3 Examples of Industry 4.0 Implementation in Transportation

Here are some examples of Industry 4.0 implementation in transportation:

- **Autonomous vehicles**: companies such as Waymo, Tesla, and Uber are leading the development of autonomous vehicles that utilize a combination of sensors, cameras, and advanced software to navigate roads without human input (Kröger, W. 2021).
- **Smart traffic management systems**: intelligent traffic management systems, implemented by cities like Singapore and Amsterdam, utilize real-time data and analytics to optimize traffic movement and reduce congestion (Deb, T., Vishwas, N., & Saha, A. 2020).
- **Connected vehicles**: connected vehicles are equipped with internet-connected sensors and software that enable communication with other vehicles and the transportation infrastructure.
- **Drones for goods delivery**: drones are being developed for goods delivery in urban areas, enabling faster and more efficient delivery of goods. Companies like Amazon and UPS are testing drone delivery systems that use autonomous drones to deliver packages to customers (Mohamed, N.,et al., 2019).
- **Hyperloop**: it is a proposed high-speed transportation system by using vacuum-sealed tubes for the movement of humans and goods up to the speed of 700 miles/ hour. (Kale, S. R., et al., 2019).

16.5.4 Benefits and Drawbacks of Industry 4.0 in Transportation

Figure 16.9 shows some of the benefits of Industry 4.0 in transportation.

Here are Industry 4.0 benefits and drawbacks in transportation.

Benefits:

- **Increased efficiency**: Industry 4.0 technologies can optimize transportation systems, reducing congestion and improving travel times.
- **Cost savings**: Industry 4.0 toolsshow a significant role in reducing transportation costs, especially in logistics and supply chain management.
- **Environmental sustainability**: Industry 4.0 technologies can reduce emissions and improve environmental sustainability by optimizing transportation systems and reducing waste.
- **Innovation**: Industry 4.0 is indeed driving innovation in the transportation industry by enabling the adoption of new technologies and business models.

  Figure 16.10 shows some of the drawbacks of Industry 4.0 in transportation.

Drawbacks:

- **Job displacement**: the utilization of Industry 4.0 technologies, including autonomous vehicles and drones, may indeed result in job displacement for human workers in the transportation industry.

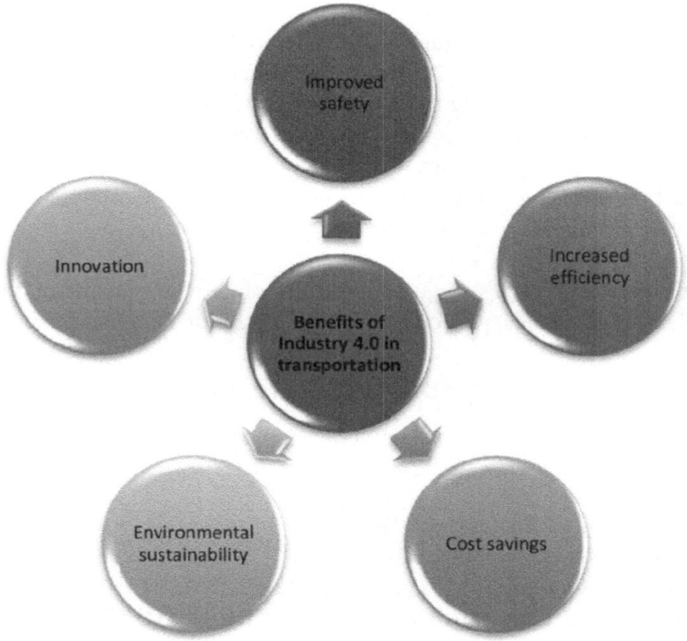

FIGURE 16.9 Benefits of Industry 4.0 in Transportation.

FIGURE 16.10 Drawbacks of Industry 4.0 in Transportation.

- **Cybersecurity risks**: Industry 4.0 technologies are indeed susceptible to cyber-attacks, posing risks of disruption to transportation systems and potential harm.
- **Cost of implementation**: the implementation of Industry 4.0 technologies can indeed be costly, particularly for small and medium-sized enterprises (SMEs), owing to the significant upfront investment required for acquiring and deploying advanced machinery, sensors, data analytics systems, and connectivity.
- **Infrastructure challenges**: implementing Industry 4.0 technologies in transportation requires significant infrastructure investments, particularly in terms of sensors and connectivity.
- **Regulatory challenges**: Industry 4.0 technologies in transportation may face regulatory challenges, particularly around issues of safety and liability.

16.5.5 Future Trends and Potential Developments

Here are some future trends and potential developments in Industry 4.0 for transportation (Laiton-Bonadiez, C., et al., 2022):

- Autonomous and electric vehicles
- Mobility-as-a-service (MaaS)
- Hyperloop and other high-speed transportation systems
- Blockchain and other distributed ledger technologies (DLT)
- Smart cities and infrastructure.

16.6 INDUSTRY 4.0 IN HEALTHCARE

Industry 4.0 is reforming the healthcare industry by enhancing the efficiency, effectiveness, and personalization of care through various implementation in healthcare (Aceto, G., et al., 2019):

- **Digital health platforms**: big data analytics and other business 4.0 technologies, AI, and the IoT are being used to develop digital health platforms that enable remote patient monitoring, personalized treatment, and real-time data analytics.
- **Wearable and implantable medical devices**: the use of such types of devices, enabled by Industry 4.0 technologies like sensors and connectivity, is expanding rapidly. These devices can observe vital signs and track medication observance.
- **Telemedicine**: telemedicine, or the use of remote communication technologies to deliver healthcare services, is becoming more popular now a days. This is enabled by Industry 4.0 technologies such as video conferencing, remote monitoring, and data analytics.
- **3D printing**: 3D printing is employed to create customized medical devices, prosthetics, and even human tissue to improve patient recovery rate, reduce costs, and increase access to care.
- **Smart hospitals**: smart hospitals are being developed that incorporate Industry 4.0 tools such as AI, IoT, and robotics to improve patient outcomes,.

16.6.1 Overview of Traditional Healthcare Systems and Challenges

Traditional healthcare systems have been primarily focused on providing reactive care, which means that healthcare providers only intervene after a patient becomes sick or develops a medical condition. This has resulted in a healthcare system that is often overburdened, expensive, and inefficient (Karatas, M.,et al., 2022). Additionally, traditional healthcare systems have faced several challenges, including:

- **Fragmentation**: healthcare systems are often fragmented, with different providers, insurers, and government agencies working in silos. This fragmentation can result in gaps in care, duplication of services, and inefficient use of resources.
- **Cost**: healthcare costs have been rising steadily, with a significant portion of healthcare spending going towards treating chronic diseases. This has resulted in an unsustainable financial burden on both patients and healthcare systems.
- **Limited access**: many people around the world lack access to basic healthcare services, with some estimates.
- **Quality of care**: the quality of healthcare services can vary widely, with some patients receiving substandard care due to factors such as inadequate resources or training.
- **Aging population**: the global population is aging, with an increasing number of people living with chronic diseases that require ongoing care and support.

16.6.2 Advanced Medical Devices, Electronic Health Records, and Telemedicine

Industry 4.0 technologies transform the healthcare industry by enabling more personalized, proactive, and efficient care (Aceto, G., Persico, V., & Pescapé, A. 2019). Some of the key Industry 4.0 technologies that are being used in healthcare include:

- **Advanced medical devices**: healthcare is indeed being transformed by advanced medical devices such as wearable sensors and remote monitoring devices, enabling healthcare providers to collect real-time patient data and remotely monitor individuals, resulting in improved accuracy in diagnoses, earlier interventions, and reduced hospitalization.
- **Electronic health records**: digitized medical records, known as electronic health records (EHRs), offer numerous benefits to healthcare providers and patients by enhancing the accuracy and accessibility of medical information, reducing paperwork, and facilitating more informed decisions about patient care.
- **Telemedicine**: healthcare services are delivered remotely through telemedicine, utilizing technology such as video conferencing and remote monitoring, thereby enhancing access to care, reducing costs, and enabling closer monitoring of patients.

16.6.3 Examples of Implementation of Industry 4.0 in Healthcare

Here are particular examples of Industry 4.0 implementation in healthcare (Paul, S., et al., 2021):

- **Robotics-assisted surgery**: robotic surgical systems are being used to perform less persistent surgeries with better precision and accuracy.
- **Personalized medicine**: Industry 4.0 technologies, such as genome sequencing and machine learning algorithms, help doctors create personalized treatments based on a patient's unique genetic profile. This can lead to better treatment and better patient outcomes.
- **Digital health platforms**: digital health platforms, such as mobile apps and patient portals, are enabling patients to access healthcare services remotely, communicate with healthcare providers, and monitor their own health. These platforms can improve patient engagement and enable healthcare providers to provide more proactive and personalized care.
- **3D printing**: 3D printing technology is being used to develop customized medical implants, prosthetics, and anatomical models. This technology enables healthcare providers to create patient-specific solutions that are more effective and comfortable for the patient.

16.6.4 Advantages and Limitations of Industry 4.0 in Healthcare

Figure 16.11 shows some of the benefits of Industry 4.0 in healthcare.

Here are some benefits and drawbacks of Industry 4.0 in healthcare (Aceto, G., et al., 2019):

Benefits:

- **Improved patient outcomes**: Industry 4.0 tools improves treatment plans, remote patient monitoring, and more accurate diagnoses.
- **Increased efficiency**: electronic health records and digital health platforms can streamline healthcare operations, reduce administrative tasks, and increase efficiency.
- **Cost savings**: Industry 4.0 technologies can reduce healthcare costs by improving operational efficiency, reducing errors, and preventing unnecessary procedures.
- **Enhanced patient experience**: telemedicine, patient portals, and other digital health platforms can improve patient engagement and enable patients to have more control over their healthcare experience.

Figure 16.12 shows some of the drawbacks of Industry 4.0 in healthcare.

Drawbacks:

- **Workforce disruption**: Industry 4.0 technologies in healthcare may require significant changes in the healthcare workforce, including new skill sets and job roles.

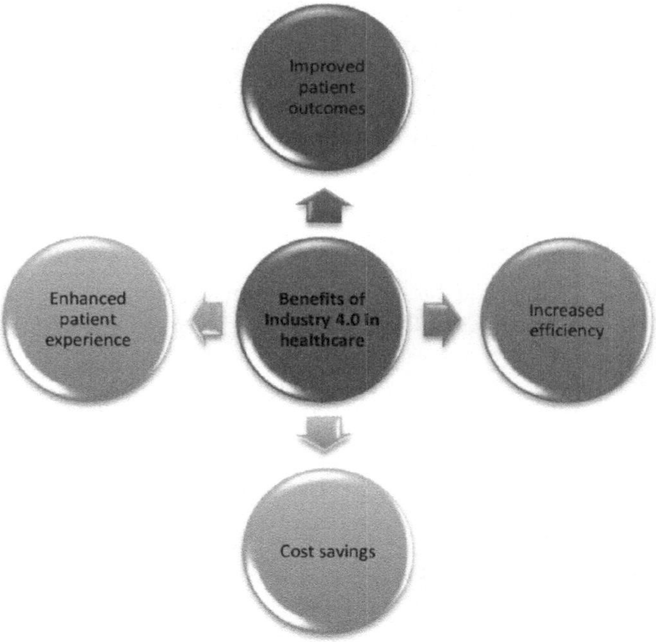

FIGURE 16.11 Benefits of Industry 4.0 in Healthcare.

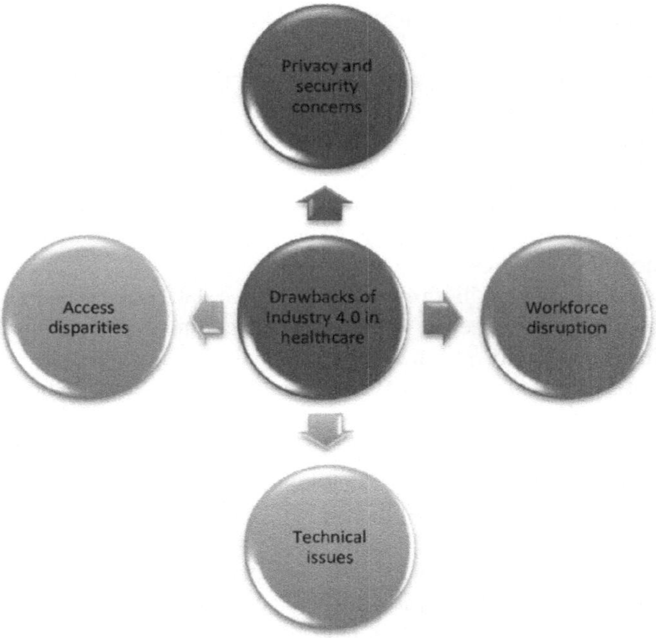

FIGURE 16.12 Drawbacks of Industry 4.0 in Healthcare.

- **Technical issues**: Industry 4.0 technologies can be complex and require significant technical expertise to implement and maintain. Technical issues can also result in system downtime or errors that can negatively impact patient care.
- **Access disparities**: the adoption of Industry 4.0 technologies may exacerbate existing disparities in access to healthcare, as these technologies may be more readily available in certain regions or healthcare systems.

16.6.5 Future Trends and Potential Developments

Here are some future trends and potential developments in Industry 4.0 in healthcare (Mohamed, N., & Al-Jaroodi, J. 2019):

- Artificial intelligence and machine learning
- Wearable technology
- 5G networks
- Blockchain
- Robotics
- Precision medicine
- Augmented reality and virtual reality.

16.7 CHALLENGES AND RISK OF INDUSTRY 4.0

While the advent of Industry 4.0 presents numerous advantages and prospects, it also introduces several hurdles and vulnerabilities that demand attention (He, Y., et al., 2023). Here are some of the challenges and risks associated with Industry 4.0:

- **Cybersecurity**: with the increasing interconnectivity and reliance on data within Industry 4.0 systems, they become susceptible to cyber threats and breaches.
- **Workforce displacement**: the integration of Industry 4.0 technologies may lead to job displacement or require extensive upskilling and reskilling of the workforce. This could potentially create a shortage of skilled labor and exacerbate income disparities.
- **Privacy and ethical dilemmas**: the extensive data generated by Industry 4.0 technologies raises concerns about privacy and ethical considerations regarding the collection and utilization of personal data.
- **Compatibility issues**: the amalgamation of various technologies and systems may result in compatibility challenges, making it difficult to share data and collaborate across different platforms.
- **Infrastructure demands**: implementing Industry 4.0 technologies necessitates significant investments in infrastructure, including high-speed connectivity, robust data storage, and enhanced computing capabilities. This could pose barriers to entry for small and medium-sized enterprises.
- **Regulatory and legal complexities**: the adoption of Industry 4.0 technologies may require the establishment of new regulatory frameworks to address issues such as data ownership, liability, and intellectual property rights.

16.7.1 Ethical and Societal Concerns

Industry 4.0 has the potential to revolutionize industries and transform society in positive ways. However, it also raises several moral and societal affects that require to be focused on (Morrar, R., Arman, H., & Mousa, S. 2017). Here are some of the main ethical and societal concerns of Industry 4.0:

- **Job displacement and earnings unfairness**: the deployment of these technologies may also require reskilling and upskilling of the workforce, which can be a challenge for some workers.
- **Privacy and data protection**: the collection and use of personal information must be transparent, and individuals must have control over their data.
- **Bias and discrimination**: Industry 4.0 technologies are only as objective as the data used to train them. If the data benefited to educate these technologies is biased, it can lead to discriminatory outcomes, reinforcing social inequalities.
- **Autonomy and accountability**: as Industry 4.0 technologies become more autonomous, there is a need for transparency and accountability. It is essential to understand how decisions are made and who is responsible when things go wrong.
- **Environmental impact**: the deployment of Industry 4.0 technologies has the potential to decrease carbon emissions and upgrade resource efficiency. However, the production and disposal of these technologies can also have a negative environmental impact.

16.7.2 Cybersecurity Risks and Data Privacy Issues

In addition to the ethical and societal concerns, cybersecurity risks and data privacy issues are significant challenges for Industry 4.0. With the grown connectivity and data sharing in Industry 4.0 systems, the risk of cyber-attacks and data breaches increases (Süzen, A. A., 2020). Here are some of the main cybersecurity risks and data privacy issues in Industry 4.0:

- **Vulnerabilities in the system**: with the increasing number of connected devices and systems, there are more potential liabilities for cyber-attacks. Cyber criminals can exploit these liabilities to increaseillegal access to sensitive data and systems.
- **Data privacy**: Industry 4.0 technologies spawn massive quantities of information, including personal information. This data is often shared across different systems and networks, raising concerns about data privacy.
- **Lack of standardization**: the deficit of calibration in Industry 4.0 technologies can create security vulnerabilities.
- **Insider threats**: the increasing use of connected devices and the internet of things (IoT) also increase the risk of insider threats.
- **Third-party risks**: as more companies collaborate and share data, third-party risks increase. Vendors, suppliers, and partners can become a weak link in the security chain if they do not have adequate security measures in place.

16.7.3 Potential Job Displacement and Skill Gaps

Industry 4.0 technology has the capability to enhance and improve many activities previously performed by humans, which could lead to job losses and skill losses. (Sima, V., et al., 2020). Here are some of the main challenges in this area:

- **Job displacement**: automation and optimization of tasks may lead to job displacement in certain industries, especially in manufacturing and transportation. Jobs that involve repetitive and routine tasks are most at risk.
- **Skill gaps**: Industry 4.0 technologies need a new set of skills that may not be present in the current workforce. For example, data analytics, machine learning, and programming skills are in high demand in industries adopting Industry 4.0 technologies.
- **Training and education**: to address the skill gaps, there is a need for training and teaching programs that equip workers with the skills needed to work with Industry 4.0 technologies.
- **Social implications**: the impact of job displacement and skill gaps may go beyond the economic implications. It may also affect the social structure and well-being of individuals and communities.

16.7.4 Regulatory and Legal Challenges

Industry 4.0 technologies raise several regulatory and legal challenges (Preuveneers, D., & Ilie-Zudor, E., 2017), such as:

- **Intellectual property rights**: the use of Industry 4.0 tools such as big data, artificial intelligence, and the internet of things (IoT) may raise questions about intellectual property rights.
- **Cybersecurity**: Industry 4.0 technologies are highly connected and generate a vast amount of data, which increases the risk of cyberattacks.
- **Liability**: with the enhancing automation and autonomy of Industry 4.0 technologies, questions may arise about liability in case of accidents or errors..
- **Standards and interoperability**: Industry 4.0 technologies require standards and interoperability.
- **Labor laws**: the increasing use of automation and artificial intelligence may raise questions about labor laws, such as minimum wage laws and working conditions.

16.8 CONCLUSION AND FUTURE OUTLOOK

Industry 4.0 signifies a transformative technological shift that is reshaping industries and fostering new opportunities for growth and innovation across various sectors. This evolution encompasses smart factories, advanced manufacturing technologies, autonomous vehicles, and telemedicine, fundamentally altering the way we live and work.

While there are several tasks and risks allied with Industry 4.0, such as ethical and societal concerns, cybersecurity risks, and potential job displacement, these can be addressed through appropriate regulations, standards, and responsible adoption of technologies.

16.8.1 Summary of Key Points

Here is a summary of the key points covered in this discussion on Industry 4.0:

- Industry 4.0 is a technological shift that is transforming industries, creating new opportunities for growth and innovation.
- Industry 4.0 has the probable to benefit a wide range of industries, involving manufacturing, communication, transportation, and healthcare.
- In manufacturing, Industry 4.0 is enabling the development of bright plants and advanced industrial technologies.
- In communication, it is enabling the development of advanced communication technologies and networks.
- In healthcare, it is enabling the development of advanced medical devices, electronic health records, and telemedicine.
- Industry 4.0 presents several challenges and risks, including ethical and societal concerns, cybersecurity risks, potential job displacement, and regulatory and legal challenges.
- To address these challenges and risks, appropriate regulations, standards, and responsible adoption of technologies are necessary.
- Looking ahead, the future of Industry 4.0 is promising, with ongoing developments in areas such as quantum computing, blockchain, and edge computing.

16.8.2 Directions for Future Research and Development

As Industry 4.0 continues to evolve and transform various industries, there are several directions for future research and development:

- **Integration of AI and ML**: while Industry 4.0 has already enabled the use of AI and ML in various industries, there is still potential for further integration and optimization of these technologies to enhance performance and productivity.
- **Standardization and interoperability**: as Industry 4.0 technologies continue to develop, there is a need for standardization and interoperability to ensure seamless integration of different systems and devices.
- **Addressing ethical and social implications**: as Industry 4.0 continues to advance, there is a need to address ethical and social implications, such as privacy concerns, job displacement, and access to technology.
- **Sustainability and environmental impact**: Industry 4.0 technologies have the prospective to significantly degrade waste and emissions in various industries. Future research should focus on the development of sustainable and environmentally friendly technologies.

REFERENCES

Aceto, G., Persico, V., & Pescapé, A. (2019). A survey on information and communication technologies for industry 4.0: State-of-the-art, taxonomies, perspectives, and challenges. *IEEE Communications Surveys & Tutorials*, *21*(4), 3467–3501.

Akyildiz, I. F., Brunetti, F., &Blázquez, C. (2008). Nanonetworks: A new communication paradigm. *Computer Networks*, *52*(12), 2260–2279.

Badri, A., Boudreau-Trudel, B., & Souissi, A. S. (2018). Occupational health and safety in the industry 4.0 era: A cause for major concern?. *Safety Science*, *109*, 403–411.

Barreto, L., Amaral, A., & Pereira, T. (2017). Industry 4.0 implications in logistics: An overview. *Procedia manufacturing*, *13*, 1245–1252.

Bhat, T. P. (2020). India and Industry 4.0. Working Paper. ISID. www.isid.org.in/wp-content/uploads/2020/07/WP218.pdf

Chen, B., Wan, J., Shu, L., Li, P., Mukherjee, M., & Yin, B. (2017). Smart factory of industry 4.0: Key technologies, application case, and challenges. *IEEE Access*, *6*, 6505–6519.

Culot, G., Fattori, F., Podrecca, M., & Sartor, M. (2019). Addressing industry 4.0 cybersecurity challenges. *IEEE Engineering Management Review*, *47*(3), 79–86.

Dalzochio, J., Kunst, R., Pignaton, E., Binotto, A., Sanyal, S., Favilla, J., & Barbosa, J. (2020). Machine learning and reasoning for predictive maintenance in Industry 4.0: Current status and challenges. *Computers in Industry*, *123*, 103298.

Deb, T., Vishwas, N., & Saha, A. (2020). A comparative study on different approaches of road traffic optimization based on big data analytics. In *Performance Management of Integrated Systems and its Applications in Software Engineering*, Millie Pant, Tarun K. Sharma, Sebastián Basterrech, & Chitresh Banerjee (Eds.), 119–126. Springer.

Dellermann, D., Ebel, P., Söllner, M., & Leimeister, J. M. (2019). Hybrid intelligence. *Business & Information Systems Engineering*, *61*, 637–643.

Ernst, F., & Frische, P. (2015). Industry 4.0/industrial internet of things-related technologies and requirements for a successful digital transformation: An investigation of manufacturing businesses worldwide. Available at SSRN 2698137.

Gardan, J. (2017). Additive manufacturing technologies: State of the art and trends. In *Additive Manufacturing Handbook*, Adedeji B. Badiru, Vhance V. Valencia, Adedeji B. Badiru, David Liu, & Carl R. Hartsfield (Eds.), 149–168. CTC Press.

He, Y., He, J., & Wen, N. (2023). The challenges of IoT-based applications in high-risk environments, health and safety industries in the Industry 4.0 era using decision-making approach. *Journal of Innovation & Knowledge*, *8*(2), 100347.

Hozdić, E. (2015). Smart factory for industry 4.0: A review. *International Journal of Modern Manufacturing Technologies*, *7*(1), 28–35.

Huang, P. Y., Ouyang, T. H., Pan, S. L., & Chou, T. C. (2012). The role of IT in achieving operational agility: A case study of Haier, China. *International Journal of Information Management*, *32*(3), 294–298.

Jadeja, Y., & Modi, K. (2012, March). Cloud computing-concepts, architecture and challenges. In *2012 International Conference on Computing, Electronics and Electrical Technologies (ICCEET)* (pp. 877–880). IEEE.

Kale, S. R., Laghane, Y. N., Kharade, A. K., & Kadus, S. B. (2019). Hyperloop: Advance mode of transportation system and optimize solution on traffic congestion. *International Journal of Applied Science & Engineering Technology*, *7*, 539–552.

Karatas, M., Eriskin, L., Deveci, M., Pamucar, D., & Garg, H. (2022). Big Data for Healthcare Industry 4.0: Applications, challenges and future perspectives. *Expert Systems with Applications*, 200, 116912.

Kenett, R. S., & Bortman, J. (2022). The digital twin in Industry 4.0: A wide-angle perspective. *Quality and Reliability Engineering International*, *38*(3), 1357–1366.

Khayyam, H., Javadi, B., Jalili, M., &Jazar, R. N. (2020). Artificial intelligence and internet of things for autonomous vehicles. In *Nonlinear Approaches in Engineering Applications: Automotive Applications of Engineering Problems*, Reza N. Jazar & Liming Dai (Eds.), 39–68. Springer.

Kröger, W. (2021). Automated vehicle driving: Background and deduction of governance needs. *Journal of Risk Research*, *24*(1), 14–27.

Laiton-Bonadiez, C., Branch-Bedoya, J. W., Zapata-Cortes, J., Paipa-Sanabria, E., & Arango-Serna, M. (2022). Industry 4.0 technologies applied to the rail transportation industry: A systematic review. *Sensors*, *22*(7), 2491.

Lee, J., Davari, H., Singh, J., & Pandhare, V. (2018). Industrial Artificial Intelligence for industry 4.0-based manufacturing systems. *Manufacturing Letters*, *18*, 20–23.

Li, B. H., Hou, B. C., Yu, W. T., Lu, X. B., & Yang, C. W. (2017). Applications of artificial intelligence in intelligent manufacturing: A review. *Frontiers of Information Technology & Electronic Engineering*, *18*, 86–96.

Liu, Y., & Xu, X. (2017). Industry 4.0 and cloud manufacturing: A comparative analysis. *Journal of Manufacturing Science and Engineering*, *139*(3). https://doi.org/10.1115/1.4034667

Lom, M., Pribyl, O., & Svitek, M. (2016, May). Industry 4.0 as a part of smart cities. In *2016 Smart Cities Symposium Prague (SCSP)* (pp. 1–6). IEEE.

Madakam, S., Holmukhe, R. M., & Jaiswal, D. K. (2019). The future digital work force: Robotic process automation (RPA). *JISTEM-Journal of Information Systems and Technology Management*, *16*. https://doi.org/10.4301/S1807-1775201916001

Mohamed, N., & Al-Jaroodi, J. (2019, April). The impact of industry 4.0 on healthcare system engineering. In *2019 IEEE International Systems Conference (SysCon)* (pp. 1–7). IEEE.

Morrar, R., Arman, H., & Mousa, S. (2017). The fourth industrial revolution (Industry 4.0): A social innovation perspective. *Technology Innovation Management Review*, *7*(11), 12–20.

Munirathinam, S. (2020). Industry 4.0: Industrial internet of things (IIOT). In *Advances in Computers* 117(1), 129–164).

Paul, S., Riffat, M., Yasir, A., Mahim, M. N., Sharnali, B. Y., Naheen, I. T., ... & Kulkarni, A. (2021). Industry 4.0 applications for medical/healthcare services. *Journal of Sensor and Actuator Networks*, *10*(3), 43.

Piccarozzi, M., Aquilani, B., & Gatti, C. (2018). Industry 4.0 in management studies: A systematic literature review. *Sustainability*, *10*(10), 3821.

Pilloni, V. (2018). How data will transform industrial processes: Crowdsensing, crowdsourcing and big data as pillars of industry 4.0. *Future Internet*, *10*(3), 24.

Preuveneers, D., & Ilie-Zudor, E. (2017). The intelligent industry of the future: A survey on emerging trends, research challenges and opportunities in Industry 4.0. *Journal of Ambient Intelligence and Smart Environments*, *9*(3), 287–298.

Rao, S. K., & Prasad, R. (2018). Impact of 5G technologies on industry 4.0. *Wireless Personal Communications*, *100*, 145–159.

Rodrigue, J. P. (2020). *The Geography of Transport Systems*. Routledge.

Rosa, D. I. D. S. (2022). *Industry 4.0 and the global automobile context: The german situation and the specific BMW's case until the BMW and Daimler's Joint Venture* (Doctoral dissertation).

Sima, V., Gheorghe, I. G., Subić, J., & Nancu, D. (2020). Influences of the industry 4.0 revolution on the human capital development and consumer behavior: A systematic review. *Sustainability*, *12*(10), 4035.

Süzen, A. A. (2020). A Risk-Assessment of Cyber Attacks and Defense Strategies in Industry 4.0 Ecosystem. *International Journal of Computer Network & Information Security*, *12*(1), 1–12.

Thoben, K. D., Wiesner, S., & Wuest, T. (2017). "Industrie 4.0" and smart manufacturing-a review of research issues and application examples. *International Journal of Automation Technology*, *11*(1), 4–16.

Tseng, M. L., Tran, T. P. T., Ha, H. M., Bui, T. D., & Lim, M. K. (2021). Sustainable industrial and operation engineering trends and challenges Toward Industry 4.0: A data driven analysis. *Journal of Industrial and Production Engineering*, *38*(8), 581–598.

Vaidya, S., Ambad, P., & Bhosle, S. (2018). Industry 4.0–A glimpse. *Procedia Manufacturing*, *20*, 233–238.

Wollschlaeger, M., Sauter, T., & Jasperneite, J. (2017). The future of industrial communication: Automation networks in the era of the internet of things and industry 4.0. *IEEE Industrial Electronics Magazine*, *11*(1), 17–27.

Xu, L. D., Xu, E. L., & Li, L. (2018). Industry 4.0: State of the art and future trends. *International Journal of Production Research*, *56*(8), 2941–2962.

Zhong, R. Y., Xu, X., Klotz, E., & Newman, S. T. (2017). Intelligent manufacturing in the context of industry 4.0: A review. *Engineering*, *3*(5), 616–630.

17 Advancing IoT Anomaly Detection through Dynamic Learning

A.V. Kalpana, T.C. Jermin Jeaunita, Vijay Anandh R, T. Chandrasekar, and Thilagavathy A

17.1 INTRODUCTION

The internet of things (IoT) represents a dynamic situation, where devices and sensors (to measure) are interconnected over the internet, reshaping the landscape of technology. This pervasive network has gained significant traction in diverse sectors, ranging from academia to business, offering the promise of "smart living". The concept revolves around leveraging the connectivity and data generated by IoT devices to make intelligent decisions that enhance efficiency and optimize resources. A tangible example is evident in smart cities, where IoT-enabled streetlights adjust their lighting levels based on sensor data, contributing to energy conservation.

Cisco's projections highlight the monumental growth of IoT-generated data, anticipated to reach a staggering 847 Zettabytes by 2021. Despite this data abundance, the sheer volume presents challenges, particularly in identifying abnormal patterns or anomalous behaviors within the datasets. Anomaly detection emerges as a critical facet of IoT data analysis, involving the identification of deviations from expected patterns. This capability is vital across diverse applications, spanning intrusion detection, fraud detection applications, as well as health monitoring.

One prominent participant in the field of anomaly identification in the internet of things is machine learning procedures. These methods could be utilised in an unsupervised or supervised manner. Identified or labelled data is used in the framework of supervised learning to categorise information according to normal and pathological classifications. Unsupervised learning, on the other hand, deals with unlabelled data by finding commonalities within the dataset, when it is unclear how to distinguish between normal and pathological groups.

This chapter uses dynamic learning approaches to navigate the complexities of identifying anomalies in the internet of things systems. It provides a thorough summary of the available options, classifies the available anomaly detection methods, and examines

DOI: 10.1201/9781003480860-17

the datasets used in this field. The investigation seeks to promote classification, give scholars a more nuanced grasp of anomaly detection methods, and highlight potential problems and directions for further study. In order to provide a definitive viewpoint on this emerging topic, the ensuing sections examine the historical context of machine learning methods for detection, evaluate current anomaly detection designs, analyse datasets and related difficulties, and suggest future research possibilities.

The January 2023 edition of IoT Analytics' Worldwide IoT Corporate Spend Tracking states that the commercial internet of things, also known as IoT, market hit $201 billion in 2022. This expansion was noteworthy. Even while this rise was less than the 23% predicted, it was nonetheless noteworthy. It is projected that the market's pace will continue to slow down in 2023. At first, the prediction was more upbeat, calling for a quicker recovery of the world economy, robust supply chains, and ongoing investments in new technology to address the labor crisis. Nevertheless, the most recent evaluation, which was completed nine months after the last update, requires that the 2023 projection be lowered to 19%, as seen in Figure 17.1.

IoT Analytics predicts that between 2022 and 2027, the IoT market would grow at a CAGR (compound annual growth rate) approximately 19.4%, reaching a total value of $483 billion. With a CAGR of 22% over the same period, the Asia-Pacific (APAC) area is expected to outperform other worldwide regions under this trajectory. North America is anticipated to expand at a quicker rate than Europe, and this is predicted to increase at a CAGR of 16% by 2027, with an anticipated compound annual growth rate (CAGR) of 20%. This nuanced forecast takes into account how the enterprise IoT landscape is changing due to factors including geographical growth rates, economic dynamics, and technology improvements.

Anomaly detection methods play a pivotal role in identifying unconventional behaviors that deviate from expected patterns [1]. These methods leverage machine

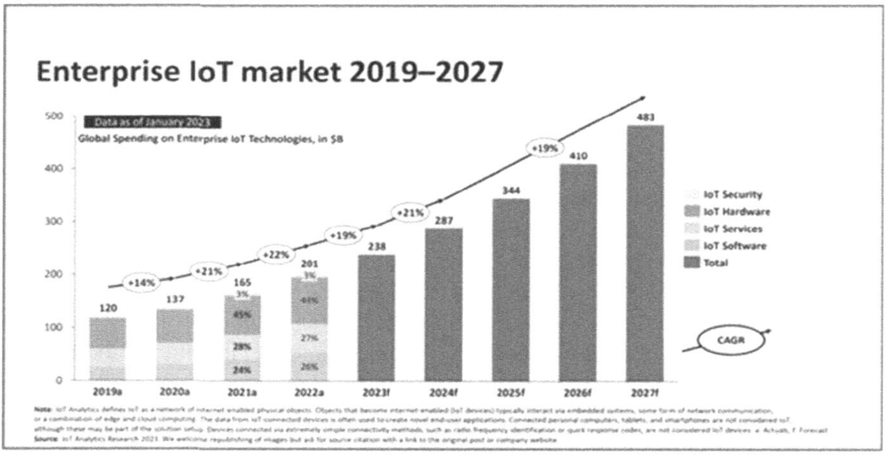

FIGURE 17.1 Global IoT Market.

Source: https://iot-analytics.com/iot-market-size/

learning algorithms to assign class labels (normal/anomaly) to data, with model construction influenced by various factors, including datasets, learning procedure types, feature selection as well as some of the evaluation systems.

The dataset serves as the basis for anomaly detection algorithms and is made up of information gathered from IoT devices. In essence, the dataset is a matrix, where columns indicate properties or features and rows represent observations, samples, or occurrences. In this study, these terms will be used interchangeably.

One important part, the learning algorithm, can work in an unsupervised or supervised manner. While unsupervised learning uses unlabeled data to find patterns without established classifications, supervised learning uses labelled data to classify observations. In order to carefully select a subgroup of characteristics or features to build the typical model, feature selection is used. Enhancing model performance and concentrating on the most pertinent data points require this phase. Machine learning algorithms generate findings, which are validated and evaluated through the use of evaluation procedures. These methods guarantee the efficacy and dependability of the detection of anomalies. Adding cutting-edge methods like deep learning as well as active training to the workflow for machine learning for identifying anomalies with-in the internet of things, also known as IoT, improves the model's performance.

Deep neural network is a sub-domain of machine learning utilizing multiple-layered artificial neural networks, or "deep architectures", to support hierarchical representation learning. By using this technique, the exemplary is able to recognize complicated patterns in the records, mimicking the intricacy of the workings of the human brain. Deep learning architectures like neural networks with deep layers (DNNs), have attracted a lot of attention recently because of their remarkable performance in a variety of fields, such as the recognition of images, retrieving, search engine retrieval of data, and natural language processing, in contrast to traditional machine learning.

Active learning introduces a strategic element to the learning process, where the algorithm actively selects the utmost enlightening occurrences for labeling. This interactive learning paradigm minimizes the need for extensive-labeled data, a common challenge in anomaly detection systems. By intelligently choosing data points that contribute the most to refining the model, active learning optimizes the learning process, making it more efficient and adaptable to evolving IoT environments.

Figure 17.2, illustrating the machine learning workflow, now encompasses the integration of deep learning and active learning. After data is gathered from IoT devices, deep learning techniques, specifically deep neural networks, are employed for detecting the anamolies, allowing the model to capture intricate patterns within the data. The active learning component strategically selects data points for labelling, optimizing the learning process and overcoming challenges associated with limited-labeled data.

This improved process highlights the anomaly detection model's resilience and adaptability in the ever-changing internet of things, demonstrating the assistances of combining the use of active deep learning, plus a conventional machine learning methods for more efficient anomaly identification.

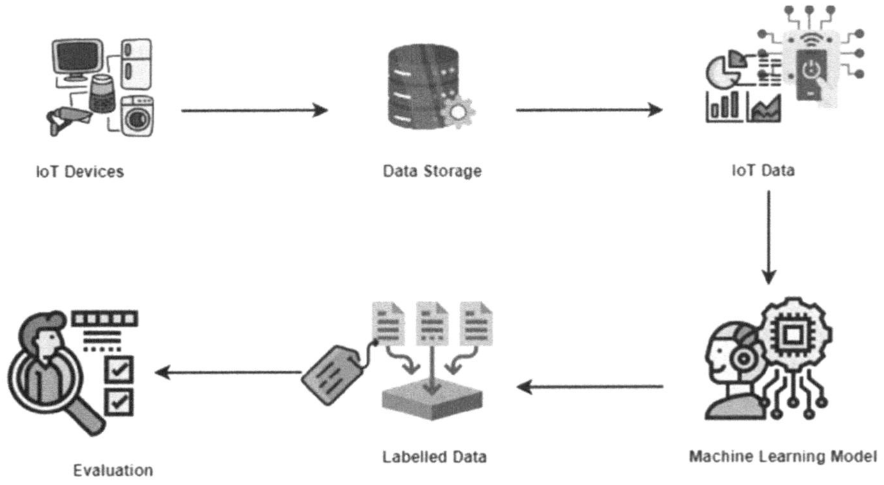

FIGURE 17.2 Integration of Machine Learning and Active Learning.

17.2 DATASETS FOR IOT ANOMALY DETECTION

The most crucial element in the algorithm's learning process is the data. As a result, we examine the existing IoT data archives for anomaly detection in this part. A range of IoT datasets that were taken from networks or perception layers were used in the earlier research that were reviewed [1]. Network information, such as satellite nets, communication protocols, and communication networks, is included in the data at the network layer. Conversely, the information gathered from tangible devices, such sensors or GPS, is stored in the perceptual layer.

A few IoT datasets are accessible to the general public, such as the following:

- The satellite Landsat [2]. Unlike the prior multi-class challenge for this dataset, Golden provides a new form with a dual classification challenge for anomaly recognition. In total, there are 36 characteristics, with values between 0 and 255. These characteristics map to four spectral bands with nine values apiece.
- The Irish Social Science Data Archive Centre (ISSDA) has made accessible data on smart meter usage collected from 2009 to 2011, accessible through ISSDA [3]. This dataset comprises three attributes.
- Aarhus, Denmark's Citypulse experiment provides pollution data that consists of 17,568 observations in addition to eight parameters.
- MIMIC consists of a over-all of 12 physiological markers [4, 5].
- DS2OS. Artificial records were collected using IoT traffic suggestions by producing a simulated IoT setting containing 357,952 instances as well as 13 attributes that encompassed both normal and pathological classes. The anomaly group comprising of eight different attack types which includes the denial-of-service assault [6].

- The Intel Lab reported the collection of 2.3 million data points from 54 sensors measuring eight variables, such as temperature, daylight, humidity as well as voltage [7].
- KDD Cup 99 contains network data by means of 24 distinct attack types and 41 attributes that identify this information as normal or abnormal [8]. A dataset called NSL-KDD [9] solves a few of the issues with KDD Cup 1999 .
- San Francisco parking. Actual records obtained using sensors installed in San Francisco parking lots between August 13 and August 26, 2013, utilizing eight measures [10].
- LWSNDR Dataset. Real sensors positioned as both indoor and outdoor sensors contributed to the creation of the LWSNDR dataset. The dataset comprising of four features are the temperature, the humidity, the Mote ID, and the reading number [11].
- CFDR uses a supercomputer to collect four types of environmental data: temperature, the voltage, date, as well as ID number [12].
- Despite being a complex system, SWaT was constructed using all available network information, sensor as well as actuator data, and 36 different forms of attacks [13].
- Kyoto 2006+ is the network traffic of Kyoto University [14].

Much of the public's access to perceptual-level datasets, such those related to smart cities or healthcare, is un-labeled. Furthermore, a few researches [15–19] have gathered data by modeling the sensors; however, none of these investigations have been published or made publicly accessible. Research articles that have made use of these datasets or dataset repositories have provided the information on the datasets.

17.3 IOT-BASED ANAMOLY DETECTION

An IoT anomaly recognition technique can be categorized into four groups based on various criteria, drawing from classifications proposed by Fahim as well as Sillitti [20]. These categories encompass the approach to the problem, the application method, the sort of method employed, and the expectancy of the system. Figure 17.3 provides a visual exemplification of these groups, offering an illustrative overview. The subsequent segment provides a concise description of each category and highlights some conventional approaches commonly employed in IoT anomaly detection.

17.3.1 By Method

Geometrical methods rely on distance- and density-based strategies to separate expected and anomalous data. These methods assume that anomalies appear in sparse regions of a dataset. They typically use static or dynamic thresholds on estimated distances to classify anomalies.

Mathematical distributions are used by statistical techniques, including the minimal volumes approach, to model normal data. The minimal volume strategy, for example, seeks to maximise real-time information points and minimise volume while

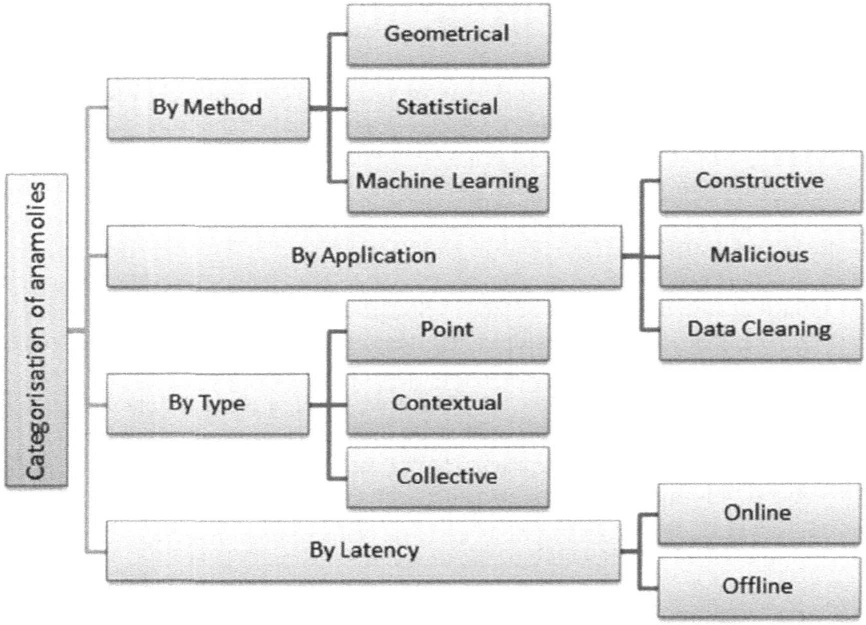

FIGURE 17.3 Categorization of Anamolies.

forming a simplex surrounding the data cloud. Exponential smoothing is another statistical technique that predicts future data points by using previous ones as a basis. Statistical techniques identify anomalies that depart from the accepted model.

Machine learning plus deep learning models have gained popularity, offering flexibility for anomaly detection. The selection of a model depends on the nature of the records. Models such as transformer as well as long short-term memory (LSTM), have favored for sequential data inputs such as time series. Conversely, non-sequential data sorts, like images, suit models like convolutional neural networks (CNN) and autoencoders (AE). These models establish decision boundaries to differentiate normal and anomalous behavior, employing techniques like support vector machines (SVM) or LSTM networks for future forecasting.

17.3.2 By Application

Anomalies can be categorized by application into constructive, destructive, and data cleaning. Constructive applications add value, such as fall prevention for the elderly or smart farming techniques. Destructive applications aim to disrupt daily operations for malicious purposes, requiring research into prevention or detection methods. Data cleaning applications, on the other hand, remove unwanted noise or spikes from input signals.

17.3.3 By Anomaly Type

Anomalies are often classified as point, collective or contextual. Point irregularities happen when a sole record point diverges from expected behavior, such as detecting credit card fraud. Contextual anomalies are anomalous in specific contexts, considering both contextual and behavioral characteristics. Collective anomalies involve anomalies across the entire dataset, like monitoring heart conditions using electrocardiograms.

17.3.4 By Latency

Detection algorithms can be categorized by latency, determining whether they operate online during data collection or offline after complete data access. Online algorithms process information serially, often using distance-based or density-based techniques. Examples include IoT-Keeper and ensemble approaches. Offline procedures have admittance to comprehensive data and employ more sophisticated techniques, completing the classic training process in offline and deploying the model in online.

17.4 ARTIFICIAL INTELLIGENCE FOR IOT ANAMOLY DETECTION

The convergence of artificial intelligence (the AI) with the internet of things (the IoT) yields robust solutions for anomaly recognition, which is vital for guaranteeing the reliability and safety of IoT systems. Anomalies, or departures from typical behavior, in IoT networks might indicate a number of problems, including broken sensors, security lapses, or unusual environmental circumstances.

Through the use of AI methods like deep learning, machine learning, and recognition of patterns, IoT systems are capable of recognising irregularities in the enormous quantities of data created by networked devices. In order to detect trends and abnormalities in both historical and real-time data, machine learning algorithms enable pre-emptive reactions to possible risks or irregularities.

Deep learning, a branch of artificial intelligence, uses neural networks to autonomously recognise intricate connections and patterns in internet of things data, providing enhanced capabilities for anomaly identification. Deep learning models are appropriate to real-time anomaly detection for active IoT contexts because they can identify anomalies substantially high accuracy, even in big and complicated datasets. Furthermore, AI-driven anomaly detection in IoT systems can facilitate predictive maintenance, enabling organizations to anticipate and prevent equipment failures before they occur. By continuously monitoring IoT data streams and detecting anomalies, organizations can optimize operations, reduce downtime, and enhance overall system reliability.

However, integrating AI-based anomaly detection into IoT environments poses several challenges, including data quality issues, scalability concerns, and computational resource constraints. Addressing these challenges requires careful consideration of data pre-processing techniques, model optimization strategies, and deployment architectures tailored to the specific requirements of IoT applications.

The examination of research articles on identifying anomalies in the IoT, was conducted using a thorough and organised method to group them according to their scope. Both systematic as well as non-systematic methods were used in the search.

The remarkable increase in IoT device dissemination in recent years has resulted in a large increase in the capacity of records created by these associated devices. The investigation of this sizable data is essential for anomaly identification since it facilitates the recognition of odd patterns or behaviors. Consequently, this review of the literature delivers a summary of the prior research on active learning-based methods for detecting abnormalities in internet of things systems. In this topic, numerous approaches were additionally examined.

According to Ahmad et. al, employing attack comprising of specific features instead of IoT-specific traits could advance the efficiency and likelihood of a machine learning-based security structure across attack as well as anomaly detection phases [21]. The NSL-KDD as well as UNSW-NB15 databases were used by the researchers in this analysis. Numerous criteria, comprising recall, accuracy, precision, and f-score, in addition to testing and training timeframes, are used to evaluate a system's performance. The radio frequency (RF) have an F1-score of 0.99, but support vector machines (also known as SVMs) possess F1-score with the value of 0.63. The major objective of this learning was to classify different categories of attacks. The outcomes established a low rate of false alarms as well as a high degree of precision in identifying various attack types when employing the obtained characteristics for this purpose.

Bukhari et.al carried out a study on anomaly detecting method in the context of strengthening cyber-attacks in smart cities. Using ANN, KNN, DT, logistic regression and RF, among other techniques, the investigators improved the overall security and reduced possible dangers of the country's smart city infrastructures. Their work looks at group techniques to increase the detection of the security framework by using bagging and boosting [22]. The two datasets UNSW-NB15 as well as CICIDS 2017 are used in their study. The findings exhibited that the SVM accomplished an accuracy rate of 91.50%. The ANN achieved an accuracy rate of 80.5% in its classification, while the boosting strategy and the stacking technique yielded accuracy rates of 99.6% & 98.8%, respectively. Thus, in the internet of things framework of smart cities, the investigational outcomes achieved with the UNSW-NB15 data set can function as an essential source of data for identifying uncommon assaults.

Zhou et al. projected an extensive structure for employing the variational long short-term memory (V-LSTM) classic model, which uses both estimating and compressing networks [23]. The investigators established a V-LSTM knowledge framework for intelligence anomaly recognition. With uneven industrial big data (IBD) data sets, this technique uses reconstructed feature demonstrations to report the trial of finding a compromise between reduced dimensionality and feature retention. With a 0.907 F1 score, and a value of 0.117 false alarm rate (the FAR), and a 0.895 areas under curve (AUC) measure, the experimental results – which included the V-LSTM policy – out-performed six other methods on the test dataset. As compared to baseline methods, the results demonstrate that their system can accurately identify real abnormalities using data on baseline network traffic in addition to dramatically lesser the false anomaly percentage for detection.

Furthermore, a novel intrusion recognition model that might be used with fog nodes was described by Alzahrani et al. This model identified unnecessary IoT device traffic using UNSW-NB15 attributes. The tabulated transformer model, which outperforms machine learning, is presented in this study. The accuracy of said model is 98.35% when separating regular from irregular traffic. But in many courses, their method accurately predicts attacks with 97.22% [24]. As per the review by Aliyu et al. the model provided new paths for studying fog node anomalies [25]. Furthermore, a range of intrusion detection methods were described by Kocher and Kumar [26]. Their work used the UNSW-NB15 dataset to train ML classifiers. To detect intrusion, the study examined LR, RF, KNN, and Naive Bayes (NB). To find the model's precision, recall, F1-score, accuracy, the mean squared error (MSE), the false positive rate (FPR), and the true positive rate (TPR), tests were run both with and lacking feature selection processes. We also looked at these machine learning classifiers, and the findings indicated that the RF algorithm can reach 99.5% accuracy when all available data is used, and 99.6% accuracy when only certain characteristics are used. To construct precise IDSs, the authors created the XGBoost technique as a feature selection method in combination with other machine learning techniques, such as by ensemble DT, KNN, SVM, LR, and ANN [27]. The UNSW-NB15 dataset was exploited by the investigators to associate methods that are used for anomaly detection. The investigational outcomes prove that techniques such as DT can achieve a test accurateness increase in binary classification between 88.12% and 90.86% by implementing the XGBoost enhanced feature selection strategy.

By leveraging attributes within the UNSW-NB15 dataset, Parra et al. [28] identified distinct sets of features pertaining to network flow, the transmission control protocol (TCP), and the message queuing telemetry transport protocol (MQTT). They addressed overfitting, the curse of dimensionality, and dataset imbalances. Supervised machine learning techniques such as ANN, SVM, and RF were employed to train the clusters. The authors achieved a multi-class classification accuracy of 97.38% and a binary classification accuracy of 98.67% based on RF. They employed cluster-based methodologies, utilising MQTT features, TCP features, RF on flow, and the most effective characteristics from both the clusters, to reach accuracy rates for classification of 96.74%, 91.96%, as well as 96.56%, respectively. They demonstrated that the proposed feature clusters surpass other state-of-the-art supervised machine learning methods in both accuracy and training time.

Recent advancements in IoT security have illuminated the potential of distributed deep learning techniques in identifying IoT threats (Parra et al., 2020) [29]. Moreover, a thorough review by Himeur et al. [29] underscores artificial intelligence's pivotal role in anomaly detection within building energy consumption. Bovenzi et al. [30] introduced a hierarchical hybrid intrusion detection system tailored for IoT applications, underscoring the expanding landscape of IoT security protocols. Cheng et al. [31] demonstrated the efficacy of semi-supervised hierarchical stacking temporal convolutional networks in anomaly detection for IoT connections. Additionally, Pajouh et al. [32] developed a two-tier classification system with two-layer dimension reduction for intrusion detection based on anomalies in IoT infrastructure networks, underscoring the imperative for robust security measures in the IoT ecosystem.

The various approaches put out to address anomaly detection in internet of things systems have been looked at and assessed in this section on literature reviews. Furthermore, because IoT environments are dynamic and ever-changing, traditional methodologies are frequently required to help accommodate the different appearances that anomalies may present over time. In conclusion, our review of the literature has highlighted the significance of the many approaches used to find abnormalities in IoT systems. Because our suggested active learning techniques do not require labeling and allow for flexibility in dynamic IoT contexts, they may enhance the efficiency and accuracy of anomaly detection. This research article's main goal is to significantly add to the body of knowledge already available in this particular topic. By doing this, a strong foundation for future expansions in the arena of IoT anomaly detection will be created, with a specific emphasis on the application of active learning methods.

17.5 SINGLE AND ENSEMBLE MODELS

Single models provide a targeted approach to analysis by using a single learner to categorize data. For example, Ford et al. [33] used data gathered by the Irish Social Science Data Archive (ISSDA) to simulate energy consumption behavior (ECB) using neural networks (NNs) [3]. With a noteworthy true positive rate (TPR) of 93.75%, their study proved the effectiveness of NNs in ECB simulation. Similar to this, Cañedo and Skjellum [15] improved IoT system security by using NNs and were able to achieve over 99% accuracy in forecasting with low false negatives in a variety of input scenarios.

In order to categorize genuine Citypulse project records for Aarhus and Denmark cities, the authors – Jain and Shah [34] – adopted a more comprehensive strategy, deploying several learners such as NNs, binary support vector machine (the b-SVMs), and multi-class SVMs. SVMs mostly surpassed NNs in accuracy by utilizing different kernels, including radial basis functions (RBF), polynomial, and linear. In particular, multiclass SVMs using a polynomial kernel performed better than alternative kernels, and binary SVMs using the RBF kernel were more accurate than linear kernels.

Using feed-forward neural networks (NNs with three distinct layers, Hodo et al. [16] presented a standalone intrusion detection algorithm on IoT networks in order to identify distributed denial of service of service (DDoS/DoS) assaults. With a remarkable 99.4% accuracy, their model was validated on a fictitious internet of things network. Additionally, Pachauri & Sharma [35] investigated ensemble models using random forest (RF) in addition to single models such as J48 decision trees and k-nearest neighbors (KNNs). They used receiver operating characteristic (ROC) curve analysis to show that RF outperformed other chosen learners using the MIMIC datasets from the PhysioNet database [4].

Hasan et al. [36] conducted a comprehensive study on attack in addition to anomaly detection, employing various machine learning techniques on the DS2OS traffic traces dataset. Through rigorous analysis, the researchers found that RF, DT, and NNs outperformed other methods, showcasing an impressive accuracy rate of

99.4%. This highlights the efficacy of these machine learning approaches in accurately detecting and mitigating attacks and anomalies within the DS2OS traffic traces dataset, emphasizing their potential for enhancing cyber-security measures in IoT environments.

A unique two-tier anomaly detection approach called TDTC was presented by Pajouh et al. [32] and is intended primarily for detecting malicious activity occurring within the internet of things backbone. This model runs in two steps: two-layer dimension reduction and then two-tier classification. It is used on the NSL-KDD dataset. Principal component analyses (PCA) in addition to linear discriminant analysis (LDA) are used in the dimension lessening phase to reduce the complexity of the dataset into a more understandable format. The data then go through a two-tier classification process using the k-nearest neighbors (KNNs) and Naive Bayes (NB) algorithms. Because of its ability to classify data, NB is initially used to identify unusual behavior. The authors next train another certainty-factor version of the KNN (CF-KNN) utilizing normal observations in order to more accurately classify data into normal versus anomaly classes, taking into account NB's limitations as being a weak learner. The effectiveness of the TDTC model and shows its excellent accuracy performance, with an amazing 84.86%. This novel method demonstrates how dimension reduction methods and numerous classifiers may be effectively integrated into a two-tier structure to provide reliable anomaly recognition in IoT networks.

Using the Intel Lab dataset, Alghuried [37] presented a two-tier anomaly identification system that made use of the inverted weight clustering (IWC) as well as C4.5 algorithms [7, 38]. IWC was initially used to classify examples into normal as well as anomaly clusters. The classified data was then put into the algorithm known as C4.5 for additional training. The model demonstrated an impressive 97% accuracy and 98.3% recall rate.

Principal component analysis (PCA) was utilized by Zhao et al. [39] to decrease the dimensionality of the record and assessed the effectiveness of k-nearest neighbors (KNNs) as well as softmax regression techniques on the KDD Cup 99 dataset [8]. Following PCA, the performance parameters of accuracy, the rate of false positives (FPR), and rate of detection showed similarities between the two techniques. Conversely, there was a notable contrast in computing complexity: KNNs required 5072.32 seconds, whereas softmax regression completed in just 5.72 seconds.

Using Intel Lab data [7], Emadi and Mazinani [40] developed the spatial clustering based on density of applications using noise (DBSCAN) in conjunction with support vector machines (SVMs). The model performed two crucial actions after selecting the features for voltage, humidity, and temperature. Coefficient correlation (CC) was first used to establish the ideal Epsilon and MinPts values, which made it easier to categorize the data into standard as well as anomaly groups based on density. After being trained on the usual data, SVMs were able to consistently achieve above 94% accuracy in a variety of studies.

Using Intel Lab data [7], Hosseini and Borojeni developed a hybrid model that included k-means clustering technique with sequential minimum optimization (the SMO) classification [41]. Data was first divided into normal as well as anomaly clusters by means of k-means clustering, and then it was sent to SMO for training.

Remarkably, the suggested model achieved 0% false positive rate, 100% accuracy, and positive detection rate.

A brand-new hybrid learning anomaly detection model (the HLMCC) for unlabeled internet of things records was presented in [42]. This model combines the decision tree (DT) classification system with the hierarchical affinity propagation (the HAP) clustering method. The DT algorithm is trained using the normal and aberrant clusters that are created by HAP clustering. Enhancement in performance parameters such as FPR, recall, accuracy, and the area under the accuracy-recall curve (AUCPR) over current models was shown by applying HLMCC to LWSNDR as well as Landsat satellite datasets [3].

Finally, a paradigm for imputation-based handling of missing values from datasets was put forth by Vangipuram et al. [43]. The model consists of two stages: classification and imputation. The imputation algorithm uses incremental clustering and marginal conditional probability to build an imputed dataset by classifying data into two groups: G1 (instances without missing values) along with G2 (instances with incomplete values). Then, the classification method makes use of the imputed dataset to classify data, and it performs significantly better than previous models that utilize k-means imputation, with remarkable accuracy that ranges from 99 to 100%.

Unsupervised learning makes use of unlabeled data to identify trends and clusters in the dataset. When the process of manual labeling is problematic or there is no prior understanding of the data structure, this strategy is especially helpful. For this, a variety of unsupervised techniques are frequently used, including principal component analysis, or PCA, and hierarchical clustering (HCA).

Zheng et al. [44] used machine learning algorithms on San Francisco parking data [10] to recognize parking spots with great occupancy as well as potential abnormalities in the data. Several techniques were used in the study, such as one-class support vector data description (SVM) clustering, maximization of expectation (EM) clustering, and farthest first (FF) clustering (SVDD). While FF clustering revealed extensively frequented places, SVDD efficiently identified both high and low occupancy zones. EM clustering made it easier to identify operational problems such as external effects or sensor failures.

Morrow et al.'s [45] work used techniques including k-mean Gaussian estimation of kernel density, and DB-SCAN to rank and find abnormalities in sensor data using supercomputing systems. Evaluation using a Cray supercomputing dataset retrieved from the Usenix and computer failure data repository (the CFDR) [13], for instance demonstrated the superiority of DB-SCAN in detecting abnormalities with high accuracy and minimizing false positives, particularly when location density is taken into account.

Using street data gathered in Barcelona, Garcia-Font et al. [18] carried out a comparative analysis among 4 machine learning algorithms namely: Maha-lanobis distance (the MD), the local outliers factor (LOF), the HCA, and one-class SVM. Metrics including TPR, which stands FPR and F-score, were assessed, and several feature selection techniques were used. Under more restricted circumstances, OC-SVM using feature vector 2 (the FV2) demonstrated encouraging results, attaining over 75% TPR.

A model combining OC-SVM with the YASA segmentation method was proposed by Martí et al. [19] to handle large amounts of unlabeled sensor data. To combat noise and inconsistencies in the data, YASA divided time series data blocks, which were then utilized to train the OC-SVM. The model's advantage over OC-SVM alone as well as statistical confidence intervals (CIs) was proven by the McNemar test after evaluation on information collected from operational turbomachines.

For the secured water treatment (SWaT) dataset, Inoue et al. [46] introduced OC-SVM and deep neural networks (DNNs) for unsupervised learning [13]. Although DNNs fared better in terms of precision and F-scores than OC-SVM, the latter did better in terms of recall.

A novel distance calculation method for PCA was introduced by Hoang and Nguyen [19] with the goal of lowering computational overhead. When the model was utilized onto the Kyoto dataset 2006+ [14], it showed respectable accuracy, TPR, and FPR performance, along with a shorter processing time than in previous distant examples.

White and Legg [47] advocated for an unsupervised learning-based anomaly detection model using OC-SVM trained solely on normal instances. Their study on network data from IoT devices like Amazon Echo showcased the model's effectiveness in distinguishing normal and anomalous behaviors across different devices.

While supervised learning relies on manually labeled data and predefined class labels, unsupervised learning excels in scenarios where labeled data is scarce or unavailable. However, evaluating unsupervised algorithms poses challenges due to the absence of ground truth. Despite its limitations in accuracy compared to supervised learning, unsupervised learning remains a valuable tool for exploring data and uncovering hidden patterns, particularly through clustering techniques.

17.6 FUTURE EXPLORATION AVENUES WITH ACTIVE LEARNING INTEGRATION

Regardless of their categorization as supervised or unsupervised, machine learning algorithms offer effective means to identify, categorize, and unveil anomalies within the IoT environment. The intrinsic advantage of learning from the vast pool of freely available IoT data, even in the absence of classification labels (normal or anomalous), is a hallmark of both approaches. However, the scarcity of labeled data poses a challenge for exclusively relying on supervised learning methods to construct anomaly detection models. Additionally, the rapid growth of IoT data demands models capable of anticipating abnormalities in newly collected and future data, a limitation practically addressed by unsupervised learning techniques. To enhance and refine these methodologies further, integrating active learning into the research agenda presents an exciting prospect. Active learning, with its capability to iteratively query unlabeled instances for labeling, can strategically improve model performance.

The following areas are proposed for future research, considering the integration of active learning:

- **Augmenting labeled internet of things data:** active learning can play a pivotal role in intelligently selecting instances for labeling, maximizing the model's learning potential. By strategically choosing the most informative data points for labeling, active learning ensures efficient utilization of labeling resources, addressing the challenge of scarce labeled data in IoT anomaly detection.

- **Tackling the unlabeled data conundrum:** the iterative nature of active learning aligns seamlessly with the evolving landscape of IoT data, providing a mechanism to iteratively improve categorization. As unlabeled IoT data accumulates swiftly, active learning can strategically query instances for labeling, enhancing the categorization process and addressing challenges associated with unlabeled information within the IoT.
- **IoT benchmarking with active learning:** integrating active learning into benchmarking studies ensures a dynamic and adaptive evaluation process. By actively selecting instances for labeling during benchmarking evaluations, researchers can obtain a nuanced understanding of how various algorithms perform under different conditions, making the benchmarking process more insightful and reflective of real-world IoT scenarios. Future research should explore the synergies between active learning and IoT benchmarking to advance the understanding and performance of anomaly detection models. Incorporating active learning into the future research agenda promises to introduce adaptive and strategic elements to the anomaly detection process in IoT, addressing key challenges related to data scarcity and un-labeled information.

17.7 CONCLUSION

In the realm of IoT, data collection is susceptible to anomalies, which can manifest as sensor malfunctions or security breaches. Anomaly detection is crucial across various domains, including fault and fraud detection, aiming to pinpoint unusual patterns diverging from expected behaviors. normal and abnormal data classes. This process involves training the algorithm to recognize typical behaviors within the dataset, allowing it to flag deviations that may indicate anomalies or outliers. By analyzing various features and patterns, machine learning algorithms can effectively differentiate between expected and unexpected data instances, enabling the identification of potential anomalies in diverse contexts such as fault detection, fraud detection, and health monitoring. This study primarily delves into machine learning methodologies for detecting anomalies in IoT settings and the associated challenges with existing datasets. We've scrutinized prevalent machine learning algorithms, both supervised and unsupervised, for IoT anomaly detection, along with their inherent challenges. Moreover, we've proposed and explored potential avenues for future research, particularly focusing on dataset-related considerations.

REFERENCES

[1] Diro, Abebe, et al. "A comprehensive study of anomaly detection schemes in IoT networks using machine learning algorithms." *Sensors* 21.24 (2021): 8320.
[2] Goldstein, Markus, and Seiichi Uchida. "A comparative evaluation of unsupervised anomaly detection algorithms for multivariate data." *PloS One* 11.4 (2016): e0152173.
[3] Commission for Energy Regulation (CER). *CER Smart Metering Project – Electricity Customer Behaviour Trial, 2009–2010* [dataset]. 1st Edition. Irish Social Science Data Archive (ISSDA). (2012). www.ucd.ie/issda/data/commissionforenergyregulationcer/

[4] Goldberger, Ary L., et al. "PhysioBank, PhysioToolkit, and PhysioNet: components of a new research resource for complex physiologic signals." *Circulation* 101.23 (2000): e215–e220. https://doi.org/10.1161/01.CIR.101.23.e215

[5] Garg, Prachi, and Ajeet Sharma. "Detection of normal ECG and arrhythmia using artificial neural network system." *International Journal of Engineering Research and Science & Technology* 4.1 (2015): 1–13.

[6] Pahl, Marc-Oliver, and François-Xavier Aubet. "All eyes on you: Distributed Multi-Dimensional IoT microservice anomaly detection." *2018 14th International Conference on Network and Service Management (CNSM)*. IEEE. (2018).

[7] Bodik, Peter, et al. "Intel Berkeley research lab data." (2004). http://db.csail.mit.edu/labdata/labdata.html

[8] KDD CUP 1999 Data. The UCI KDD Archive. (1999). http://kdd.ics.uci.edu/databases/kddcup99/kddcup99.html

[9] Tavallaee, Mahbod, et al. "A detailed analysis of the KDD CUP 99 data set." *2009 IEEE Symposium on Computational Intelligence for Security and Defense Applications*. IEEE. (2009).

[10] Bock, Fabian, and Sergio Di Martino. "On-street parking availability data in San Francisco, from stationary sensors and high-mileage probe vehicles." *Data in Brief* 25 (2019): 104039.

[11] Suthaharan, Shan, et al. "Labelled data collection for anomaly detection in wireless sensor networks." *2010 Sixth International Conference on Intelligent Sensors, Sensor Networks and Information Processing*. IEEE. (2010). https://doi.org/10.1109/ISSNIP.2010.5706782

[12] Schroeder, Bianca, and Garth A. Gibson. "The computer failure data repository (CFDR)." Workshop on Reliability Analysis of System Failure Data (RAF'07), Microsoft Research, Cambridge, UK. (2007).

[13] Goh, Jonathan, et al. "A dataset to support research in the design of secure water treatment systems." *Critical Information Infrastructures Security: 11th International Conference, CRITIS 2016, Paris, France, October 10–12, 2016, Revised Selected Papers 11*. Springer International Publishing. (2017).

[14] Song, Jungsuk, et al. "Statistical analysis of honeypot data and building of Kyoto 2006+ dataset for NIDS evaluation." *International Workshop on Building Analysis Datasets and Gathering Experience Returns for Security* (2011).

[15] Canedo, Janice, and Anthony Skjellum. "Using machine learning to secure IoT systems." *2016 14th Annual Conference on Privacy, Security and Trust (PST)*. IEEE. (2016). https://doi.org/10.1109/PST.2016.7906930

[16] Hodo, Elike, et al. "Threat analysis of IoT networks using artificial neural network intrusion detection system." *2016 International Symposium on Networks, Computers and Communications (ISNCC)*. IEEE. (2016). https://doi.org/10.1109/ISNCC.2016.7746067

[17] Garcia-Font, Victor, Carles Garrigues, and Helena Rifà-Pous. "A comparative study of anomaly detection techniques for smart city wireless sensor networks." Sensors 16.6 (2016): 868. https://doi.org/10.3390/s16060868

[18] Martí, Luis, et al. "Anomaly detection based on sensor data in petroleum industry applications." *Sensors* 15.2 (2015): 2774–2797. https://doi.org/10.3390/s150202774

[19] White, Jonathan, and Phil Legg. "Unsupervised one-class learning for anomaly detection on home IoT network devices." *2021 International Conference on Cyber Situational Awareness, Data Analytics and Assessment (CyberSA)*. IEEE. (2021). https://doi.org/10.1109/CyberSA52016.2021.9478248

[20] Fahim, Muhammad, and Alberto Sillitti. "Anomaly detection, analysis and prediction techniques in IoT environment: A systematic literature review." *IEEE Access* 7 (2019): 81664–81681.
[21] Ahmad, Muhammad, et al. "Intrusion detection in internet of things using supervised machine learning based on application and transport layer features using UNSW-NB15 data-set." *EURASIP Journal on Wireless Communications and Networking* 2021 (2021): 1–23.
[22] Bukhari, Owais, et al. "Anomaly detection using ensemble techniques for boosting the security of intrusion detection system." *Procedia Computer Science* 218 (2023): 1003–1013.
[23] Zhou, Xiaokang, et al. "Variational LSTM enhanced anomaly detection for industrial big data." *IEEE Transactions on Industrial Informatics* 17.5 (2020): 3469–3477.
[24] Alzahrani, Abdullah I.A., et al. "Anomaly detection in fog computing architectures using custom tab transformer for internet of things." *Electronics* 11.23 (2022): 4017.
[25] Aliyu, Farouq, et al. "Human immune-based intrusion detection and prevention system for fog computing." *Journal of Network and Systems Management* 30.1 (2022): 11.
[26] Kocher, Geeta, and Gulshan Kumar. "Analysis of machine learning algorithms with feature selection for intrusion detection using UNSW-NB15 dataset." (2021). https://doi.org/10.2139/ssrn.3784406
[27] Kasongo, Sydney M., and Yanxia Sun. "Performance analysis of intrusion detection systems using a feature selection method on the UNSW-NB15 dataset." *Journal of Big Data* 7.1 (2020): 105.
[28] Parra, Gonzalo De La Torre, et al. "Detecting Internet of Things attacks using distributed deep learning." *Journal of Network and Computer Applications* 163 (2020): 102662.
[29] Himeur, Yassine, et al. "Artificial intelligence based anomaly detection of energy consumption in buildings: A review, current trends and new perspectives." *Applied Energy* 287 (2021): 116601.
[30] Bovenzi, Giampaolo, et al. "A hierarchical hybrid intrusion detection approach in IoT scenarios." *GLOBECOM 2020-2020 IEEE Global Communications Conference*. IEEE. (2020).
[31] Cheng, Yongliang, et al. "Leveraging semisupervised hierarchical stacking temporal convolutional network for anomaly detection in IoT communication." *IEEE Internet of Things Journal* 8.1 (2020): 144–155.
[32] Pajouh, Hamed Haddad, et al. "A two-layer dimension reduction and two-tier classification model for anomaly-based intrusion detection in IoT backbone networks." *IEEE Transactions on Emerging Topics in Computing* 7.2 (2016): 314–323.
[33] Ford, Vitaly, Ambareen Siraj, and William Eberle. "Smart grid energy fraud detection using artificial neural networks." *2014 IEEE Symposium on Computational Intelligence Applications in Smart Grid (CIASG)*. IEEE. (2014). https://doi.org/10.1109/CIASG.2014.7011557
[34] Jain, Raj, and Hitesh Shah. "An anomaly detection in smart cities modeled as wireless sensor network." *2016 International Conference on Signal and Information Processing (IConSIP)*. IEEE. (2016). https://doi.org/10.1109/ICONSIP.2016.7857445
[35] Pachauri, Girik, and Sandeep Sharma. "Anomaly detection in medical wireless sensor networks using machine learning algorithms." *Procedia Computer Science* 70 (2015): 325–333. https://doi.org/10.1016/j.procs.2015.10.026
[36] Hasan, Mahmudul, et al. "Attack and anomaly detection in IoT sensors in IoT sites using machine learning approaches." *Internet of Things* 7 (2019): 100059.
[37] Alghuried, Ahod. "A model for anomalies detection in internet of things (IoT) using inverse weight clustering and decision tree." (2017). https://doi.org/10.21427/D7WK7S

[38] Guestrin, Carlos, et al. "Distributed regression: an efficient framework for modeling sensor network data." *Proceedings of the 3rd International Symposium on Information Processing in Sensor Networks*. (2004). http://db.csail.mit.edu/labdata/labdata.html

[39] Zhao, Shengchu, et al. "A dimension reduction model and classifier for anomaly-based intrusion detection in internet of things." *2017 IEEE 15th International Conference on Dependable, Autonomic and Secure Computing, 15th International Conference on Pervasive Intelligence and Computing, 3rd International Conference on Big Data Intelligence and Computing and Cyber Science and Technology Congress (DASC/PiCom/DataCom/CyberSciTech)*. IEEE. (2017). https://doi.org/10.1109/DASCPICom-DataCom-CyberSciTec.2017.141

[40] Saeedi Emadi, Hossein, and Sayyed Majid Mazinani. "A novel anomaly detection algorithm using DBSCAN and SVM in wireless sensor networks." *Wireless Personal Communications* 98 (2018): 2025–2035. https://doi.org/10.1007/s11277-017-4961-1

[41] Hosseini, Mostafa, and Hamid Reza Shayegh Borojeni. "A hybrid approach for anomaly detection in the Internet of Things." *Proceedings of the International Conference on Smart Cities and Internet of Things*, Association for Computing Machinery, New York, NY. (2018). https://doi.org/10.1145/3269961.3269975

[42] Alghanmi, Nusaybah, Reem Alotaibi, and Seyed M. Buhari. "HLMCC: A hybrid learning anomaly detection model for unlabeled data in Internet of Things." *IEEE Access* 7 (2019): 179492–179504. https://doi.org/10.1109/ACCESS.2019.2959739

[43] Vangipuram, Radhakrishna, et al. "A machine learning approach for imputation and anomaly detection in IoT environment." *Expert Systems* 37.5 (2020): e12556. https://doi.org/10.1111/exsy.12556

[44] Zheng, Yanxu, Sutharshan Rajasegarar, and Christopher Leckie. "Parking availability prediction for sensor-enabled car parks in smart cities." *2015 IEEE Tenth International Conference on Intelligent Sensors, Sensor Networks and Information Processing (ISSNIP)*. IEEE. (2015). https://doi.org/10.1109/ISSNIP.2014.6827618

[45] Morrow, Adam, Elisabeth Baseman, and Sean Blanchard. "Ranking anomalous high performance computing sensor data using unsupervised clustering." *2016 International Conference on Computational Science and Computational Intelligence (CSCI)*. IEEE. (2016). https://doi.org/10.1109/CSCI.2016.0124

[46] Inoue, Jun, et al. "Anomaly detection for a water treatment system using unsupervised machine learning." *2017 IEEE International Conference on Data Mining Workshops (ICDMW)*. IEEE. (2017). https://doi.org/10.1109/ICDMW.2017.149

[47] Hoang, Dang Hai, and Ha Duong Nguyen. "A PCA-based method for IoT network traffic anomaly detection." *2018 20th International conference on Advanced Communication Technology (ICACT)*. IEEE. (2018).

Index